RIVER SEDIMENTATION

PROCEEDINGS OF THE 13TH INTERNATIONAL SYMPOSIUM ON RIVER SEDIMENTATION, ISRS 2016, STUTTGART, GERMANY, 19–22 SEPTEMBER 2016

River Sedimentation

Editors

Silke Wieprecht, Stefan Haun, Karolin Weber,
Markus Noack & Kristina Terheiden
*Institute for Modelling Hydraulic and Environmental Systems,
University of Stuttgart, Germany*

CRC Press
Taylor & Francis Group
Boca Raton London New York Leiden

CRC Press is an imprint of the
Taylor & Francis Group, an **informa** business

A BALKEMA BOOK

CRC Press/Balkema is an imprint of the Taylor & Francis Group, an informa business

© 2017 Taylor & Francis Group, London, UK

Typeset by MPS Limited, Chennai, India
Printed and bound in Great Britain by CPI Group (UK) Ltd, Croydon, CR0 4YY

All rights reserved. No part of this publication or the information contained herein may be reproduced, stored in a retrieval system, or transmitted in any form or by any means, electronic, mechanical, by photocopying, recording or otherwise, without written prior permission from the publishers.

Although all care is taken to ensure integrity and the quality of this publication and the information herein, no responsibility is assumed by the publishers nor the author for any damage to the property or persons as a result of operation or use of this publication and/or the information contained herein.

Published by: CRC Press/Balkema
P.O. Box 11320, 2301 EH Leiden, The Netherlands
e-mail: Pub.NL@taylorandfrancis.com
www.crcpress.com – www.taylorandfrancis.com

ISBN: 978-1-138-02945-3 (hardback + USB)
ISBN: 978-1-315-62320-7 (eBook PDF)

Table of contents

Preface	XVII
Committees	XIX
Sponsors	XXI

Keynote papers

Form, function and physics: The ecology of biogenic stabilization *D.M. Paterson, J.M. Kenworthy & J.A. Hope*	3
Mixed cohesive and noncohesive sediment transport: A state of the art review *W. Wu*	9

A. Integrated sediment management at the river basin scale

Erosion on irregular slope surface: A full N-S equation based numerical study *Y. An & Q.Q. Liu*	21
Impacts of recent climate and land use dynamics on spatial and temporal changes of sediment budget and reservoir siltation in small agricultural catchments of the European Russia *V. Belyaev & A. Malyutina*	22
Variability of total sediment supply of the Chao Phraya River, Thailand *B. Bidorn, S.A. Kish, J.F. Donoghue, W. Huang & K. Bidorn*	23
Responsible management of alpine rivers: The Arly Basin/Savoie, France *P. Ergenzinger & C. de Jong*	24
Suspended sediment yield transportation by rivers of the Kamchatsky Krai into the Pacific Ocean *L.V. Kuksina & N.I. Alexeevsky*	25
Study on strategy of wide floodplain training in the Lower Yellow River *J. Li, E. Jiang & X. Zhang*	26
Study on sediment regulation approaches of Liu River *L. Lin, X. Guan & T. Yu*	27
The International Sediment Initiative (ISI) and its case studies *C. Liu, D.E. Walling, M. Spreafico, J. Ramasamy, H.D. Thulstrup & A. Mishra*	28
Contribution in the study of sediment transport in northern Algeria *M. Meddi*	29
Application of airborne gamma-ray imagery to assist soil survey in the upper Pasak basin, Thailand *R. Moonjun, D.P. Shrestha & K. Duangkamol*	30
More bed load in rivers. Achieving a sediment balance close to the natural state in the canton of Bern *M. Pauli, L. Hunzinger & O. Hitz*	31
Van Deemter's analysis of drainage to incised ditches in lowland areas *M.J.M. Römkens*	32
An overview of hydro-sedimentological characteristics of intermittent rivers in Kabul region of Kabul river basin *N. Sadid, S. Haun & S. Wieprecht*	33
Analyses on trends and reasons of runoff and sediment load of Yellow River stem *H.L. Shi, C.H. Hu, A.J. Deng & Q.Q. Tian*	34

Identifying the sources of fine sediment to quantify the success of sustainable flood risk strategies 35
S. Twohig & I. Pattison

Stream flow modeling for a karst basin using coupled hydrological-hydrodynamic models:
Case study of Lijiang River, China 36
Q.F. Wu, Y. Cai, S.G. Liu, Y.M. Jiang, A.N. Makhinov & A.F. Makhinova

Mechanical effects of vegetation in soil conservation and soil erosion reduction 37
P.Q. Xiao, W.Y. Yao, Z.Z. Shen & C.X. Yang

Impact of sediment management on integrated water resources management 38
F. Yazdandoost & F. Farahani

Integrated sediment transport modelling for rivers feeding lakes and wetlands 39
F. Yazdandoost & N. Khorami

Erosion risk mapping for Hulu Langat River basin 40
R. Zainal Abidin, N. Yusoff, M.S. Sulaiman & T. Mohamed Mustafa

The impact of sub-branch direction on sediment deposition in the condition of reciprocating mainstream flow 41
M. Zhu

Development of sediment control structure for dam sedimentation counter measurement approach 42
J. Zulfan, N.S. Slamet & A. Prasetyo

B. Sediment transport

Erosion and seepage failure around sheet-pile using two-phase WC-SPH model 45
A.M. Abdelrazek, I. Kimura & Y. Shimizu

Investigation on sandy riverbank failure eroded by water level rising 46
R. Arai, K. Ota, T. Sato & Y. Toyoda

Integrated investigation of space-time variability in bed load transport rates using remote sensing 47
M. Bakker & S.N. Lane

Application of Euler-Euler method in estimation of hydraulic structures scour 48
Sh. Basirat & S.A.A. Salehi Neyshabouri

Reliability analysis of a 2D sediment transport model: An example of the lower river Salzach 49
F. Beckers, M. Noack & S. Wieprecht

Contraction rate of the flow, velocities, river bed stratification impact on the scour at the guide banks 50
B. Gjunsburgs & M. Bizane

Study progress of bottom-block scour on Yellow River 51
Y. Cao, E. Jiang, J. Li & Q. Zhang

Quartz silt deposition 52
S. Capapé, J.P. Martín-Vide & F. Colombo

Analysis of the sediment hydrography by modeling hyper-concentrated flow and sediment transport 53
Y.Y. Chiu & K.C. Yeh

Flow and riverbed erosion-deposition simulation around submerged water intake 54
Z.F. Cui & D.C. Hu

Effect on bed load transport discharge of Chongqing reach by backwater of
Three Gorges Reservoir in upper Yangtze River 55
X. Fu, S. Yang, Y. Chen, J. Hu, S. Tong & Y. Xiao

Influence of high Paraná's River dunes variability in Itaipu's reservoir sedimentation 56
P.E. Gamaro, L.H. Maldonado, J.L. Castro & V.P. Bastolla

Suspended load monitoring for sustainable hydropower development 57
M. Guerrero, A. Antonini, N. Rüther & S. Stokseth

Study on sedimentation velocity in transition zone 58
Y. Guo & L. Gao

Estimation of maximum local scour depth around submerged spur-dike 59
S.Y. Hao, Y.F. Xia & H. Xu

Mathematical description of flow memory effects on graded bedload 60
K. Hassan & H. Haynes

Evolution of Modaomen Bar at Pearl River Estuary 61
Y. He, C. Lu, J. Deng, Y. Yang & L. Yang

Modeling of non-capacity bed load transport in swash zone 62
Z. He, L. Tan & P. Hu

Influence of groundwater flow intensity on sediment motion 63
O. Herrera-Granados

Effects of bed-load on flow resistance and stability in step-pool systems 64
B. Hohermuth & V. Weitbrecht

State of the art on remote sensing methods for suspended sediment concentration in inland
and coastal waters 65
A.E. Holdefer & K. Formiga

Study on the incipient velocity of biofilm-coated sediment 66
L. Huang, H.W. Fang, Q.Q. Shang, Y.S. Chen & G.J. He

Riverbank erosion rates prediction incorporating soil erodibility and soil properties
relationship: Bernam River, Malaysia case study 67
S.L. Ibrahim, J. Ariffin & A. Saadon

Development of a bedload sensor for continuous measurement and its applicability 68
T. Itoh, T. Nagayama, R. Utsunomiya, M. Fujita, D. Tsutsumi, S. Miyata & T. Mizuyama

Distributional characteristics of sediment concentration in the lotus-root-shape compound channels 69
Z. Ji, C.H. Hu & F. He

Discharge coefficients derived from sediment concentration to estimate discharge across a Sabo dam 70
K. Kawaike, H. Nakagawa, N. Kim & H. Zhang

Near-bed turbulence characteristics in unsteady hydrograph flows over mobile and immobile
gravel beds 71
J. Kean, A. Cuthbertson & L. Beevers

Numerical simulation of local scour around three circular cylinders in staggered array 72
H.S. Kim, M. Park, I. Kimura, Y. Shimizu & M. Nabi

Simulation of sediment hyper concentration in the lower Yellow River using variational data
assimilation method 73
R. Lai, M. Wang, M. Wang & H. Wang

Experimental study on local scour protection of piers for Hangzhou Bay Bridge by using
twisted duplex blocks 74
Z. Li, Y. Shi, R. Wang, J. Zhang & J. Wang

Effect of proportion of wash load to suspended load on river erosion and deposition 75
C.T. Liao, K.C. Yeh, G.H. Liu & K.W. Wu

Experimental investigation on local shear stress and turbulence intensity over a rough bed
with and without sediment using Laser Doppler Anemometry and Particle Image Velocimetry 76
P. Lichtneger, C. Sindelar, H. Habersack, J. Kitzhofer & E.A. Prager

Analysis on transport characteristic of bimodal bed load & a new calculation model for
non-uniform bed load 77
M.X. Liu, G.D. Li, D.P. Sun & L.Q. Han

Sediment transport along the Deepwater Navigational Channel of Changjiang Estuary, China 78
S. Lou, S.G. Liu, G.H. Zhong & G.F. Ma

Study on the bed coarsening and limit scour depth of the lower reaches of the Three Gorges Reservoir 79
J.-X. Mao & X. Geng

Observation evaluating water and sediment runoff at Sabo dam in Kiso river basin 80
S. Matsuda, T. Nagayama, T. Ikeshima, K. Goto, Y. Nishi & T. Itoh

Turbulent flow and its characteristics over submerged obstacle marks 81
B.S. Mazumder & H. Maity

Highly seasonal suspended sediment and bed load transport dynamic in tropical mountain catchments 82
S.B. Morera, A. Crave & J.L. Guyot

Prediction of bed load transport rate using Shiono and Knight model 83
N. Movahedi, A.R. Zahiri & A.A. Dehghani

River embankment failure and resultant flood and sediment inflow discharges due to
overtopping river flow 84
H. Nakagawa, H. Mizutani, Y. Wang, K. Kawaike, O. Kitaguchi & H. Zhang

Comparison of acoustic backscatter to turbidity for suspended sediment estimation in
the Sacramento-San Joaquin Delta in California 85
M. Ozturk & P.A. Work

Proposing BEHI-NBS method for the estimation of river bank erosion on a river in Nepal 86
S. Pakuwal & S. Panthee

Turbulent hydrodynamics through cross-sections at upstream, interior and downstream of
sparse vegetation patch in open channel flow 87
D. Pal, S. Maji, P.R. Hanmaiahgari, M.D. Bui & P. Rutschmann

Experimental investigation of a cantilever failure for cohesive riverbanks 88
S. Patsinghasanee, I. Kimura & Y. Shimizu

Bedload monitoring in a steep alpine stream: Results from the 2014 measurement campaign 89
R. Rainato, L. Picco & L. Mao

Monitoring topography of laboratory fluvial dike models subjected to breaching based on a
laser profilometry technique 90
I. Rifai, S. Erpicum, P. Archambeau, D. Violeau, M. Pirotton, K. El Kadi Abderrezzak & B. Dewals

A data-driven fuzzy approach to simulate the critical shear stress of cohesive sediments 91
A. Schäfer Rodrigues Silva, M. Noack, D. Schlabing & S. Wieprecht

Discussion of the impact of pressure fluctuations on local scouring 92
W. Schanderl, M. Manhart & O. Link

Processes and effects of reversing currents on the erosion stability of wide-graded grain material 93
A. Schendel, N. Goseberg & T. Schlurmann

Control of bridge abutment scour using triangular vanes 94
M. Shafai-Bejestan & N. Raee

Estimation for the riverbank collapse volume with sandy-riverbank in the desert reach
of the upper Yellow River 95
A. Shu, X. Zhou, G. Duan, F. Li & S. Wang

Incipient motion for gravel particles in cohesive mixture of clay-silt-gravel 96
U.K. Singh, Z. Ahmad & A. Kumar

Assessment of sediment rating curves for mountain rivers in Malaysia 97
S.K. Sinnakaudan, M.R. Shukor, S.I.H. Ismail, A.Z. Abdul Razad & M.N. Ahmad

Super Deltaflex – Advanced development of transit time acoustic flow measurement 98
W. Stedtnitz & T. Schott

Transport of sediment in the presence of stable clast 99
M.S. Sulaiman, R. Zainal Abidin & S.K. Sinnakaudan

Research of bed load distribution density based on image recognition technology 100
D.P. Sun, H. Chen, Y. Sun, A. Gao, M.J. Dong, L.Q. Han & M.X. Liu

Combination of permeable and impermeable spur dikes to reduce local scour and to create diverse river bed   101
A. Tominaga & S.H. Sadat

Estimated response of Nieuwe Waterweg Rotterdam to deepening of the navigation channel   102
A.P. Tuijnder, L.M. Perk, R.C. Steijn, B.T. Grasmeijer, J. Adema, N. Geleynse, J. Cleveringa & L.C. van Rijn

Variation in total sediment rate prediction using different fall velocity methods   103
S.I. Waikhom & S.M. Yadav

Characteristics of sediment movement and river-bed morphology at mountainous stream confluence region   104
X.K. Wang, E. Huang, X.N. Liu, X.F. Yan & H.F. Duan

Experiment for bed erosion focusing on combination of horizontal distance and overlapping height between main and counter Sabo dam   105
H. Watabe, K. Kaitsuka, M. Sugiyama, T. Itoh, H. Muramatsu, T. Nagayama, H. Ogawa, T. Miike, A. Miyamoto, Y. Yamada & T. Mizuyama

Homogeneous two-dimensional Poissonian model applied to the suspended movement of pollutant and non uniform fine sediment in open channel flow   106
G. Wilson Júnior & C.S.G. Monteiro

Experimental study on energy dissipation and beach protection effects of a new type of penetrating frames   107
Y.F. Xia, H. Xu, Z.M. Fu, K.H. Chen & F. Chen

Suspended sediment dynamics of an allogenic dryland river channel   108
G.A. Yu, M. Disse & Z.W. Li

Sediment transport in the middle reach of the Huaihe River   109
B. Yu, J. Ni, H. Zhou, J. Sui, P. Wu & R. Juepner

Sediment transport and silting in Zarzis commercial harbor (Tunisia)   110
M. Zelleg, I. Said, E. Hamdi & Z. Lafhaj

The influences of water-sediment conditions on the sediment delivery rate of the Three Gorges Reservoir   111
D. Zhandi, L. Qin, H. Haihua & J. Zuwen

The response of riverbed erosion and deposition adjustment to the flow and sediment process in the Lower Yellow River   112
X. Zhang, D.P. Sun, Y. Sun & M.X. Liu

Comparison of capacity and non-capacity sediment transport models for dam break flow over movable bed   113
J. Zhao, I. Özgen, R. Hinkelmann, F. Simons & D. Liang

Numerical modelling of scour – the influence of small scale morphological processes   114
L. Zhou & R.J. Perkins

Experimental study on scouring characteristics of cohesive bank soil in the Middle Yangtze River   115
Q.L. Zong, J.Q. Xia & Y. Zhang

C. River morphodynamics

Two-dimensional river bed configuration analysis of the Hii River and diversion channel flood in September 2013   119
R. Akoh, S. Maeno, S. Hirashita, K. Yoshida & T. Matsumoto

Morphodynamic modelling of a meandering sand bed river using Delft3D   120
M.S. Banda, A. Dittrich & J. Pervez

Laboratory experiments on the influence of the length of a sediment replenishment applied with alternated geometrical configuration   121
E. Battisacco, M.J. Franca & A.J. Schleiss

Application of 2D numerical modelling to determination of sediment transport in a Mexican river   122
G. Cardoso-Landa

Historical and current uses of the Morvan's Rivers (central France): Impacts on bedload
transport and fluvial morphology   123
L. Gilet, F. Gob, E. Gautier & C. Virmoux

Sediment transport and evolution at Pearl River estuary   124
J. Deng & H. Deng

Conceptual modeling of bank retreat process in the Upper Jingjiang Reach   125
S.S. Deng, J.Q. Xia, M.R. Zhou & J. Li

How fast evolve the river-bottom profile and grain-size composition at basin scale   126
G. Di Silvio, M. Franzoia & M. Nones

A cellular automata model for riverbed evolvement   127
M.J. Dong

Laboratory experiments on gravel deposit erosion   128
F. Friedl, V. Weitbrecht & R.M. Boes

Landslide dam breach during 2015 earthquake in Nepal: Computational modelling of
hydraulic and morphological effects   129
S. Giri, M. Nabi, J.D. Bricker, B.R. Adhikari & W. Schwanghart

Analysis of the interaction between the Yangtze River and Poyang Lake, China based on Chaos theory   130
J. Hu, Z.L. Wang & Y. Lu

Dynamics of sediment storage in non-alluvial channels   131
C.S. James

Bed variation during floods in the Chikugo River estuary with complex structures of bed layers   132
Y. Kaneko & S. Fukuoka

Study for restoring bank protection functions of longitudinal dikes existing in the river
with alternate bars   133
S. Kato, T. Gotoh & S. Fukuoka

Long-term numerical investigations of the effects of training structures in a river reach with
ongoing river bed deepening   134
A. Kikillus, L. Seitz, S. Haun & S. Wieprecht

Computations on bedform by DEM-URANS coupling with two-way approach   135
I. Kimura, K. Horiuchi & Y. Shimizu

Numerical assessment of the interactions between hydrodynamics, bed morphodynamics
and bank erosion   136
E.J. Langendoen, M.E. Ursic, A. Mendoza, J.D. Abad, R. Ata, K. El Kadi Abderrezzak & P. Tassi

A novel engineering desilting measure – "auto-desilting gallery"   137
S. Li, Q. Yi, W. Cheng & Q. Liu

Critical discharge of erosion-deposition process of mid-channel bar head in anabranching channel   138
Z.W. Li, G.A. Yu & C.D. Zhang

Study on the flow around the Baguazhou Island in the lower reach of the Yangtze River   139
D. Liang, X. Wang, P. Yu & H. Tang

Experiment study on sediment control function of river narrow-section   140
C.H. Lin, C.L. Shieh, C.J. Liu, S.H. Lin & Y.J. Tsai

Experimental study on velocity pattern and bed morphology around a model patch of vegetation   141
C. Liu, D. Wang, K. Yang & X.N. Liu

Effect of the Three Gorges Dam and other upstream factors on the hydrological conditions
of Yichang reach, Yangtze River   142
H. Liu & Y. Lu

Numerical modelling of the Danube river channel morphological development at the
Slovak–Hungarian river section 143
M. Lukac & K. Holubová

Bed-slope-related diffusion of an erodible hump 144
S. Maldonado, M.J. Creed & A.G.L. Borthwick

Restoration of the Eggrank bend at the Thur River in Andelfingen ZH 145
M. Mende, M. Müller, P. Sieber & M. Oplatka

Formation of river dunes by measurement, linear stability analysis and simulation with Bmor3D 146
P. Mewis

Dynamic state of river-mouth bar in the Yuragawa River and its control under flood flow conditions 147
H. Miwa, K. Kanda, T. Ochi & H. Kawaguchi

Computational modelling of secondary flow on unstructured grids 148
M. Nabi, W. Ottevanger & S. Giri

Morphological development of tidal tributaries in relation to turbidity and sediment concentration
of the main estuary river 149
E. Nehlsen & P. Fröhle

Numerical modeling of antidune formation and propagation 150
N.R.B. Olsen

On the effect of different upstream schemes on the simulation of the antidunes propagation 151
E. Rademacher & A. Malcherek

Quasi-three dimensional computations for flows and bed variations in curved channel with
gently sloped outer bank 152
T. Sasaki & S. Fukuoka

Relation between Ishikari River mouth stability and construction of the Ishikari Bay New Port 153
M. Takezawa, H. Gotoh, R. Hanada, O. Ishikawa, M. Tanaka & T. Yamamoto

A look to valley types developed along the Göksu River (between Mut and Silifke: Southern Turkey) 154
A. Turan

Annual change of water environment and topographic feature at urban river mouth 155
K. Uno & S. Kishimoto

Numerical simulation of gravel deposit erosion 156
L. Vonwiller, D.F. Vetsch & R.M. Boes

Characteristics of flow and sediment at the confluences of mainstream and tributary of the upper
reaches of the Yangtze River 157
P.Y. Wang, L.F. Han, C.Y. Yang & T. Yu

Bedrock channel morphological modeling on the river in Taiwan 158
K.W. Wu, K.C. Yeh, C.T. Liao & Y.G. Lai

Recent channel adjustments in the Jingjiang Reach controlled by various boundary conditions 159
J.Q. Xia, M.R. Zhou, S.S. Deng & J.Y. Lu

A comparison of two total sediment transport models for rivers 160
V.K. Yadav, S.M. Yadav & S.I. Waikhom

Experiments on the channel plane form with nodes and anti-nodes 161
S. Yamaguchi, Y. Watanabe & K. Sumitomo

Braided channel evolution in the middle and lower reaches of the Yangtze River after operation
of the Three Gorgers Reservoir 162
S. Yao, G. Qu & H. Wang

Coastline change of the Yellow River Delta since 1855 163
S. Yu & S. Tian

A physically-based model of individual step-pool stability in mountain streams 164
C.D. Zhang, Z.L. Wang & Z. Li

Sensitivity of deposition and erosion to bed composition in the Iffezheim reservoir, Germany *Q. Zhang, T. Speckter, R. Hinkelmann, G. Hillebrand, T. Hoffmann & H. Moser*	165
Features of recent scouring and silting of the river channel of the Jingjiang River downstream of the Three Gorges Project *Y.H. Zhu, X.H. Guo, G. Qu, F. Tang & L.H. Gu*	166

D. Hydromorphology meets ecology

Effects of sediment bypass tunnels on sediment grain size distribution and benthic habitats *C. Auel, S. Kobayashi, T. Sumi & Y. Takemon*	169
River restoration in sand-dominated lowland streams – a comparison of morphodynamic impacts and response *V. Berger, A. Niemann & C.K. Feld*	170
Application of the hydromorphological assessment framework Valmorph to evaluate the changes in suspended sediment distribution in the Ems estuary *C. Borgsmüller, I. Quick & Y. Baulig*	171
Mechanics of biofilm-coated sediment transport *H.W. Fang, H.M. Zhao, W. Cheng, M. Fazeli, Y.S. Chen, Q.Q. Shang, G.J. He & L. Huang*	172
Heavy metal concentrations and enrichment of sediment cores: Correlation between geochemistry and geoaccumulation index *F. Fernandes & C. Poleto*	173
Microbial biostabilization and flocculation – what can we learn for sediment transport modelling? *S.U. Gerbersdorf, H. Schmidt, M. Thom & S. Wieprecht*	174
Compensatory measures at a Heavily Modified Waterbody (HMWB) improve the hydromorphological quality, a practical example from the Moselle *D. Gintz & Y. Baulig*	175
The analysis of sediment diameter with biofilm *G.J. He, H.W. Fang, Q.Q. Shang, F. Mahede & L. Huang*	176
Coarse sand as a specific problem for aquatic ecosystems in granite-dominated landscapes *S. Höfler, C. Gumpinger & C. Hauer*	177
Current status, sources and effects of fine sediments in Upper Austrian streams *S. Höfler, C. Scheder, C. Gumpinger, B. Piberhofer & C. Hauer*	178
Reconnection of the Danube floodplain channels as a vital step to restore river morphology and fluvial dynamics *K. Holubová, M. Čomaj & K. Mravcová*	179
Modeling of nutrient dynamics and vegetation succession, and comparison between with and without riverbed geomorphological simulation *H. Itoh, S. Yamauchi, M. Sugano, K. Sanjaya & T. Asaeda*	180
Flow patterns, turbidity and sediment size distribution on the Luneplate tidal polder, Lower Weser *E. Kemayou Tchamako, B. Koppe & U. von Bargen*	181
Explicitly salinity and sediment concentration on flocculation processes in estuaries *A. Mhashhash, B. Bockelmann-Evans & S. Pan*	182
River restoration: The need for a better monitoring agenda *M. Nones*	183
Reconciling the debate on the impact of vegetation density on river channel braiding *I. Pattison & R. Roucou*	184
Analysing sediment characteristics of the alpine River Brixentaler Ache (Austria) including *in-situ* measurements of dissolved oxygen *L. Seitz, M. Noack, S. Haun, R. Reindl, G. Senn & M. Schletterer*	185
Correlation between the shelter of juvenile salmonids and bed substrate *M. Szabo-Meszaros, N. Rüther & K. Alfredsen*	186

Analysis of tidal effects on heavy metal transport in coastal aquifers   187
A. Tao, S.G. Liu, S. Lou, C.M. Dai, B. Tan, R.S. Chalov & S.R. Chalov

The role of surface adhesion in biostabilization processes   188
M. Thom, H. Schmidt, S. Wieprecht & S.U. Gerbersdorf

Characterizing natural riparian plant stands for modeling of flow and suspended sediment transport   189
K. Västilä & J. Järvelä

The effect to the river environmental preservation of artificial flood in Satsunai River   190
Y. Watanabe, K. Sumitomo, S. Yamaguchi & H. Yokohama

Study on influence of waterway regulation engineering to fish habitat   191
Y.F. Geng & Z.L. Wang

Impact of biofilm on the sediment properties and its environmental effects   192
H.M. Zhao, W.H. Cao, L.Q. Tang, C.H. Wang, Y.H. Wang, D.B. Liu, C.S. Guo, J. Lu & Y.F. Zhang

E. Reservoir sustainability

Impact of a single dam on sediment transport continuity in large lowland rivers   195
Z. Babiński & M. Habel

Experiences of controlled sediment flushing from four alpine reservoirs   196
M.L. Brignoli, P. Espa, S. Quadroni, G. Crosa, G. Gentili & R.J. Batalla

On the vertical turbulent interaction of non-Newtonian fluid mud   197
O. Chmiel, A. Malcherek & M. Naulin

Improving the RESCON approach   198
N. Efthymiou, S. Palt, P. Pintz, P.K. Thapa, G.W. Annandale & P. Karki

Designing reservoir sediment management alternatives with automated concentration constraints in a 1D sediment model   199
S. Gibson & P. Boyd

Reservoir sedimentation issues in India as a part of Dam Rehabilitation and Improvement Project (DRIP): Field reconnaissance and modelling   200
S. Giri, M. Nabi, P. Cleyet-Merle & B.R.K. Pillai

Sedimentation in rivers and reservoirs following the eruptions of Kelut Volcano, Indonesia   201
F. Hidayat, P.T. Juwono, A. Suharyanto, A. Pujiraharjo, D. Sisinggih & D. Legono

Modelling deposition, consolidation and erosion of cohesive sediments in the Upper Rhine   202
T. Hoffmann, G. Hillebrand & M. Noack

The aging of Japan's dams: Innovative technologies for improving dams water and sediment management   203
S.A. Kantoush & T. Sumi

Improvement of a bedload transport rate measuring systems in sediment bypass tunnels   204
T. Koshiba, C. Auel, D. Tsutsumi, S.A. Kantoush & T. Sumi

Development of a management strategy based on in-situ observation for Agondian Reservoir   205
C.C. Li, Y.J. Tsai, T.H. Wu & H.C. Tai

Density driven underflows with suspended solids in Lake Constance   206
S. Mirbach & U. Lang

Development of oblique flow in barrages due to shoal formation   207
K. Mishra

Hydrodynamic instabilities in shallow reservoirs: Implications for sediment management   208
Y. Peltier, A. de Cuyper, S. Erpicum, P. Archambeau, M. Pirotton & B. Dewals

Long term simulation of reservoir sedimentation with turbid underflows   209
G. Petkovšek

Controlling sediment flushing to mitigate downstream environmental impacts   210
S. Quadroni, G. Crosa, S. Zaccara, P. Espa, M.L. Brignoli, G. Gentili & R.J. Batalla

Economic assessment of the effects of sediment replenishment to rivers and the effectiveness of sediment management 211
K. Tomita, T. Homma & T. Sumi

The monitoring of empty flushing operation at Agondian Reservoir, Kaohsiung, Taiwan 212
Y.J. Tsai & C.C. Li

Study on sediment desilting operation mode and structure layout of Pakistan Karot hydropower project 213
J. Zhao, X.N. Liu, B. Fan, G. Wei, M. Wang & Z. Jin

F. Social, economic and political aspects of sediment management

Sediment management at Sukkur Barrage – How competing needs and uses of the structure impact the design 217
S. Aziz, M. Roca-Collell & I. Heijne

Overlooked costs of dams: Barrier to sustainability 218
M. George & R. Hotchkiss

River sand and gravel mining in Iran 219
S. Norouzi, J. Habibi, E. Zrdchoghai & A. Zardchoghai

SS 1 Hydropower and sediment management

Challenges facing Atbara Dam Complex (ADC) operation management 223
A.A. Ahmed

Field calibration of bedload monitoring system in a sediment bypass tunnel: Swiss plate geophone 224
I. Albayrak, M. Hagmann, C.R. Wyss & R.M. Boes

Measuring sediment fluxes in periglacial reservoirs using water samples, LISST and ADCP 225
D. Ehrbar, L. Schmocker, D.F. Vetsch, R.M. Boes & M. Döring

Sensitivity analysis of measured sediment fluxes in a reservoir 226
S. Haun & L. Lizano

HPP Vrhovo operation under reservoir sediment management 227
L. Javornik, A. Kryžanowski & M. Mikoš

Sediment management for sustainable hydropower development 228
M. Omelan, J. Visscher, N. Rüther & S. Stokseth

Flow field and sediment flux measurements at alpine desanding facilities 229
C. Paschmann, J.N. Fernandes, D.F. Vetsch & R.M. Boes

Ensuring sediment continuity through a reservoir: Challenges and methodology applied to define favorable hydraulic scenarios in the case study of the Champagneux run-of-river dam on the Rhône River, France 230
C. Peteuil, D. Alliau, T. Frétaud, M. Decachard, S. Roux, S. Reynaud, N. Boisson, A. Vollant & Y. Baux

Experimental analysis of the interaction between hydroelectric sluice gates and sediment transport 231
G.R. Pisaturo, M. Righetti, F. Amante & E. Bigliotti

3D fully coupled numerical modelling of local sediment flushing scour at dam bottom outlets for sustainable hydropower operation 232
O. Sawadogo & G.R. Basson

Integrative monitoring approaches for the sediment management in alpine reservoirs: Case study Gepatsch (HPP Kaunertal, Tyrol) 233
M. Schletterer, B. Hofer, R. Obendorfer, A. Hammer, M. Hubmann, R. Schwarzenberger, M. Boschi, S. Haun, M. Haimann, P. Holzapfel, H. Habersack, B. Brock, B. Schmalzer & C. Hauer

Sediment management of reservoirs: Sediment discharge in dependence on the suspended load concentration in the run-off water – Theoretical foundations and practical experiences 234
F. Sollerer & G. Gökler

SS 2 Navigation and river morphology

Analysis of sedimentation of the Yangtze Estuary channel, China — 237
X.P. Dou, Z.X. Jiao, X.Y. Gao, L. Ding & J. Jiao

Evolution characteristics of the north branch of the Yangtze Estuary — 238
X.Y. Gao, X.P. Dou, L. Ding, Z.R. Gao & J. Jiao

Design of bank protection for inland waterways with GBBSoft+ — 239
C. Gesing, B. Söhngen & K. Kauppert

Tension between bridge and waterway in the middle of Yangtze River with its countermeasures — 240
D. Li & L. Chen

Adaptability of numerical model for siltation in the Yangtze Estuary channel — 241
T.L. Li, L.M. Chen, X.Z. Zhang, W.Y. Zhang & X.Y. Gao

The potential of alternative technical-biological bank protection measures on federal waterways – an applied research approach — 242
K. Schmitt & L. Symmank

Study on sediment transport of silt coast by wave and tidal current — 243
J. Mu & C. Yin

Back siltation in Bach Dang navigation channel, Nam Trieu Estuary, Vietnam — 244
V.T. Nguyen, M.D. Do & M.T. Vu

Scour geometry and flow velocities induced by an experimental ship propeller jet — 245
F. Núñez-González, K. Koll, B. Söhngen & D. Spitzer

German guidelines for designing alternative bank protection measures — 246
B. Söhngen, P. Fleischer & H. Liebenstein

Sediment budget of the Rhine River as basis for optimizing navigation along the Mittelrhein waterway — 247
S. Vollmer, G. Hillebrand, T. Hoffmann & S. Schriever

River Rhine between Mainz and Bingen – Morphodynamic analysis of a navigational bottleneck — 248
S. Wurms

Turbulence based approach for the transported particle size concerning ship induced propulsion flux — 249
R. Zimmermann, J. Stamm, T. Beck & B. Söhngen

SS 3 Innovative measurement techniques

Bathymetry of Zipingpu Reservoir by earthquake and flood induced turbidity currents — 253
A. Ruidong, L. Jia & Y. Zhongluan

Feasibility tests to airborne gravelometry for prealpine rivers — 254
M. Detert, L. Kadinski & V. Weitbrecht

Combining *in-situ* laser diffraction (LISST) and vibrating tube densimetry to measure low and high suspended sediment concentrations — 255
D. Felix, I. Albayrak & R.M. Boes

UAV based determination of grain size distribution at River Jachen, Germany — 256
C. Haas, P. Thumser & L. Seitz

Experimental study on development and migration of sand waves in a flume — 257
C. Liu, W.H. Cao, L. Xu, J. Lu & L. Liu

Comprehensive measurement techniques of water flow, bedload and suspended sediment in large river using Acoustic Doppler Current Profiler — 258
S. Okada, A. Yorozuya, H. Koseki, S. Kudo & K. Muraoka

Estimation of sediment deposition in Koyna Reservoir by integrated bathymetric survey — 259
R.A. Patil & R.V. Shetkar

Densitometric probe based on non-differential pressure: A monitoring technique for high suspended sediment concentrations — 260
D. Petrovic, A. Marescaux, J.-P. Vanderborght & M.A. Verbanck

Suspended sediment measurements with multi-frequency backscatter acoustics *J. Skripalle, T. Hies & H.H. Nguyen*	261
Continuous grid monitoring to optimize sedimentation management *T. Van Hoestenberghe, R. Vanthillo, M. De Paepe, N. Dezillie & N. Van Ransbeeck*	262

SS 4 SEDITRANS – Sediment transport in fluvial, estuarine and coastal environment

Failure by overtopping of earth dams: Novel methods to determine the breach effluent hydrograph *S. Amaral, T. Viseu, J.E. Santos, A. Lopes, A.M. Bento, L. Caldeira, R. Cardoso & R.M.L. Ferreira*	265
A particle counter prototype and video imaging techniques for calculation of bedload fluxes *F. Antico, P. Sanches, L. Mendes, R. Aleixo & R.M.L. Ferreira*	266
Coupling of large eddy simulations with the level-set method for flow with moving boundaries *F. Kyrousi, A. Leonardi, F. Zanello & V. Armenio*	267
River morphodynamics under the effect of flow variability *B. Oliveira & R. Maia*	268
Numerical investigation on the effect of suspended sediment load on flow field around a cylinder *T. Paone, R.M.L. Ferreira, A.H. Cardoso & V. Armenio*	269
Impact of placer mining on suspended sediments in rivers of the Kamchatka Peninsula (Russian Federation) and the Selenga River basin (Mongolia) and its modeling *E. Promakhova & N.I. Alexeevsky*	270

SS 5 Sustainable land management

Research-praxis integration in South China – the rocky road to implement strategies for sustainable rubber cultivation in the Mekong Region *T. Aenis, J. Wang, S. Hofmann-Souki, T. Lixia, G. Langenberger, G. Cadisch, K. Martin, M. Cotter, M. Krauss & H. Waibel*	273
Managing rubber plantations towards improved water protection *G. Langenberger, H. Liu, S. Blagodatskiy, G. Cadisch, M. Krauss, J. Wang, T. Aenis, S. Min & H. Waibel*	274
Dynamics of soil erosion in rubber plantations and its mitigation by herbicide management *H. Liu, S. Blagodatskiy & G. Cadisch*	275
Mitigation of forest to rubber change impact on soil erosion and stream quality by integrated land management *H. Liu, X. Yang, S. Blagodatskiy, C. Marohn & G. Cadisch*	276
Reduction of fine sediment infiltration into rivers by implementing riparian buffer strips in an agricultural dominated area in Southwest China *L. Seitz, S. Wieprecht, M. Krauss, N. Azizi & H. Steinmetz*	277
Author index	279

Preface

Worldwide the majority of water bodies (rivers, reservoirs and lakes) are no longer in a natural morphodynamic status due to anthropogenic interventions in the catchment area as well as in the river itself. Alterations in the catchment, e.g. uprooting, may lead to high sediment intrusions in river systems even during relatively short precipitation events. The effect of high suspended load concentration and the intrusion of fine sediments (clogging) result in a reduction of habitat quality for many aquatic species. Furthermore, the low water quality can negatively affect both macroinvertebrates as well as microbial organisms on the sediment bed due to increased turbidity as well as sediment smothering that alter the light climate and oxygen penetration, respectively. Thus, in the long term, fine material intrusion also provokes a reduction of biodiversity. Consequently, rivers lose their capability to provide several ecosystem functions and services. Additionally, river modifications due to hydraulic engineering disturb the sediment continuity and natural sediment transport dynamics. The construction of dams and reservoirs, e.g. for drinking water purposes, flood mitigation or energy production results worldwide in a reduction of sediment availability in rivers of 20×10^9 ty^{-1}. This discontinuity reduces the storage potential of the reservoirs and thus causes economic losses but also impairs the whole aquatic ecosystem sustainably. As a result, ongoing river bed erosion downstream of dams and thus so called 'hungry rivers' can be observed which coincide mostly with a reduction of the groundwater level and with local bank erosion and bridge foundation scour. From an ecological perspective, the uncontrolled and diffusive inflow of nutrients and pollutants into rivers also changes the environmental conditions and severely impacts the natural system. For instance, an additional input of nutrients and phosphate might stimulate the growth of cyanobacteria and increase the secretion of harmful toxins. As a consequence affected rivers and lakes can no longer provide drinking or irrigation water or serve for recreation purposes.

In the context of the growing number of hydraulic mega projects and the associated effects on the sediment transport and budget of concerned river systems, in 1980 the first time the International Symposium on River Sedimentation (ISRS) was organized and established as a triennial event. Under the auspices of UNESCO-IRTCES (International Research and Training Centre on Erosion and Sedimentation), the ISRS symposia have been successfully held in Beijing, China (1980), Nanjing, China (1983), Jackson, USA (1986), Beijing, China (1989), Karlsruhe, Germany (1992), New Delhi, India (1995), Hong Kong, China (1998), Cairo, Egypt (2001), Yichang, China (2004), Moscow, Russia (2007), Stellenbosch, South-Africa (2010) and Kyoto, Japan (2013). Since the foundation of the "World Association for Sedimentation and Erosion Research" (WASER) during the ninth symposium in 2004 the ISRS serves as official symposia series of WASER.

The symposium aims at providing a platform for scientists, engineers and opinion leaders for an in-depth and stimulating exchange of information. The objectives are to develop sustainable revitalization and management strategies that address the ongoing negative effects of anthropogenic activities whilst improving river systems towards a healthy ecological status. The fundamental research and understanding of interactive processes between water and sediments is as important as the sharing and exchange of knowledge in applied projects.

Topics of nature-orientated use of water bodies with a focus on the connection between ecological needs and the structural and hydraulic requirements are a main theme of the symposium. Besides the spatial scale (from the initiation of motion of single grains to processes occurring at the catchment scale), the temporal scale plays an important role. While typical morphological structures are formed over years, hydrodynamic considerations may concentrate on processes occurring at short timescales (seconds). For example, eddy structures can be defined in very high-resolutions, which is one pre-condition for a detailed description of the interaction between water and sediments.

During the ISRS2016 six main topics are addressed:

A. Integrated sediment management at the river basin scale
B. Sediment transport
C. River morphodynamics
D. Hydromorphology meets Ecology
E. Reservoir sustainability
F. Social, economic and political aspects of sediment management

Furthermore, special topics are discussed which are in the focus of research in Germany as well as worldwide:

SS 1 Hydropower and sediment management
SS 2 Navigation and river morphology
SS 3 Innovative measurement techniques
SS 4 SEDITRANS – sediment transport in fluvial, estuarine and coastal environment
SS 5 Sustainable land management

The Local Organizing Committee of ISRS2016 received 303 abstracts from 51 countries and regions, of which 269 full papers were requested after the first round of peer review by 70 members of the Review Board. The full papers submissions were subsequently subjected to a second peer review by 67 experts from around the world. Based on the comments of the reviewers, the Board finally selected 185 papers to be included in this book. The accepted contributions cover the broad spectrum of river sediment related issues. All papers included in this book are presented by the corresponding authors during the four day symposium either in oral or poster presentations.

We are proud to also introduce three keynote lectures representing the wide spectrum of sediment transport. Worldwide leading experts in their field report about their actual scientific results. Bruce W. Melville from New Zealand talks about "Local scour at hydraulic structures", David M. Paterson from Great Britain gives an insight in "Form, function and physics" and Weiming Wu from United Stated of America shares his knowledge about "Advances and challenges in mixed cohesive/non-cohesive sediment transport research".

A special "Workshop on International Sediment Advancements" (WISA) is launched on the second day of the symposium. The workshop, under the headline of "Hydraulic, morphological and biological interactions in sediment management", is proposed by Prof. Giampaolo Di Silvio. He is the current president of WASER and full of enthusiasm and ideas to promote the topic of sediment transport and its related processes. He puts great effort in the international exchange and is a major contributor to the success of this Symposium. This inter-organizational workshop, under the auspices of UNESCO-IHP-ISI is devoted to disseminate beyond the limits of each membership the most significant progress attained by several scientific associations operating in the field of sediment research and management. We are pleased that Manfred Spreafico (ISI-UNESCO) talks about "Material fluxes in river eco-systems with special attention to reservoirs", Zhaoyin Wang (IAHR) discusses "Eco-sedimentology. A new area in sediment studies", Desmond Walling (IAHS) refers to "Changing perspectives on the suspended sediment load of rivers", Ildefonso Pla Sentis (CONSOWA) speaks about "Hydrological processes in soils of sloping lands as a basis for sediment production and sediment yield", Silke Wieprecht (LOC) shares ideas about "Influence of morphological changes on ecology: a cascade of scales" and finally Giampaolo Di Silvio (WASER) addresses the subject of "A hydro-, morpho-, bio-dynamic model for long-term, basin-scale river simulations".

We would like to express our sincere gratitude to our colleagues who submitted abstracts and papers to ISRS2016, the paper reviewers who made in-depth reviews amidst their busy schedule, the advisory, scientific and local organizing committee members who significantly promoted this event and the secretariat members who were kept busy with the preparation and arrangements of all kinds of detailed matters and always had an open ear for all kinds of requests and desires. We express our gratitude to our collaborating publisher, CRC Press/Balkema (Taylor and Francis Group) who made our proceedings more visible and professional.

Editors
Stuttgart, September 2016

Committees

LOCAL ORGANIZING COMMITTEE

- Chairperson: Silke Wieprecht, University of Stuttgart
- Andreas Dittrich, Technische Universität Braunschweig
- Stefan Haun, University of Stuttgart
- Andreas Malcherek, Universität der Bundeswehr München
- Markus Noack, University of Stuttgart
- Holger Schüttrumpf, RWTH Aachen University
- Jürgen Stamm, TU Dresden
- Karolin Weber, University of Stuttgart

INTERNATIONAL ADVISORY COMMITTEE

- Robert M. Boes, Switzerland
- Subhasish Dey, India
- Giampaolo Di Silvio, Italy
- Rollin H. Hotchkiss, USA
- Pierre Y. Julien, USA
- Bruce W. Melville, New Zealand
- Anil Mishra, France
- Mathias J.M. Römkens, USA
- Manfred Spreafico, Switzerland
- Hans D. Thulstrup, China
- Leo C. van Rijn, The Netherlands
- Desmond E. Walling, UK
- Zhaoyin Wang, China
- Sam S.Y. Wang, USA
- Ulrich C.E. Zanke, Germany

INTERNATIONAL SCIENTIFIC COMMITTEE

- Jorge D. Abad, USA
- Jochen E. Aberle, Norway
- Mustafa S. Altinakar, USA
- Aronne Armanini, Italy
- Markus Aufleger, Austria
- James E. Ball, Australia
- Kazimierz Banasik, Poland
- Alistair G.L. Borthwick, Ireland
- Benoît X. Camenen, France
- Roger A. Falconer, UK
- Norbert Fenzl, Brasil
- Rui M.L. Ferreira, Portugal
- Ana Maria Ferreira da Silva, Canada
- Naziano P. Filizola, Brasil
- Heide Friedrich, New Zealand
- Shoji Fukuoka, Japan
- Marcelo H. García, USA
- Valentin Golosov, Russian Federation
- Helmut Habersack, Austria
- Willi H. Hager, Switzerland
- Heather Haynes, UK
- Christopher S. James, South Africa
- Sameh A. Kantoush, Japan
- Pravin Karki, USA
- Rebekka Kopmann, Germany
- Stuart N. Lane, Switzerland
- Cheng Liu, China
- Gil M. Mahé, France
- Juan P. Martín-Vide, Spain
- Bijoy S. Mazumder, India
- Hajime Nakagawa, Japan
- Ali A.S. Neyshabouri, Islamic Republic of Iran
- Nils R.B. Olsen, Norway
- André Paquier, France
- Pawel M. Rowinski, Poland
- Nils Rüther, Norway
- Nicole Saenger, Germany
- Anton J. Schleiss, Switzerland
- Andreas Schmidt, Germany
- Thorsten Stoesser, UK
- Stefan Vollmer, Germany
- Roman Weichert, Germany
- Volker Weitbrecht, Switzerland
- Geraldo Wilson Júnior, Brasil
- Farhad Yazdandoost, Islamic Republic of Iran

Gold Sponsor

Silver Sponsor

Bronze Sponsors

Sponsors

Institutional Support

Keynote papers

Form, function and physics: The ecology of biogenic stabilization

D.M. Paterson & J.M. Kenworthy
Scottish Ocean Institute, School of Biology, University of St Andrews, Scotland

J.A. Hope
Institute for Modelling Hydraulic and Environmental Systems, University of Stuttgart, Germany

ABSTRACT: The effect of biological cohesion on the behaviour of sediments is gaining increasing notice. This is partly supported by ecological theory in terms of the role of organisms as "ecosystem engineers" and the associated discussion of "niche construction", suggesting an evolutionary role for habitat modification by biological action. In addition there is a strong societal and policy drive toward the "ecosystem approach" supporting an integrated examination of the functional roles of biota in selected habitats. In this context the increasing recognition of the importance of biological activity in the mediation the erosion, transport, deposition and consolidation (ETDC) cycle of sediments is important and advances in technology will improve our ability to examine these effect under their natural settings. This will shortly be combined with vastly enhanced molecular tools that will allow the discrimination of microbial biodiversity and examination of their metabolism contribution to ecosystem function. This may lead to a step-change in our ability to research the influence of microbiota on natural sediment dynamics and opens an exciting era for new interdisciplinary research.

1 INTRODUCTION

The earliest visible remnants of organismal life on Earth come from the fossil record of bacterial remains and laminated deposits preserved in early rock formations (cf 3.5 billon years BP, Altermann et al., 2003). The laminated fossils, known as stromatolites, are clear early evidence of life but also of the impact of biology on sediment dynamics (Paterson et al., 2008). Paleo-ecologists often debate what the form of ancient microbial remains, biofilms and mats tell us about the ambient conditions in the ancient environments when the material was first preserved (Noffke & Paterson, 2008). There is evidence that these early bacterial assemblages were capable of trapping and binding sediments, helping to create one of the first recognizable ecosystems on the planet. The stabilization of sediment at the interface between the bed and the water column allows biogeochemical gradients to become established, initiating the difference in local conditions (niche separation). This helps to promote specialization among bacteria that, in combination with ability "fix" those changes (improved fitness), drives the evolutionary process. The ability of organisms to affect their environment is well-known and often described as "ecosystem engineering" (Jones et al., 1994). However, the examples of ecosystem engineers that are given are often larger charismatic species (elephants, beavers, otters etc.) while organisms such as bacteria, protists and algae, which are equally if not often more important, are less often described (Boogert et al., 2006, Gerbersdorf et al., 2009). There is a further interesting twist to the "ecosystem engineering" debate. Some evolutionary theorists consider that the alteration of the environment by organisms should in itself be recognized as an evolutionary pressure. The development of this theory, termed "niche construction" (Laland et al., 2009), is somewhat controversial (Wright & Jones, 2006) but seems a logical extension from the acceptance of the widespread nature of the "ecosystem engineering" process. However well-developed the theories of "ecosystem engineering" and "niche construction", the practical implications are of more concern to environmental engineers who are engaged with the fundamental question:

> "To what extent must the impact of biological processes be taken into account when predicting the erosion, transport, deposition and consolidation cycle (ETDC) of natural sediments"?

Efforts to answer this question have been accelerating in recent years since the early work of Manzenreider (1983), showing that natural sediments, complete with their compliment of living and dead organic material, did not conform to the predictions of the Shields curve (Black et al., 2002). There was not much immediate follow up to this research, but as modeling became more sophisticated and questions concerning the variation between the predicted sediment behaviour and actuality emerged more frequently, the importance of understanding more about the moderation of the ETDC cycle by biota became clear. This was additionally driven by the need to understand the distribution of material or chemicals often associated with the fine particulate fraction of the bed matrix. This includes nutrients, xenobiotic compounds, heavy metals, and even pathogens such as *E. coli* (Gerbersdorf et al., 2007; Pachepsky, and Shelton, 2011) which can be released several years after the pollutants were trapped (Gerbersdorf et al., 2007). Any changes to the management of an area, be it dredging, reservoir management or natural flood events can mobilise these deeper contaminated sediments. Therefore understanding the natural dynamics, location and re-location of sediments is of increasing environmental importance, especially for fluvial and coastal systems (Gerbersdorf et al., 2015).

2 THE ECOSYSTEM APPROACH

Examination of natural sediment systems can be characterised under three broad approaches: laboratory experimentation, field experimentation and field observation. All have required a change in ethos and advances in technology. The change in ethos is important. Across many spheres of environmental management the concept of the "ecosystem approach" has been gaining ground. The "ecosystem approach" has many definitions (Apitz et al., 2006) but fundamentally requires a more holistic understanding of the interactions that drive the ecological dynamics of a given habitat. This means including, and understanding, the interactions between the biology, physics, biogeochemistry and ecology of the system as these are part of the combination of factors and processes that delivers the goods and services (Beaumont et al., 2007) valued by humankind. An appreciation of the complexity of the natural environment and the real difficulty of recreating natural conditions in the laboratory suggests a requirement, in keeping with the ecosystem approach, to make more measurements/observations of sediment erosion and dynamics in the field. Of course, fieldwork cannot, and should not, replace all laboratory studies but there is a balance to be struck and at present there is still a dearth of high quality fieldwork.

2.1 *Laboratory observations*

Part of the original problem of determining the effect of natural populations on the dynamics of sediment behaviour was that any effect of biology was traditionally removed from test systems. Relationships such as the Shields curve were based on experiments conducted in laboratories with sediments that had been cleaned and sorted. A few researchers such as Nowell et al. (1981), Manzenrieder (1983) and Parchure & Mehta (1985) attempted to examine the influence of biology on the erosional characteristics of natural sediments but their work was peripheral to the main quest to develop a theoretical understanding of sediment dynamics. One of the problems of this effort was the lack of suitable methodologies to apply to natural systems and unwillingness to bring the field into the laboratory. This has changed in the last few years with some large scale experiments either taking natural sediments or creating model "sediment systems" designed to test biological effects (Malarkey et al., 2015, Parsons et al., 2016). One remarkable effort (Moller et al., 2014) was the dissection, transport and re-assembling of an area of salt marsh in what the authors claim is the largest open access flume system in the world (310 m long by 5 m wide and 7 m in depth). The result demonstrated the importance of vegetation in reducing wave energy and the resilience of the marsh despite shearing of individual plant elements. These examples illustrate the recent move towards larger and more complex experiments more closely replicating natural conditions using the traditional laboratory flume approach. While this marks a welcome development, the cost and logistical effort behind this work is immense and so unlikely to be widely repeated.

2.2 *Field observations*

Development of *in situ* technology and data capture now allow the analysis of water motion, particle tracking, floc dynamics and bedform properties to be remotely assessed in real time. This combined approach has allowed detailed, high resolution analysis of the changing nature of natural beds but also makes possible the correlation of other factors (biological and biogeochemical) collected at the same time with the properties of the bed and water column. For example, the NERC funded COHesive BEDforms (COHBED) project (NE/1027223/1), combined multidisciplinary expertise and techniques to investigate the influence of physical, chemical and biological processes on the formation and migration of bedforms. High resolution ripple profiles, laser scans and ADV data were combined with biogeochemical samples from the bed, allowing the biological cohesion to be related to the morphological changes in bedforms (Lichtmann et al., in prep). Field campaigns were complimented with further simplified flume studies to allow the effect of biostabilisation to be determined for both cohesive (Parsons et al., 2016) and non-cohesive sediments (Malarkey et al., 2015). Increasingly, it is acknowledged that biological effects, often through the accumulation extracellular polymeric substances (EPS), influence not only the biostabilisation of sediment beds, but mediate the transport and deposition of fine material once it is resuspended. Imaging techniques (LabSFloc–2 system, Manning et al., 2010) allowed several characteristics of the sediment dynamics, for example floc settling velocity and effective density, to be monitored *in situ*. Floc characteristics were then related to their organic content, quantifying cytochemically-stained acidic polymeric substances (APS) and assessing temporal variation in both (Hope, 2016; Manning et al., in prep). The results of this study suggested that a particular fraction of the organic content (APS) was a significant predictor of the microfloc effective density (Figure 1) while particle loading (suspended particulate matter – SPM) was not. Variation in the floc effective density at low SPM concentrations (Figure 1) may be due to variation in the primary particle size but this was difficult to quantify without large volumes of sample. APS and other more specific organic quantification methods may provide a good predictor variable for floc behaviour whereas a gross organic content of the material may not correlate quick as good. This is attributed to the sticky nature of. The results undoubtedly have important implications for future models, since SPM concentration and turbulence are often considered to mediate flocculation and depositional behaviour, while biological properties are, more often than not, excluded.

2.3 Field experiments

While both laboratory and field observations are effective techniques for monitoring the environment, they cannot always be used to successfully predict the implications of environmental change. Sedimentary habitats are subject to multiple stressors of both anthropogenic and climatic origin, the intensities of which are likely to increase in the coming decades (Halpern et al., 2008). The effect of these changes on sediment dynamics is difficult to predict. Laboratory studies can and have been used effectively to analyze the impacts of stressors (e.g. Hicks et al., 2011), however they do this in isolation from natural habitats and natural communities. These experiments fail to take into consideration environmental and biological contexts; the roles these attributes play in mediating stressor impacts is not fully understood (Bishop & Kelaher, 2013). Natural environments tend to be heterogeneous whereas mesocosm or laboratory studies tend to be controlled homogeneous environments (Dyson et al., 2007). This heterogeneity can buffer against stressor effects (Godbold et al., 2011) or negate patterns that would otherwise be observed due to stressor effects under laboratory conditions (Bulling et al., 2008).

Stressors to ecological systems rarely occur singularly (Crain et al., 2008; Halpern et al., 2008) and may overlap in time and space. Furthermore, the impact of multiple stressors will not always be additive, as often the effects are synergistic or cumulative (Crain et al., 2008). As natural habitats are subjected to multiple stressors daily, the addition of a new perturbation will likely interact with present conditions unpredictably. For example, Christensen et al. (2006) conducted long-term observational studies in combination with mesocosm experiments in boreal lake systems. Interactions between changing climatic conditions and ocean acidification led to ecological surprises that were unpredictable based upon their additive impact. To eliminate this high degree of uncertainty, there is a need for well-designed field experimentation.

While the effects of multiple stressors on communities have received greater attention in recent years

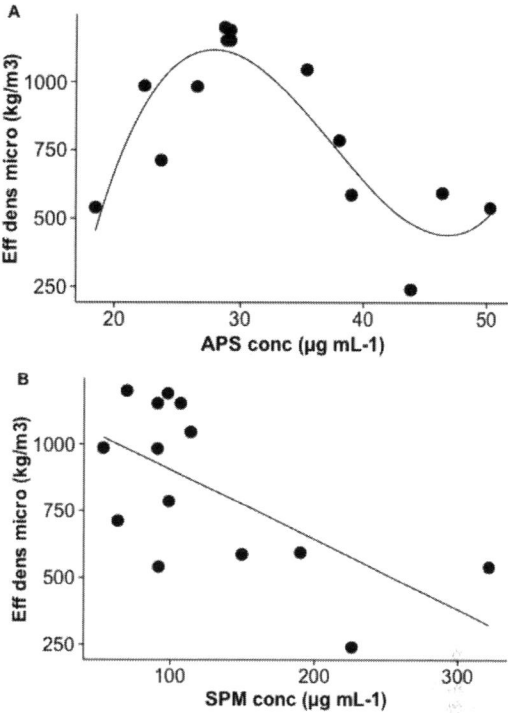

Figure 1. The Effective density of the microfloc population (Eff dens micro) as a function of A) Mean acidic polymeric substances (APS) concentration (pseudo-$r^2 = 0.79$, $P < 0.01$). B) SPM concentration and ($r^2_{adj} = 0.43$, $F_{1,12} = 10.83$, $P < 0.001$). After Hope, 2016.

(Alsterberg et al., 2014), assessing multiple stressor effects on various functions and processes the communities perform has received little attention. The complex biological effects, such as the colonization of surface sediments by microphytobenthos (MPB) and the formation of biofilms (Aspden et al., 2002; Consalvey et al., 2004) together with the grazing and bioturbating activity of macrofauna interact and impact sediment stabilisation. This necessitates that any evaluation of ecosystem service in response to multiple stressors can only meaningfully be done in the field. The imperative of addressing this question in a natural setting is compounded by the important role that physical factors play in determining sediment erosion potential, in some instances outweighing any relationship between sediment stability and the biota (Paterson et al., 2000; Defew et al., 2002). In situ techniques and devices, such as the use of a cohesive strength meter CSM (Vardy et al., 2007; Grabowski et al. 2010) allow the quantification of multiple stressors impacts on sediment stability by allowing quick and repeatable measures of erosion thresholds (Kenworthy, 2016). Through the use of high quality experimental designs, field experiments can therefore effectively be used to test the application of additional stress on a system (Figure 2). These field experiments are optimal for examining how a natural

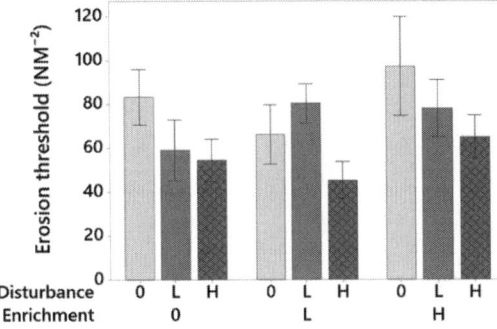

Figure 2. Effects of stressors on mean erosion thresholds (NM^{-2}: CSM, ± S.E) following 4 months of increased nutrient enrichment and physical disturbance in an Australian intertidal mudflat. Two factor ANOVA: disturbance – $F_{(2,48)} = 3.18$, $p = 0.047$; enrichment $P > 0.05$; interaction $P > 0.05$. Stress levels, $0 =$ Zero stress (natural sediment), $L =$ Low, $H =$ High. After Kenworthy, 2016.

system reacts to stressors, but often lack the precision to isolate the mechanisms by which particular stressor impacts occur (Daehler and Strong, 1996; Crane et al., 2007). For this reason, both field and laboratory approaches are required.

3 THE MICROBIAL WORK: BIOFILMS AND RESEARCH

As noted by Boogert et al. (2006), larger and charismatic organisms more often linked to the concept of ecosystem engineering and although the theory is being developed and supported (Byres et al., 2006) it is still unusual for bacterial processes to be considered. The stabilization of aquatic and fluvial sediment by plants is however recognised as an important ecosystem function but the role of microbial assemblages needs to be more strongly asserted. While individual bacterial cells are unlikely to have a significant measurable effect on system behaviour, once conditions are suitable, cells proliferate rapidly and where this growth is associated with a surface, a matrix comprising the cells and their extracellular exudates (extracellular polymeric material, (EPS)) will develop. Biofilm are a well-studied phenomenon (Characklis & Marshall, 1990) but there does seem to be a separation in the literature between the biomedical approach to the investigation of biofilms, which concentrate, on molecular interactions at the cell surface, infection and resistance and the study of natural microbial assemblage and the environmental ecology of biofilms. Understandably, the medical approach is often concerned with monocultures under relatively "clean" conditions while the complexity of natural biofilm is a much more developed (Stoodly et al 2002). The mechanism by which microbial cells influence their physical surroundings is now well-known, either through protein interactions (Cell adhesion molecules (CAMs) between the cell surface and/or an adjacent substratum (direct attachment) or through development of matrix (mainly mucopolysaccharides) creating a biofilm, permeating the sediment water interface. The role of these organic molecules in helping shape the dynamics of aquatic systems has recently been emphasised (Malarkey et al 2015, Parson et al., 2016).

The next phase of research is likely to rely on advances in molecular analysis, metagenomics (Thomas et al., 2012) and metabolomics (Nicholson & Lindon, 2006). Prokaryotes (both archaebacteria and eubacteria) have largely been treated as a "black box" since only about 2% of known bacteria can be cultured (Wade, 2002). Now metagenomic analysis of environmental DNA creates a database of operational taxonomic units (OTU, a molecular analog for species) that reveals the diversity of entire microbial assemblages in immense detail. In our own recent work, we have recorded nearly 200,000 bacterial OTUs from a laboratory incubation experiment using natural sediments (Hicks et al., submitted). Having to interpret this level of microbial biodiversity will become common in the next decade but in itself will be of less interest than the study of the processes (metabolomics) that the bacterial assemblages drive. Part of this research should be the analysis of polymer production and secretion into the environment, the medium that is probably the major factor in mediating the response of the sediment to environmental forcing. Knowledge of how bacterial metabolism changes in response to environmental challenge will become a driving force for environmental microbial ecology (Logue et al., 2015).

4 CONCLUSIONS

There is a societal move away from solving all coastal and fluvial problems by hard engineering towards a more natural and ecosystem based approach to coastal protection and flood mediation. In terms of the coast, sea defenses were being breached and new areas of saltmarsh or mudflat being encouraged while for rivers there is a recognition that channelised systems can exacerbate episodic extreme events and several projects have been initiated to reintroduce more natural pattern of flow and upstream "storage" of waters to alleviate such extremes. This is essentially a form of "geo-engineering" that is enhanced by a healthy and resilient biotic component. To fully understand how aquatic habits will adjust to change and to promote conditions that support the provision of ecosystem services, we require a holistic approach including all aspects that control the distribution of sediment and the ETDC cycle. The role of microbial assemblages is now recognised as a significant factor deserving greater attention.

ACKNOWLEDGEMENTS

JAH was support by the NERC award, COHBED (NE/1027223/1) while DMP received funding from the MASTS pooling initiative (The Marine Alliance for Science and Technology for Scotland) and their support is gratefully acknowledged. MASTS is funded by the Scottish Funding Council (grant reference HR09011) and contributing institutions.

REFERENCES

Altermann, W. Józef, K. 2003. Archean microfossils: a reappraisal of early life on Earth. *Research in Microbiology, Amsterdam, the Netherlands: Elsevier for the Pasteur Institute,* 154(9): 611–617.

Altermann, W. 2008. Accretion, Trapping and Binding of Sediment in Archean Stromatolites—Morphological Expression of the Antiquity of Life. *Space Science Reviews,* 135: 1–4.

Apitz, S.E., Elliott, M., Fountain, M., & Galloway, T.S. 2006. European environmental management: Moving to an ecosystem approach. *Integrated environmental assessment management,* 2: 80–85. doi: 10.1002/ieam.5630020114.

Aspden, R.J., Vardy S, & Paterson D.M. 2004 Salt Marsh Microbial Ecology: Microbes, Benthic Mats and Sediment Movement. In: *Coastal and Estuarine Studies. The Ecogeomorphology of Tidal Marshes (Fagherazzi S, Marani M, Blum LK eds), American Geophysical Union,* 59: 115–136.

Beaumont, N.J., Austen, M.C., Atkins, J., Burdon, D., Degraer, S., Dentinho, T.P., Derous, S., Holm, P., Horton, T., van Ierland, E., Marboe, A.H., Starkey, D.J., Townsend, M. & Zarzycki, T. 2007. Identification, Definition and Quantification of Goods and Services provided by Marine Biodiversity: *Implications for the Ecosystem Approach. Marine Pollution Bulletin* 54 253–265.

Bishop, M.J., & Kelaher, B.P. 2013. Context-specific effects of the identity of detrital mixtures on invertebrate communities. *Ecology and Evolution.* 3, 3986–3999.

Black K.S., Tolhurst T.J., Hagerthey S.E., & Paterson D.M. 2002. Working with Natural Cohesive Sediments. *Journal of Hydraulic Engineering* 128(1): 1–7.

Boogert N.J., Paterson D.M., & Laland K.N. 2006. The implications of niche construction and ecosystem engineering for conservation biology. *Biosciences,* 57(7): 570–578.

Bulling, M.T., Solan, M., Dyson, K..E., Hernandez-Milian, G., Luque, P., Pierce, G.J., Raffaelli, D., Paterson, D.M., & White, P.C.L., 2008. Species effects on ecosystem processes are modified by faunal responses to habitat composition. *Oecologia* 158, 511–520.

Byers, J.E. , Cuddington, K., Jones, C.J., Talley, T.S., Hastings, A., Lambrinos, J.G., Crooks, J.A., & Wilson, W.G. 2006. Using ecosystem engineers to restore ecological systems. *Trends in Ecology and Evolution.* 21: 9. 493–500.

Characklis WG, & Marshall KC, (ed.). 1990. *Biofilms.* New York: John Wiley & Sons; 195–231.

Christensen, M.R., Graham, M.D., Vinebrooke, R.D., Findlay, D.L., Paterson, M.J., & Turner, M. A. 2006. Multiple anthropogenic stressors cause ecological surprises in boreal lakes. *Global Change Biology.* 12, 2316–2322.

Consalvey M, Paterson D.M., & Underwood G.J.C. (2004) The ups and downs of life in a benthic biofilm: Migration of benthic diatoms. *Diatom Research* 19(2): 181–202.

Crane, M., Burton, G.A., Allen, G., Culp, J.M., Greenberg, M.S., Munkittrick, K.R., Ribeiro, R., Salazar, M.H., & St-Jean, S.D. 2007. Review of Aquatic *In Situ* Approaches for Stressor and Effect Diagnosis. *Integrated environmental assessment management.* 3, 234.

Daehler, C.C., & Strong, D.R. 1996. Can you bottle nature? The roles of microcosms in ecological research. *Ecology* 77, 663–664. doi:10.2307/2265487

Defew, E.C., Tolhurst, T.J., & Paterson, D.M. 2002. Site-specific features influence sediment stability of intertidal flats. *Hydrology and Earth System Sciences.* 6, 971–982.

Dyson, K.E., Bulling, M.T., Solan, M., Hernandez-Milian, G., Raffaelli, D.G., White, P.C.L., & Paterson, D.M. 2007. Influence of macrofaunal assemblages and environmental heterogeneity on microphytobenthic production in experimental systems. *Proceedings of the Royal Society: Biological Sciences* 274, 2547–54. doi:10.1098/rspb.2007.0922.

Gerbersdorf, S.U., Jancke, T. & Westrich, B., 2007. Sediment Properties for Assessing the Erosion Risk of Contaminated Riverine Sites. An approach to evaluate sediment properties and their covariance patterns over depth in relation to erosion resistance. First investigations in natural sediments. *Journal of Soils & Sediments,* 7(1), pp. 25–35.

Gerbersdorf S.U., Bittner R, Lubarsky H, Manz W, & Paterson D.M. 2009. Microbial assemblages as ecosystem engineers of sediment stability. *Journal of Soils & Sediments* 9(6): 640–652 doi: 10.1007/s11368-009-0142-5.

Gerbersdorf, S.U., Cimatoribus, C., Class, H., Engesser, K.-H., Helbich, S., Hollert, H., Lange, C., Kranert, M., Metzger, J.W., Nowak, W., Seiler, T.-B., Steger, K., Steinmetz, H. and S. Wieprecht. 2015. Anthropogenic Trace Compounds (ATCs) in aquatic habitats – research needs on sources, fate, detection and toxicity to ensure timely elimination strategies and risk management. *Environmental International,* 79 85–1085.

Grabowski, R.C., Droppo, I.G., & Wharton, G. 2010. Estimation of critical shear stress from cohesive strength meter-derived erosion thresholds. *Limnology and Oceanography: Methods,* 8:678–685 doi: 10.4319/ lom.2010.8.0678.

Godbold, J.A., Bulling, M.T., & Solan, M. 2011. Habitat structure mediates biodiversity effects on ecosystem properties. *Proceedings of the Royal Society: Biological Sciences.* 278, 2510–8. doi:10.1098/rspb.2010.2414.

Halpern, B.S., Walbridge, S., Selkoe, K.A., Kappel, C. V, Micheli, F., D'Agrosa, C., Bruno, J.F., Casey, K.S., Ebert, C., Fox, H.E., Fujita, R., Heinemann, D., Lenihan, H.S., Madin, E.M.P., Perry, M.T., Selig, E.R., Spalding, M., Steneck, R., & Watson, R., 2008. A Global Map of Human Impact on Marine Ecosystems. *Science.* 319, 948–952. doi:10.1126/science.1149345.

Hicks, N., Bulling, M.T., Solan, M., Raffaelli, D., White, P.C.L., & Paterson, D.M. 2011. Impact of biodiversity-climate futures on primary production and metabolism in a model benthic estuarine system. *BMC Ecology,* 11(7). doi:10.1186/1472-6785-11-7.

Hope, J. A. (2016). The biological mediation of cohesive and non-cohesive sediment dynamics. PhD thesis, University of St Andrews.

Jones, C.G., Lawton, J.H., & Shachak, M. 1994. Organisms as ecosystem engineers. *Oikos.* 69: 373–386.

Wright, J.P. & Jones, C.G. 2006. The Concept of Organisms as Ecosystem Engineers Ten Years On: Progress, Limitations, and Challenges. *BioScience,* 56:3, 203–209.

Kenworthy, J.M. 2016. Comparative estuarine dynamics: Trophic linkages and ecosystem function. PhD thesis, University of St Andrews.

Logue, J. B., Findlay, S.E.G. & Comte, J. 2015. Editorial: Microbial Responses to Environmental Changes. *Frontiers in Microbiology,* 6. 1364. doi: 10.3389/ fmicb.2015.01364.

Laland, K.N., Odling-Smee, F.J. & Feldman, M.W. 1999. Evolutionary consequences of niche construction and their implications for ecology. *Proceedings of the National Academy of Sciences*. 96: 10242–10247.

Lubarsky, H.V., Hubas, C., Chocholek, M., Larson, F., Manz, W., Paterson, D.M. & Gerbersdorf, S.U. (2010). The stabilisation potential of individual and mixed assemblages of natural bacteria and microalgae. *PloS one*, 5(11), p.e13794.

Malarkey, J., Baas, J.H., Hope, J.A., Aspden, R.J., Parsons, D.R., Peakall, J., Paterson, D.M., Schindler, R.J., Ye, L., Lichtman, I.D., Bass, S.J., Davies, A.G., Manning, A.J. & Thorne, P.D. 2015. The pervasive role of biological cohesion in bedform development. *Nature Communications*, 6:6257 doi: 10.1038/ncomms7257.

Manning, A.J., Baugh, J.V., Spearman, J.R. & Whitehouse, R.J. 2010. Flocculation settling characteristics of mud:sand mixtures. *Ocean dynamics*, 60(2), pp. 237–253. doi:

Manzenrieder, H. 1983. Retardation of initial erosion under biological effects in sandy tidal flats, Leichtweiss, Inst. Tech. Univ., Braunschweig, 469–479.

Möller, I., Kudella, M., Rupprecht, F., Spencer, T., Paul, M., van Wesenbeeck, B.K., Wolters, G., Jensen, K., Bouma, T.J., Miranda-Lange, M. & Schimmels, S., 2014. Wave attenuation over coastal salt marshes under storm surge conditions. *Nature Geoscience*, 7(10), pp. 727–731. doi: 10.1038/ngeo2251.

Nicholson, J.K., Lindon, J.C. 2008. Systems biology: Metabonomics. *Nature* **455** (7216): 1054–6.

Noffke, N & Paterson, D.M. 2008. Microbial interactions with physical sediment dynamics, and their significance for the interpretation of Earth's biological history. *Geobiology* 6: 1–4 doi: 10.1111/j.1472-4669.2007.00132.x

Nowell, A.R.M., Jumars, P.A., & Eckman. J.E. 1981. Effects of biological activity on the entrainment of marine sediments. *Sedimentary dynamics of continental shelves*, C. A. Nitrrouer, ed., Elsevier, 133–154.

Pachepsky, Y.A., & Shelton, D.R. 2011. *Escherichia Coli* and Fecal Coliforms in Freshwater and Estuarine Sediments, *Critical Reviews in Environmental Science & Technology*, 41:12, 1067–1110.

Parchure, T.M. & Mehta, A.J. 1985. Erosion of soft cohesive sediment deposits. *Journal of Hydraulic Engineering*, 111, 1308–1326.

Parsons, D.R., Schindler, R.J., Hope, J.A., Malarkey, J., Baas, J.H., Peakall, J., Manning, A.J., Ye, L., Simmons, S., Paterson, D.M., Aspden, R.J., Bass, S.J., Davies, A. G., Lichtman, I.D., & Thorne, P.D. 2016. The role of biophysical cohesion on subaqueous bed form size. *Geophysical Research Letters*, vol Early View, pp. 1–8, doi: 10.1002/2016GL067667

Paterson, D.M., Tolhurst, T.J., Kelly, J.A., Honeywill, C., de Deckere, E.M.G.T., Huet, V., Shayler, S.A., Black, K.S., de Brouwer, J., & Davidson, I. 2000. Variations in sediment properties, Skeffling mudflat, Humber Estuary, UK. *Continental. Shelf Research* 20, 1373–1396.

Paterson, D.M., Aspden, R.J., Visscher, P.T., Consalvey, M, Andres, M.S., Decho, A.W., Stolz, J., & Reid, R.P. 2008. Light-Dependant Biostabilisation of Sediments by Stromatolite Assemblages. *PLoS ONE* 3(9): e3176. doi:10.1371/ journal.pone.000317.

Shields, A. 1936. Anwendung der Ähnlichkeitsmechanik und der Turbulenzforschung auf die Geschiebebewegung; In Mitteilungen der Preussischen Versuchsanstalt für Wasserbau und Schiffbau, Heft 26.

Stoodley, P, Sauer, K, Davies, D.G., Costerton, J.W. 2002. Biofilms as complex differentiated communities. *Annual Review of Microbiology*; 56:187–209. Epub 2002 Jan 30.

Tolhurst, T.J., Black K.S., Paterson D.M, Mitchener, H, Termaat, R, Shayler, S.A. 2000. Comparison of four *in situ* sediment erosion devices. *Continental Shelf Research* 20 (10–11): 1397–1418.

Tolhurst, T.J., Black, K.S., & Paterson, D.M. 2009. Muddy Sediment Erosion: Insights From Field Studies. *Journal of Hydraulic Engineering* 135(2): 73–87 doi.org/10.1061/(ASCE)0733-9429(2009)135:2(73).

Thomas, T., Gilbert, J., Meyer, F. 2011. Metagenomics – a guide from sampling to data analysis. *Microbial Informatics & Experimentation* 20122:3 doi: 10.1186/2042-5783-2-3.

Vardy, S, Saunders, J.E., Tolhurst, T.J., Davies, P, & Paterson, D.M. 2007. Calibration of the high-pressure cohesive strength meter (CSM). *Continental Shelf Research*, 27, 1190–1199.

Wade, W. 2002. Unculturable bacteria—the uncharacterized organisms that cause oral infections. *Journal of the Royal Society of Medicine*, 95(2), 81–83.

Wingender, J. & Flemming, H.C. 2011. Biofilms in drinking water and their role as reservoir for pathogens. *International Journal of Hygiene & Environmental Health*, 214(6), pp. 417–423. doi: 10.1016/j.ijheh.2011.05.009

Mixed cohesive and noncohesive sediment transport: A state of the art review

W. Wu
Department of Civil and Environmental Engineering, Clarkson University, Potsdam, NY, USA

ABSTRACT: Noncohesive sediments move as individual particles, whereas cohesive sediments ($< \sim 0.01$ mm in size) usually transport in flocs that consist of fine particles irregularly bonded by interparticle electrostatical forces and undergo continuous, dynamic aggregation and disaggregation. When cohesive and noncohesive sediments are mixed, interactions between them play an important role, and the sediment mixture experiences much more complex transport processes. Researchers and engineers have encountered significant challenges when dealing with such mixed sediments which widely exist in estuaries, coastal inlets, reservoirs, rivers, uplands, etc. Presented in this paper is a state-of-the-art review of recent advances in experiments, formulas and models of mixed cohesive and noncohesive sediment transport.

1 INTRODUCTION

Sediments on the Earth surface consist of clay, silt, sand, gravel, and cobble. Sediment particles with size less than about 0.01 mm, i.e. clay and fine silt, are generally considered to be cohesive, while coarse silt, sand, gravel and cobble are noncohesive. Noncohesive sediments transport in individual particles. However, because of the action of electrostatical forces that are comparable to or larger than the gravity forces, cohesive sediment particles may stick together and form flocs or aggregates when they collide. The flocs may be transported by convection (due to river flows, tides, winds, and waves), turbulent diffusion, and gravitational settling, and undergo continuous aggregation and disaggregation during transport. The settled cohesive deposits may consolidate, due to gravity and the overlying water pressure.

When cohesive and noncohesive sediments are mixed, interactions between them play an important role in the transport processes. It has been recognized that when the fraction of cohesive sediments is larger than about 10% (van Rijn 1993; Torfs 1995; van Ledden 2003), the sediment mixtures exhibit cohesive properties and undergo much more complicated transport processes than pure cohesive and noncohesive sediments. Historically, studies have been concentrated largely on the behaviors of either cohesive or non-cohesive sediments. Only a limited number of studies have concerned about mixed cohesive/noncohesive sediments. In this paper, knowledge, experiments, formulas and models of mixed cohesive and noncohesive sediment transport are reviewed.

2 EXPERIMENTS ON MIXED SEDIMENT TRANSPORT

A number of laboratory experiments and field measurements investigated the transport characteristics of cohesive and noncohesive sediment mixtures. Table 1 summarizes the used facilities, flow conditions, mixture compositions, and the investigated aspects in some of these experiments. Most of the experiments considered only unidirectional flows in open flumes, closed tunnels, or annular flumes. Three sets of experiments were conducted under waves or oscillatory flow using wave flume or oscillatory tray (van Rijn and Louisse 1987; Williamson and Ockenden 1993; Panagiotopoulos et al. 1997). Most of the experiments investigated the erosion threshold and erosion rate of mixed cohesive/noncohesive sediments, four sets of experiments studied the transport mechanism, four sets for bed-form development, and one set for flocculation/settling. These experiments contribute significantly the knowledge on the bulk properties, erosion, settling, transport and bed forms of mixed sediments.

3 POROSITY OF MIXED SEDIMENTS

Bulk properties, such as porosity, of mixed cohesive and noncohesive sediments depend on how the components are mixed or packed. Because the cohesive sediment particles are much finer than the non-cohesive sediment particles, the filling of fines in the pores of coarse particles exists. When the fraction of fines

Table 1. Past experiments on transport of mixed cohesive and noncohesive sediments.

Investigators	Experiment facility	Flow condition	Clay or mud, sand content in mixture	Studied items
Murray (1976)	Open flume: 1.52 m long, 0.11 m wide 0.11 m deep	Unidirectional flow	Clay: 0, 10, 18%	Transport rate
Grissinger et al. (1981)	Potable flume	Unidirectional flow	Sand: 45–84%	Erosion threshold, rate
Kamphuis and Hall (1983)	Flume-tunnel: 9 m long, 0.15 m wide	Unidirectional flow	Sand: 1–50%	Erosion threshold
Van Rijn and Louisse (1987)	Wave flume: 17 m long, 0.3 m wide, 0.5 m deep	Regular waves	Clay: 0, 25, 75, 100%	Erosion threshold, rate, bed forms
Collins (1989)	Open flume	Unidirectional flow	Mud: <10%	Erosion rate, transport rate, bed forms
Nalluri and Alvarez (1992)	Flume with circular cross-section	Unidirectional flow	Clay: <30%	Erosion threshold
Williamson and Ockenden (1993)	Annular flume: 6.0 m in outer diameter, 0.4 m wide, 0.3 m deep	Unidirectional flow	Sand: 0, 11, 20, 50, 66%	Erosion threshold, rate
Williamson and Ockenden (1993)	Wave flume: 23 m long and 0.3 m wide	Random waves	Sand: 0, 20, 40%	Erosion threshold, rate
Torfs (1995)	Open flume: 9 m long, 0.4 m wide, 0.4 m deep	Unidirectional flow	Clay: 7–13%	Erosion threshold, rate, transport rate
Panagiotopoulos et al. (1997)	Oscillating tray	Unidirectional, Oscillatory flows	Mud: 0–50%	Erosion threshold
Gailani et al. (2001)	Water tunnel: 2 m long, 0.1 m wide, 0.02 m high	Unidirectional flow	Clay: 0–16%	Erosion rate
Barry (2003)	Open flume: 4.3 m long, 0.15 m wide, 0.2 m deep	Unidirectional flow	Clay: <14%	Erosion threshold, rate
Aberle et al. (2004)	In-situ flume: ~1 m long, 0.1 m high, 0.2 m wide	Unidirectional flow	Sand: 2–54%	Erosion rate
Banasiak and Verhoeven (2008)	Semicircular pipe with diameter 0.39 m, length 11 m	Unidirectional flow	Clay: 3–10%	Transport rate, bed forms
Jain and Kothyari (2008, 2009, 2010)	Open flume: 16 m long, 0.75 m wide, 0.5 m deep	Unidirectional flow	Clay: 10–50%	Erosion threshold, transport rate
Manning et al. (2010)	Mini-annular flume: 1.2 m in diameter, 0.1 m wide and 0.15 me deep	Unidirectional flow	Mud: 25, 50 and 75%	Flocculation and settling
Ye et al. (2011)	Closed circular channel: 1.8 m long, 0.2 m wide, 0.2 m deep	Unidirectional flow	Mud: 0–100%	Erosion threshold
Baas et al. (2013)	Open flume: 10 m long, 0.3 m wide	Unidirectional flow	Mud: <18%	Bed forms
Wang (2013)	Flume: 6.1 m long, 0.38 m wide	Unidirectional flow	Silt-kaolin mixtures: 0–100% fines	Erosion threshold, rate
Smith et al. (2015)	Sedflume: 10 cm wide, 2 cm deep conduit	Unidirectional flow	Mud: 0–100%	Erosion threshold, rate

is small, the cohesive sediments are distributed in the voids of or between noncohesive sediments. This case is called coarse packing. When the fraction of fines is large, the coarse noncohesive particles are suspended or dispersed by fine particles. This case is called fine packing. Figure 1(a) and (b) show the ideal coarse packing and ideal fine packing modes. In reality, the packing can be random, so that no packing, fine packing and coarse packing can all exist in a mixture.

Two ideal packing or filling models were considered in some early studies to estimate the porosity of sediment mixtures (e.g., Marion et al. 1992). Later studies showed that these two ideal packing modes significantly underestimate the mixture porosity (Koltermann and Gorelick 1995). Koltermann and Gorelick (1995) and Kamann et al. (2007) proposed fractional packing models for porosity in bimodal sediment mixtures, which consider both fine and coarse packing modes existing in reality and thus fit experiment data better. The weakness in these fractional packing models is that they use a minimum porosity that need to be measured.

Han et al. (1981) investigated the filling phenomenon between fine and coarse particles and proposed a random filling model for the overall dry density of the poorly-sorted sediment mixtures. One

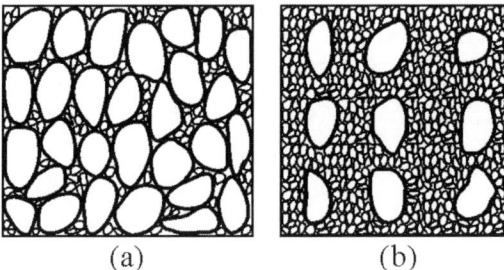

(a)   (b)

Figure 1. Packing of fine and coarse sediment particles: (a) ideal coarse packing, and (b) ideal fine packing.

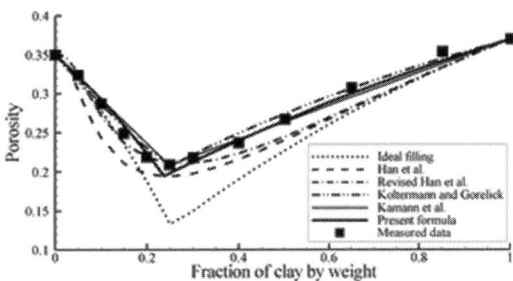

Figure 2. Porosity of sand-clay mixture in the experiment of Marion et al. (1992).

drawback in the model of Han et al. (1981) is that only a two-dimensional planar packing configuration is considered. It was improved by Wu (2015) using a three-dimensional packing configuration. Wu (2015) also proposed a new model by considering how many fine sediment particles are required and how many are available to cover the surface areas of coarse particles in the mixture. The new model performs well among the existing models, so it is briefly described below.

As discussed by Han et al. (1981) and Wu (2015), the mixture dry density has the following relation:

$$\frac{1}{\rho_d} = \frac{p_s}{\rho_{ds}}(1-B_f) + \frac{p_s}{\rho_s}B_f + \frac{p_m}{\rho_{dm}} \quad (1)$$

where ρ_d = dry density of the sediment mixture, ρ_{ds} = dry density of coarse component, ρ_{dm} = dry density of fine component, ρ_s = material density of sediment, p_s = fraction of coarse component, p_m = fraction fine component, and B_f = a filling coefficient. Wu (2015) set B_f as the ratio of the numbers of fine particles available and needed to cover the surface of a coarse particle in the mixture:

$$B_f = \min\left[R_N/(nN_c), B_{fmax}\right] \quad (2)$$

where R_N = number of fine particles available to cover a coarse sediment particle, n = number of layers of fine particles needed to fill the void of a coarse particle, N_c = number of fine particles needed to cover the entire surface of a coarse particle, and B_{fmax} = a limit factor representing the maximum level of filling that the mixture can reach.

The filling coefficient B_f has values between 0 and 1. When $B_f = 0$, no filling occurs, and then Equation (1) becomes the Colby (1963) formula. $B_f = 1$ indicates the case of ideal fine packing, in which coarse particles are dispersed in fine particles and lose their void completely. The ideal packing is difficult to reach even in laboratory settings, and thus B_f has a maximum value of B_{fmax} less than 1. Wu (2015) suggested that B_{fmax} is about 0.9 for noncohesive sediment mixtures, such as sand and gravel mixtures, and between 0.65–0.8 for the sediment mixtures of sand and clay, with 0.65 for less compacted mixtures and 0.8 for more compacted mixtures.

The parameters R_N, n and N_c in Equation (2) are determined as (Wu 2015)

$$R_N = \frac{p_m d_s^3}{p_s d_m^3} \quad (3)$$

$$n = 0.1124 \frac{d_s}{d_m} \quad (4)$$

$$N_c = \frac{\pi}{\left\{\arcsin\left[d_m/(d_s+d_m)\right]\right\}^2 \cos 30^o} \quad (5)$$

where d_s and d_m = representative diameters of coarse and fine particles, respectively.

The porosity of cohesive sediment changes in time and space due to consolidation, whereas that of non-cohesive sediment does not. Thus, the porosity of cohesive component in a mixture is difficult to measure. Equation (1) can be used to determine the dry density of cohesive component when those of the mixture and coarse component are known.

Figure 2 compares the measured porosities of sand and clay mixtures in an experiment run of Marion et al. (1992) against the predictions of different formulas. The mixtures were obtained by shaking sand and clay in a consistent manner in a sealed plastic container until the mixtures appeared visually homogeneous. The chosen experiment run was carried out at confining hydrostatic pressures of 20 MPa and pore pressure of 1 MPa. Experiments and formulas show that the sand-clay mixture porosity decreases and then increases as the clay content increases, and has a minimum value when the clay content is about 20%–30%.

4 EROSION OF MIXED SEDIMENTS

Erosion of sediments is affected by flow conditions, sediment properties, and bed configurations. Noncohesive sediments erode in disperse particles, whereas three modes have been observed for cohesive sediment erosion. The first erosion mode is surface or floc erosion, in which sediment is eroded from the bed in particles and flocs, due to the breaking of interparticle electrochemical bonds under the action of the flow that exceeds a critical shear stress. The second mode is mass erosion, in which sediment is eroded in layers, due to bed failures along planes below the bed surface when the applied shear stress exceeds the bed bulk strength. The third mode is sediment entrainment due to bed fluidization followed by the destabilization of the water-sediment interface (Mehta 1986).

Mixed cohesive/noncohesive sediments exhibit much more complex erosion mechanisms due to the interactions between the size classes. All the above erosion modes can exist. In the case of low cohesive fraction, the mixture is likely packed as shown in Figure 1(a) and the mixture behaves as noncohesive, so that erosion occurs by particles and flocs, i.e. surface erosion. In the case of high cohesive fraction, the mixture is likely packed as Figure 1(b) and behaves as cohesive, so that the typical three erosion modes of cohesive sediment appear. Kamphuis and Hall (1983) found that the size of the eroded particles tended to be larger for mixtures with higher sand contents.

The surface erosion rate of cohesive soils under unidirectional flow has been described with an excess shear stress approach (Arulanandan et al. 1980):

$$\varepsilon = k_d \left(\tau - \tau_{ce} \right)^n \tag{6}$$

where $\varepsilon =$ erosion rate, $k_d =$ erodibility coefficient, $\tau =$ hydraulic boundary shear stress, $\tau_{ce} =$ critical shear stress, and $n =$ empirical exponent usually assumed to equal 1.0 (Hanson and Cook 2004). Partheniades (1965) suggested a function of the dimensionless excess shear stress:

$$\varepsilon = M \left(\frac{\tau - \tau_{ce}}{\tau_{ce}} \right)^n \tag{7}$$

where $M =$ erodibility rate coefficient. One can see that k_d and M are related by $k_d = M/M\tau_{ce}^n$. Partheniades (1965) used $n = 1$, whereas Gailani et al. (1991) found that n is between 2 and 3 and the erodibility coefficient M is related to the deposition or consolidation time.

Raudkivi and Hutchison (1974) and Mehta et al. (1982) established exponential relations between the surface erosion rate and dimensionless excess shear stress. The exponential relations are usually valid for partly consolidated beds, whereas the linear relation (6) or (7) is valid for fully consolidated beds in which soil properties do not vary with time and over depth (Ariathurai and Mehta 1983; Mehta 1986). Sanford and Maa (2001) applied Equation (6) in the case of stratified (partially consolidated) bed with τ_{ce} varying with depth, and derived an erosion formula that can be used in depth-limited erosion and steady-state erosion:

$$\varepsilon = \rho_d \beta \left(\tau - \tau_{ce} \right) e^{-\alpha\beta(t-t_0)} \tag{8}$$

where $\rho_d =$ dry density of the bed; $\beta =$ a local parameter, with $\rho_d \beta$ being the erosion rate coefficient at $t = t_0$; $t_0 =$ time when the bed shear stress τ starts to apply; and $\alpha =$ vertical gradient of the critical shear stress.

Equations (6)–(8) may be used to determine the erosion rate of the mixture when the cohesive size class is dominant, or extended to the fractional erosion rate of the cohesive size class (Lin and Wu, 2013). In either case, the erodibility coefficient and the critical shear stress may be different from those of pure cohesive sediments, as discussed in the next two sections.

Aberle et al. (2004) measured the erosion rate of mixed cohesive/noncoheisve sediments using a benthic in-situ flume under unidirectional flow, and found that the formula of Sanford and Maa (2001) is suitable for interpretation of the measured erosion rate data. The erosion rate coefficient depends on bed material properties, such as dry bulk density, water content, organic content, sand content, and salinity.

Equations (6)–(8) were developed for cohesive sediment erosion under unidirectional flows, and may be extended to wave erosion. For example, Equation (7) has been extended to nonbreaking wave erosion by defining τ as the peak bed shear stress during the wave cycle (Mehta 1996). The erosion coefficient and critical shear stress can be different from those associated with current-induced erosion due to the effect of cyclic loading on the soil matrix (Maa 1986).

5 CRITICAL SHEAR STRESS FOR MIXED SEDIMENT EROSION

Experiments found that the clay or mud (including clay and silt) content of the mixtures correlated positively and significantly with the critical erosion shear stress and the erosion resistance of the soils. Panagiotopoulos et al. (1997) found that the rate of increase in the critical shear stress is smaller for clay mineral content <11–14% (by weight), due to increase of the internal friction angle. When the clay mineral content exceeds 11–14%, the sand particles are not in contact with each other (Figure 1(b)), so that the clay component controls the erodibility of the sediment mixture.

Van Rijn (1993), Torfs et al. (2000), Whitehouse et al. (2000), Sharif (2002), van Ledden (2003), Kothyari and Jain (2008), Ahmad et al. (2011) and Wu and Perera (2015) proposed formulas for the critical shear stress for the sediment mixture erosion as functions of clay or mud content. The formulas of van Ledden (2003), Torfs et al. (2000) and Wu and Perera (2015) are presented here as typical examples.

Van Ledden (2003) suggested a critical mud content, p_{mc}, below which the sand and mud mixture is cohesionless and above which the mixture exhibits cohesion. The prosed formula determines the critical shear stress of the cohesionless and cohesive sand-mud mixtures as follows:

$$\tau_{cem} = \begin{cases} \tau_{cr}(1+p_m)^\beta & p_m \leq p_{mc} \\ \dfrac{\tau_{cr}(1+p_{mc})^\beta - \tau_{ce}}{1-p_{mc}}(1-p_m) + \tau_{ce} & p_m > p_{mc} \end{cases} \tag{9}$$

where τ_{cem} = critical shear stress of the mixture, τ_{cr} = critical shear stress of pure sand, τ_{ce} = critical shear stress of pure mud, and β = empirical exponent.

Wu and Perera (2015) tested the formulas of van Ledden (2003) and Ahmad et al. (2011) and found both have significant errors when compared with measurement data. Wu and Perera (2015) developed a new formula for the critical shear stress of sand and mud mixtures, which is written as

$$\tau_{cem} = \tau_{ceL} + (\tau_{ce} - \tau_{ceL})\exp\left[-\alpha\left(\frac{p_s}{p_m}\right)^{1.2}\right] \quad (10)$$

where $\alpha = 0.365 e^{-3.265 d_s}$, with d_s = sand diameter in mm; and τ_{ceL} = critical shear stress for erosion of sediment mixtures with low mud contents:

$$\tau_{ceL} = \tau_{cr} + 1.25(\tau_{ce1} - \tau_{cr})\min(p_m, 0.05) \quad (11)$$

where τ_{ce1} = critical shear stress of pure mud.

In Equation (10), τ_{ce} is the critical shear stress for erosion of mud corresponding to the porosity of the mud in each sediment mixture:

$$\tau_{ce} = \tau_{ce1}\left(\frac{r}{r_1}\right)^{1.7} \text{ or } \tau_{ce} = \tau_{ce1}\left(\frac{\rho_{dm}}{\rho_{dm1}}\right)^{2.17} \quad (12)$$

where r and ρ_{dm} = solid/void volume ratio and the dry density of the mud component in the sediment mixture, respectively; and r_1 and ρ_{dm1} = solid/void volume ratio and the dry density of the pure mud corresponding to τ_{ce1}.

Note that when the mud content is low, the second term of Equation (10) is very small and thus Equation (10) reduces to Equation (11).

Torfs et al. (2000) examined erosion of mixtures of fine-grained sediments with sand based on laboratory flume experiments under steady flows of Torfs (1995), and developed the following formula:

$$\tau_{cem} = \frac{\alpha_3 \tan\phi}{\alpha_1 + \alpha_2 \tan\phi} g(\rho_{sm} - \rho)d + K'\varsigma(p_v - p_{vc})^\xi \quad (13)$$

where α_1 and α_3 = area and volume shape factors of coarse grain sediment; α_2 is equal to the product of α_1 and the lift and drag coefficient ratio; g = gravitational acceleration rate; ρ_{sm} = grain or floc density; ρ = water density; d = representative mixture diameter; ϕ = repose or internal friction angle; p_v = solid volume fraction of the fine sediment-water mixture before it is mixed with sand; p_{vc} = "threshold" value of p_v below which the bed has no shear strength; and K', ς and ξ = coefficients related to sediment size composition and consolidation level. The first term on the right-hand side of Equation (13) is coarse-grain dominated, and the second term is fine-grain dominated.

Equation (13) shows that the critical shear stress for mixture erosion, τ_{cem}, varies non-monotonically with the fine sediment weight fraction, p_c. Starting with a sandy bed, as p_c increases, at first τ_{cem} seemingly decreases from its value for sand and reaches a minimum at a weight fraction p_{cm}, then increases. It appears that p_{cm} may be a measure of the space-filling concentration of fine particles within the pores of the sandy bed matrix. Barry (2003) experimentally found that p_{cm} is about 4% (between 2–6%) for the mixtures with median sand and several clays. For a bed of 0.85 mm sand in fresh water, as p_c increases to about 13%, the critical shear stress for erosion returns to the sand critical shear stress. Barry (2003) developed a quasi-hydrodynamic model of clay lubrication, which explains well the variations of τ_{cem} with p_c observed in the experiments.

Note that the formulas of van Ledden (2003) and Wu and Perera (2015), as well as those of Kothyari and Jain (2008) and Ahmad et al. (2011), suggest that the critical shear stress of erosion increases monotonically with increasing clay or mud content, whereas the formula of Torfs et al. (2000) presents a trend of decrease first and then increase with increasing cohesive fraction. Moreover, Williamson and Ockenden's (1993) erodibility tests under unidirectional flow and waves suggested that higher sand content (<66%) increased the shear stress required for erosion and decreased the erosion rate. This might be due to that the presence of sand assisted the drainage, resulting in more rapid compaction, higher bed density and thus less erodibility. All these contradicting results warrant further clarification using more experiment data.

6 EROSION COEFFICIENT FOR MIXED SEDIMENTS

Simon et al. (2003) and several others found that the erosion coefficient k_d is related to the critical shear stress τ_{ce} in the linear erosion formula, Equation (6) with n = 1. Wu and Perera (2015) collected 22 groups of measurement data from literature to test this possible relation. Most of the data were obtained by using the jet erosion test device for different soils, and some are from cohesive strength meter, erosion function apparatus and SEDFlume. Based on these data, Wu and Perera (2015) developed a relation between k_d and τ_{ce} in the linear erosion law as follows:

$$k_d = \frac{2.286 \times 10^{-6}}{1 + 0.317\tau_{ce}^{1.2}} \quad (14)$$

where k_d is in m/s-Pa, and τ_{ce} is in Pa.

Mehta and Parchure (2000) showed that the erodibility coefficient k_d in the linear erosion law exponentially decreases with the shear strength for erosion:

$$k_d = k_{d0} \exp\left(-\chi_s \tau_{ce}^{\lambda_s}\right) \quad (15)$$

where k_{d0}, χ_s, and λ_s = sediment-specific coefficients.

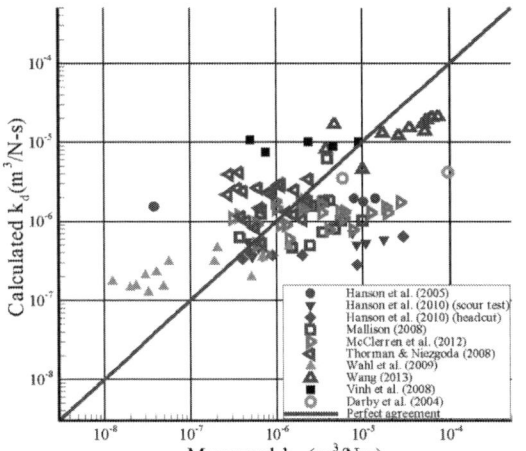

Figure 3. Comparison of SITES k_d formula and measured data ($R^2 = 0.65$).

Torfs et al. (2000) applied Equation (15) with Equation (13) for the erosion rate of sediment mixture measured in the experiments of Torfs (1995). As the fine size fraction increases, the critical shear stress increases, and in turn the erosion rate coefficient k_d decreases according to Equation (15). This decrease of k_d with increasing clay content was also observed in the experiments of Gailani et al. (2001), which showed that as bentonite (up to 16% by volume) was added, the soil erosion rate decreased rapidly.

The SITES program (USDA-NRCS 1997) determines the erosion rate of the sediment mixture using the linear Equation (6), with τ_{ce} determined using the Shields diagram and k_d (in m/hr-Pa) related to soil properties as

$$k_d = \frac{0.036\gamma}{\gamma_d} \exp\left[-0.121 c_\%^{0.406} \left(\frac{\gamma_d}{\gamma}\right)^{3.1}\right] \quad (16)$$

where $c_\%$ = ratio (fraction) of clay in the soil, γ = specific weight of water, and γ_d = dry specific weight of the soil. Equation (16) shows decrease of k_d with increasing clay content and increasing bed density.

Wu and Perera (2015) tested the SITES formula with available data. Figure 3 shows the SITES formula works relatively well, but the data scatters and the formula has room for improvement.

7 FLOCCULATION AND SETTLING OF MIXED SEDIMENTS

It is well known that cohesive sediments flocculate when the particles collide. The factors affecting flocculation include sediment size, sediment concentration, salinity, turbulence, temperature and organic matters (Wu 2007). For cohesive and noncohesive sediment mixtures, flocculation and settling are more complex due to interactions between cohesive and noncohesive sediments.

Manning et al. (2010) conducted experiments on the flocculation and settling of mixed sediments using a mini-annular flume. For pure muds, the macroflocs are regarded as the most dominant contributors to the total depositional flux. By adding more sand to a mud/sand mixture, the fall velocity of the macrofloc fraction slows and the settling velocity of microflocs quickens. Their experiments demonstrated that flocculation is an extremely important factor with regards to the depositional behavior of mud/sand mixtures, and these factors must be considered when modelling mixed sediment transport in the estuarine or marine environment.

8 TRANSPORT OF MIXED SEDIMENTS

The noncohesive size class in the mixed sediments transports as bed load and/or suspended load. Traditionally, cohesive sediment is usually considered as suspended load. However, large-sized flocs may transport near the bed in the mode of bed load while small-sized flocs in suspended load. Certainly, this needs to be clarified in the case of mixed sediment transport.

Murray (1976), Collins (1989), and Banasiak and Verhoeven (2008) conducted experiments on the transport of mixed sediments under unidirectional flow, and found the reduction of sand concentration or transport rate due to the presence of cohesive sediment. This reduction was also observed in the wave erosion experiments of van Rjin and Louisse (1987).

Jain and Kothyari (2009, 2010) measured the fractional bed-load and suspended-load transport rates for two types of sediment mixtures: one of clay and fine gravel, and the other with clay, sand and fine gravel. The clay content in the sediment mixtures was varied from 10% to 50% by weight. They proposed correction factors to modify the existing formulas of the fractional transport rates of bed load and suspended load for the sediment mixtures. The proposed correction factors are functions of clay content, cohesion, and unconfined compressive strength. Extension of these correction factors to clay contents >50% and <10% needs to be verified by experiment data.

9 BED FORMS BY MIXED SEDIMENTS

In the experiments of van Rijn and Louisse (1987), Collins (1989), Banasiak and Verhoeven (2008), and Baas et al. (2013), the bed-form development on mixed cohesive/noncohesive sediment beds was investigated. Among the four experiments, van Rijn and Louisse (1987) considered waves, and the other three used unidirectional flows. All the four experiments demonstrated strong suppression of sand ripples by the presence of cohesive sediments in the bed. Baas et al.

(2013) showed that the ripple height (h_r) in unidirectional flows had an inversely proportional relationship with the initial bed mud fraction (p_{m0}) within the range of 0–18%:

$$h_r = \begin{cases} 15.24 - 0.37 p_{m0} & p_{m0} < 13\% \\ 3.77 & p_{m0} > 13\% \end{cases} \quad (17)$$

Van Rijn and Louisse's (1987) experiments considered four runs with clay contents: 0%, 25%, 75%, and 100%, and did not derive a quantitative relationship between ripple height and clay content in wave conditions. Van Rijn (2007) suggested a correction factor to account for the effect of fine sediments on bed roughness. Validation of the correction factor by measurement data is needed.

10 COMPUTATIONAL MODELING OF MIXED SEDIMENT TRANSPORT

Traditionally, sediment is treated as either cohesive or noncohesive, particularly in singe-sized sediment transport models, and thus the interactions between cohesive and non-cohesion sediment classes are lumped into model parameters, such as erosion and deposition rates and settling velocity. This limitation has been recognized in several recent models (van Ledden 1993; Ziegler and Nisbet 1995; Lin and Wu 2013), which simulate the multiple-sized transport of noncohesive and cohesive sediment mixtures. Among these models, the model of Lin and Wu (2013) is more comprehensive and provides a general modeling framework, as briefly described below.

The model of Lin and Wu (2013) divides the sediment mixture into a suitable number of size classes based on the gradation of the dispersed sediment particles. Because it is difficult to solve the details of aggregation and disaggregation processes using multiple size classes for the flocs, only one size class is used to represent all cohesive particles in the mixture. The model describes transport of all non-cohesive and cohesive size classes in a channel using the same 1-D equation, which is given below:

$$\frac{\partial}{\partial t}\left(\frac{Q_{tk}}{\beta_{tk} U}\right) + \frac{\partial(Q_{tk})}{\partial x} = E_{bk} - D_{bk} + q_{tlk} \quad (18)$$

$$(k = 1, 2, \ldots, N)$$

where Q_{tk} = actual transport rates of the kth size class, U = section-averaged flow velocity, D_{bk} = deposition rate, E_{bk} = erosion rate, q_{tlk} = side discharge per unit channel length, β_{tk} = correction coefficient that is the ratio of the sediment and flow section-averaged velocities, k = size class index, and N = total number of size classes.

The sediment deposition rate is determined as

$$D_{bk} = B \alpha_k \omega_{sf,k} C_{tk} \quad (19)$$

where B = channel width at water surface, $\omega_{sf,k}$ = settling velocity, and C_{tk} = section-averaged sediment concentration. For non-cohesive sediment, the coefficient α_k is called the adaptation coefficient, which is determined using the formula of Armanini and di Silvio (1988) or calibrated by measurement data. For cohesive sediment, α_k is called the deposition probability coefficient and determined by the formula of Mehta and Partheniades (1975). For non-cohesive sediment, the settling velocity $\omega_{sf,k}$ is determined using Wu and Wang's (2006) formula. For cohesive sediment, $\omega_{sf,k}$ is determined by the formula proposed by Wu and Wang (2004), which takes into account the influences of sediment size, sediment concentration, salinity, and turbulence intensity on flocculation.

The sediment erosion rate is determined by

$$E_{bk} = p_{bk} E_{bk}^* \quad (20)$$

where p_{bk} = fraction of the kth size class in the surface layer of bed material, and E_{bk}^* = potential erosion rate of the kth size class. For non-cohesive sediment classes, E_{bk}^* is determined with

$$E_{bk}^* = B \alpha_k \omega_{sf,k} C_{tk}^* \quad (21)$$

where C_{tk}^* = potential transport capacity in terms of concentration of bed-material load and is determined using the formula of Wu et al. (2000). For cohesive sediment class, E_{bk}^* is determined with

$$E_{bk}^* = BM \left(\frac{\tau_b - \tau_{ce}}{\tau_{ce}}\right)^n \quad (22)$$

where n = empirical exponent that is 1 through 3.

The critical bed shear stress for surface erosion is significantly affected by consolidation. It increases as the dry bed density increases with time and depth. τ_{ce} is approximated by Nicholson and O'Connor's formula (1986) as a function of dry bed density:

$$\tau_{ce} = \tau_{ce0} + k_\tau (\rho_d - \rho_{d0})^{n_\tau} \quad (23)$$

where ρ_{d0} = initial dry bed density (kg/m^3), τ_{ce0} = initial critical shear stress (N/m^2) at ρ_{d0}, k_τ = coefficient of about 0.00037, and n_τ = exponent of about 1.5.

The critical shear stress of noncohseive sediment size classes is affected by the appearance of cohesive sediment. This has not been experimentally studied. Lin and Wu (2013) introduced two critical fractions of cohesive particles in the mixture, denoted as p_{cmin} and p_{cmax}. When the fraction of cohesive sediment, p_c, is less than p_{cmin}, the critical shear stress of noncohesive size fractions, denoted as τ_{ck}, is not affected by the cohesive size class. When $p_c > p_{cmax}$, the sediment mixture is fully dominated by cohesive sediment and τ_{ck} is equal to that of cohesive sediment. If p_c is between p_{cmin} and p_{cmax}, τ_{ck} is assumed to be a linear function of p_c:

$$\tau_{ck} = \tau_{ck,n} + (\tau_{ce} - \tau_{ck,n}) \frac{p_c - p_{cmin}}{p_{cmax} - p_{cmin}} \quad (24)$$

where $\tau_{ck,n}$ = critical bed shear stress of the same size class in the situation where only noncohesive sediments exist, and τ_{ce} = critical shear stress of cohesive size class or that of the mixture when $p_c > p_{cmax}$. However, Equation (24) was only indirectly tested through numerical modeling, because no direct measurement data is available.

The rate of change in bed sediment mass, $\partial M_{bk}/\partial t$, due to transport of the kth size class is determined by

$$\frac{\partial M_{bk}}{\partial t} = \rho_s \left(D_{bk} - E_{bk} \right) \quad (25)$$

The bed material sorting is simulated using a multiple-layer approach. The bed is divided into a suitable number of layers in the vertical to account for the heterogeneity of bed material. The first or top layer is the mixing or active layer, which directly exchanges with the moving sediments. The bed material size composition in the mixing layer is determined as

$$\frac{\partial (M_m p_{bk})}{\partial t} = \frac{\partial M_{bk}}{\partial t} + p_{bk}^* \left(\frac{\partial M_m}{\partial t} - \frac{\partial M_b}{\partial t} \right) \quad (26)$$

where M_m = mass of bed material in the mixing layer per unit channel length (kg/m); $\partial M_b/\partial t$ is the total bed mass change; and p_{bk}^* is equal to the fraction of the kth size class of bed material in the mixing layer if $\partial M_b/\partial t - \partial M_m/\partial t > 0$, and that contained in the second bed layer if $\partial M_b/\partial t - \partial M_m/\partial t < 0$.

In addition to bed change caused by sediment transport, the bed lowers due to consolidation. The change of each bed layer thickness is determined by

$$\frac{\partial}{\partial t}\left(\delta_j \rho_{dj} \right) = 0 \quad (27)$$

where δ_j, ρ_{dj} = thickness and dry density of the jth layer of bed material. The change of dry bed density is determined by using the modified Hayter (1983) formula and the Lane and Koelzer (1953) formula.

11 CONCLUSIONS

The transport of mixed cohesive and noncohesive sediments is a very complex phenomenon that has not been well understood. The internal structure of the mixture is the key factor that affects the transport behaviors. In the case of low cohesive fraction less than about 10%, the mixture behaves as noncohesive. As the cohesive fraction increases, the mixture becomes cohesive.

The porosity of the sediment mixture is dependent on the packing or filling between fine and coarse particles. When the fine fraction is low, the mixture undergoes coarse packing. When the fine fraction is high, the mixture undergoes fine packing. Several packing or filling models exist in literature and explain that the porosity has a minimum when the fine component is about 20–30% of the mixture by weight.

The erosion of the sediment mixture is still often modeled using the linear or nonlinear law as used for cohesive sediments. However, the critical shear stress for erosion and the erosion coefficient are found to be functions of mud or clay fraction, as well as other parameters such as dry density of mud component.

Only a few studies have concerned about the flocculation and settling of the sediment mixture. It has been demonstrated that the flocculation of cohesive sediment component is affected by the noncohesive component.

Experiments showed that noncohesive sediment in the mixture may still transport in bed load and suspended load, and the cohesive component is usually suspended but large-sized flocs may move near the bed as bed load. The transport rate of noncohesive component is reduced by adding the cohesive component.

Bed forms can occur on mixed sediment beds. The height of bed forms is suppressed by increasing the cohesive fraction.

Finally, several numerical models for mixed cohesive and noncohesive sediment transport have been reported in literature. The latest one has considered to some extent the interactions between cohesive and noncohesive sediments. Further development is needed with better understanding of the involved physical and chemical processes and more reliable laboratory and field data.

REFERENCES

Aberle, J., Nikora, V. & Walters, R. 2004. Effects of bed material properties on cohesive sediment erosion. *Marine Geology* 207, 83–93.

Ahmad, M.F., Dong, P., Mamat, M., Wan Nik, W.B. & Mohd, M.H. 2011. The critical shear stresses for sand and mud mixture. *Applied Mathematical Sciences* 5(2): 53–71.

Ariathurai, R. & Mehta, A.J. 1983. Fine sediments in waterway and harbor shoaling problems. *Proc. Int. Conf. on Coastal and Port Eng. in Developing Countries*, Colombo, Sri Lanka, 1094–1108.

Armanini, A. & di Silvio, G. 1988. A one-dimensional model for the transport of a sediment mixture in non-equilibrium conditions. *J. Hydraulic Res.* 26(3): 275–292.

Arulanandan, K., Gillogly, E. & Tully, R. 1980. Development of a quantitative method to predict critical shear stress and rate of erosion of natural undisturbed cohesive soils. *Technical Rep. GL-805*, US Army of Engineers' Waterway Experiment Station, Vicksburg, Mississippi.

Baas, J.H., Davies, A.G. & Malarkey, J. 2013. Bedform development in mixed sand–mud: The contrasting role of cohesive forces in flow and bed. *Geomorphology* 182, 19–32.

Banasiak, R. and Verhoeven, R. (2008). "Transport of sand and partly cohesive sediments in a circular pipe run partially full." *J. Hydraulic Eng.* 134(2): 216–224.

Barry, K.M. 2003. The effect of clay particles in pore water on the critical shear stress of sand. PhD Dissertation, University of Florida, Gainesville, FL.

Colby, B.R. 1963. Discussion of "Sediment transportation mechanics: Introduction and properties of sediment", Progress report by the Task Committee on Preparation of Sediment Manual of the Committee on Sedimentation of the Hydraulics Division, V. A. Vanoni, Chmn., *J. Hydr. Div.*, ASCE, 89(1): 266–268.

Collins, M. 1989. The behavior of cohesive and non-cohesive sediments. *Proc. the 17th Int. Seminar on the Environmental Aspects of Dredging Activities*, Nov. 27-Dec. 1, Nantes.

Gailani, J.Z., Jin, L., McNeil, J. & Lick, W. 2001. "Effects of bentonite on sediment erosion rates." DOER Technical Notes Collection (ERDC TN-DOER-N9), U.S. Army Engineer Research and Development Center, Vicksburg, MS. www.wes.army.mil/el/dots/doer

Gailani, J., Ziegler, C.K. & Lick, W. 1991. Transport of suspended solids in the Lower Fox River. *J. Great Lakes Res.* 17(4): 479–494.

Grissinger, E.H., Little, W.C. & Murphey, J.B. 1981. Erodibility of streambank materials of low cohesion. *ASAE*, 24(3): 624–630.

Han, Q.W., Wang, Y.C. & Xiang, X.L. 1981. Initial specific weight of deposits. *J. Sediment Res.*, No. 1, pp. 1–13 (in Chinese).

Hanson, G.J. & Cook, K.R. 2004. Apparatus, test procedures, and analytical methods to measure soil erodibility in-situ. *Applied Engineering in Agriculture*, ASAE, 20(4): 455–462.

Hayter, E.J. 1983. Prediction of cohesive sediment movement in estuarial waters. PhD thesis. University of Florida.

Jain, R.K. & Kothyari, U.C. 2009. Cohesion influences on erosion and bed load transport. *Water Resour. Res.* 45(W06410), doi:10.1029/2008WR007044.

Jain, R.K. & Kothyari, U.C. 2010. Influence of cohesion on suspended load transport of nonuniform sediments. *J. Hydraulic Res.* 48(1): 33–43.

Kamann, P.J., Ritzi, R.W., Dominic, D.F. & Conrad, C.M. 2007. Porosity and permeability in sediment mixtures. *Ground Water* 45(4): 429–438, doi: 10.1111/j.1745-6584.2007.00313.x.

Kamphuis, J.W. & Hall, K.R. 1983. Cohesive material erosion by unidirectional current. *J. Hydraul. Eng.*, ASCE, 109(1): 49–61.

Koltermann, C.E. & Gorelick, S.M. 1995. Fractional packing model for hydraulic conductivity derived from sediment mixtures. *Water Resour. Res.* 31(12): 3283–3297.

Kothyari, U.C. & Jain, R.K. 2008. Influence of cohesion on the incipient motion condition of sediment mixtures. *Water Resour. Res.* 44(W04410), doi:10.1029/2007WR006326.

Kramer, H. (1935). "Sand mixtures and sand movement in fluid models." Trans., ASCE, Vol. 100, Paper No. 1909, 798–878.

Lane, E.W. & Koelzer, V.A. (1953). Density of sediments deposited in reservoirs. Report No. 9, A study of Methods Used in Measurement and Analysis of Sediment Loads in Streams, Engineering District, St. Paul, Minnesota.

Lin, Q. & Wu, W. 2013. A one-dimensional model of mixed cohesive and non-cohesive sediment transport in open channels. *J. Hydraulic Res.* 51(5): 506–517, doi: 10.1080/00221686.2013.812046.

Maa, J.P.-Y. 1986. Erosion of soft mud by waves. PhD Thesis, University of Florida, Gainesville, FL.

Manning, A.J., Baugh, J.V., Spearman, J.R., & Whitehouse, J.S. 2010. Flocculation settling characteristics of mud/sand mixtures. *Ocean Dynamics* 60: 237–253, doi: 10.1007/s10236–009–0251–0.

Marion, D., Nur, A., Yin, H. & Han, D. 1992. Compressional velocity and porosity in sandy-clay mixtures. *Geophysics* 57(4): 554–563.

Mehta, A.J. 1986. Characterization of cohesive sediment properties and transport processes in estuaries. *Estuarine Cohesive Sediment Dynamics*, A. J. Mehta (ed.), Springer-Verlag, pp. 290–325.

Mehta, A.J. 1996. Interaction between fluid mud and water waves. *Environmental Hydraulics*, V.P. Singh and W.H. Hager (eds.), Kluver, Dordrecht, The Netherlands, 153–187.

Mehta, A.J. & Parchure, T.M. 2000. Surface erosion of fine-grained sediment revisited. *Muddy Coast Dynamics and Resource Management*, B.W. Flemming, M.T. Delafontaine, and G. Liebezeit, eds., Elsevier, Oxford, UK, 55–74.

Mehta, A.J., Parchure, T.M., Dixit, J.G. & Ariathurai, R. 1982. Resuspension potential of deposited cohesive sediment beds. *Estuarine Comparisons*, V. S. Kennedy (ed.), Academic Press, New York, pp. 591–609.

Mehta, A.J. & Partheniades, E. 1975. An investigation of the depositional properties of flocculated fine sediment. *J. Hydraulic Res.* 13(4): 361–381.

Murry, W.A. 1976. Erodibility of coarse sand/clayey silt mixtures. *Report No. 411.2*, Fritz Engineering Laboratory, http://preserve.lehigh.edu/engr-civil-environmental-fritz-lab-reports/2180

Nalluri, C. & Alvarez, E.M. 1992. The influence of cohesion on sediment behavior. *J. Water Sci. Tech.* 25(8): 151–164.

Nicholson, J. & O'Connor, B.A. 1986. Cohesive sediment transport model. *J. Hydraulic. Eng.*, ASCE, 112(7): 621–639.

Panagiotopoulos, I., Voulgaris, G. & Collins, M.B. 1997. The influence of clay on the threshold of movement of fine sandy beds. *Coastal Eng.* 32: 19–43.

Partheniades, E. 1965. Erosion and deposition of cohesive soils. *J. Hydr. Div.*, ASCE, 91(HY1): 105–139.

Raudkivi, A.J. & Hutchison, D.L. 1974. Erosion of kaolinite clay by flowing water. *Proc. Royal Society, London*, England, Series A, 337: 537–544.

Sanford, L. & Maa, J.P.-Y. 2001. A unified erosion formulation for fine sediments. *Marine Geology* 179: 9–23.

Sharif. A.R. 2002. Critical shear stress and erosion of cohesive soils. PhD Thesis, SUNY Buffalo, NY.

Simon, A., Collison, A. J. & Layzell, A. 2003. Incorporating bank-toe erosion by hydraulic shear into the ARS bank-stability model: Missouri River, Eastern Montana. *Proc. of World Water and Environmental Resources Congress*.

Smith et al. 2015. Private Communication.

Torfs, H. 1995. Erosion of mud/sand mixtures. PhD Thesis, Katholieke Univ. Leuven, Belgium.

Torfs, H., Jiang, J. & Mehta, J. 2000. Assessment of the erodibility of fine/coarse sediment mixtures. *Coastal and Estuarine Fine Sediment Processes*, Elsevier, 3: 109–123.

U.S. Department of Agriculture, Natural Resources Conservation Service (USDA-NRCS). 1997. Chapter 51: Earth spillway erosion model. *Part 628 Dams, National Engineering Handbook*.

Van Ledden, M. 2003. Sand and mud segregation. PhD thesis, Delft University of Technology, Delft, The Netherlands.

Van Rijn, L.C. & Louisse, C.J. 1987. The effect of waves on cohesive bed surfaces. *Proc. The 2nd Int. Conf. on Coastal and Port Engineering in Developing Countries*, Beijing, 1518–1532.

Van Rijn, L.C. 1993. *Principles of Sediment Transport in Rivers, Estuaries and Coastal Seas*. Aqua Publications, The Netherlands.

Van Rijn, L.C. 2007. Unified view of sediment transport by currents and waves, part I: initiation of motion, bed roughness, and bed-load transport. *J. Hydraulic Eng.*, ASCE, 133(6): 649–667.

Wang, Y.C.B. 2013. Effects of physical properties and rheological characteristics on critical shear stress of fine sediments. Ph.D. thesis, Georgia Institute of Technology.

Whitehouse, R., Soulsby, R., Roberts, W. & Mitchener, H. 2000. *Dynamics of Estuarine Muds: A Manual for Practical Applications*. Thomas Telford Publishing, London.

Williamson, H.J. & Ockenden, M.C. 1993. Laboratory and field investigations of mud and sand mixture. *Proc. First Int. Conf. on Hydro-Science and Engineering*, Washington D.C., 622–629.

Wu, W. 2007. *Computational River Dynamics*, Taylor & Francis, UK, 494 p.

Wu, W. 2015. Porosity of bimodal sediment mixtures. Under review by *J. Hydraulic Engineering*.

Wu, W. & Perera, C. 2015. Erosion of mixed cohesive and noncohesive sediments. Technical Report, Clarkson University, NY.

Wu, W. & Wang, S.S.Y. 2004. Depth-averaged 2-D calculation of tidal flow, salinity and cohesive sediment transport in estuaries. *Int. J. Sediment Res.* 19(3): 17–2190.

Wu, W. & Wang, S.S.Y. 2006. Formulas for sediment porosity and settling velocity. *J. Hydraulic Eng.* 132(8): 858–862.

Wu, W., Wang, S.S.Y. & Jia, Y. 2000. Nonuniform sediment transport in alluvial rivers. *J. Hydr. Res.* 38(6): 427–434.

Ye, Z., Cheng, L. & Zang, Z. 2011. Experimental study of erosion threshold of reconstituted sediments. *Proc. the ASME 30th International Conference on Ocean, Offshore and Arctic Engineering*, ASME, pp. 973–983.

Ziegler, C.K. & Lick, W. 1988. The transport of fine-grained sediment in shallow waters. *Environ. Geol. Water Sci.* 11(1): 123–132.

A. Integrated sediment management at the river basin scale

Erosion on irregular slope surface: A full N-S equation based numerical study

Y. An & Q.Q. Liu
Institute of Mechanics, Chinese Academy of Sciences, Beijing, China

ABSTRACT

Soil erosion by overland flow on hillslopes is one of the most crucial fundamental sediment yield processes. Generally, Saint-Venant equation based overland flow model is widely used to model this special flow, and in turn, the sheet erosion. However, it becomes more and more clear that the overland flow, which generally involves irregular micro landform, could behave different characteristics from river flow in which classical depth-averaged model suits well. The key issue is that the characteristic length of the irregular micro landform is considerable large comparing with the overland flow depth, which leads to non-uniform velocity profile in the depth direction. Thus, this paper employed full Navier-Stokes equation to study the overland flow on partially inundated irregular micro-relief and discuss the self-enhanced sheet erosion processes.

To understand the self-enhanced sheet erosion processes, we first generate irregular micro-relief (0.4 m × 0.2 m) using Monte Carlo method basing on statistics data obtained from 3D scanner. Next, an automatic mesh generation script is utilized to generate valid CFD meshes (2 mm spatial resolution and minimum 0.05 mm in height direction with dynamic adaption technique). On this mesh, the flow is solved with a CFD code using VOF method and SIMPLE scheme on the "Yuan" super computer of CAS. Finally, we extract the surface shear stress and flow structure from simulation results. Potential erosion intensity is estimated with empirical equation of shear stress. As the sheet erosion induced landform change is much slower than flow solving time steps, we assume that the surface morphology change could be decoupled from flow solving. Thus, by designing different micro-relief to reflect the influence of morphology change, the self-enhanced sheet erosion processes could be discussed. In the simulation, the adaption refine technology is employed in the VOF calculation to reduce possible shear stress errors around the interface between water surface and landform for partially inundated cases. And the *low-Re k-ε* model is chosen to simulate the viscosity of the overland flow to balance the accuracy and compute efficient.

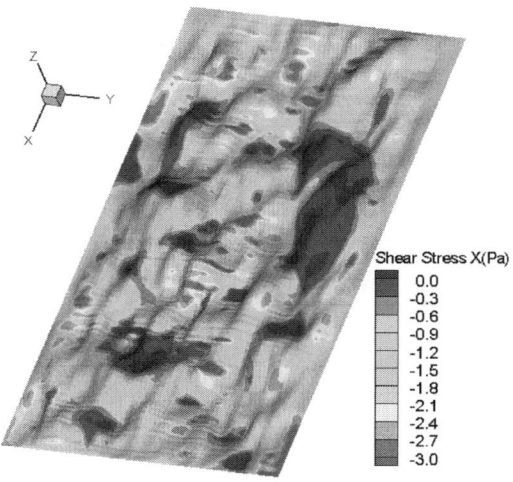

Figure 1. Typical shear stress distribution on an irregular hillslope surface (note that the blue area is not inundated).

It is note that the inundated situation is the critical factor determining the sheet erosion development, which has also been discussed by Dunkerley (2004) and Emilio et al. (2012). When the slope is mild, the inundate area would not change a lot when flow discharge increases, and the sheet erosion distribution is relatively wide on the surface. While for the steep cases, the mean shear stress is strong and the sheet erosion is localized around the lower zones near obstacles. The erosion process in the latter situation is developed more rapidly and might lead to rill erosion.

REFERENCES

Dunkerley, D. 2004. Flow threads in surface run-off: Implications for the assessment of flow properties and friction coefficients in soil erosion and hydraulics investigations. Earth Surf Proc Land., 29(8):1011–1026.

Emilio, R.C., Yolanda, C., Chamizo, S., Afana, A. & Solé-Benet, A. 2012. Effects of biological soil crusts on surface roughness and implications for runoff and erosion. Geomorphology, 145–146(1):81–89.

Impacts of recent climate and land use dynamics on spatial and temporal changes of sediment budget and reservoir siltation in small agricultural catchments of the European Russia

V. Belyaev & A. Malyutina
Laboratory of Soil Erosion and Fluvial Processes, Faculty of Geography, Lomonosov Moscow State University, Moscow, Russia

ABSTRACT

It is generally accepted that both climatic fluctuations and changes of land use patterns can exert substantial influence on spatial and temporal characteristics of sediment redistribution in a fluvial system. However, in most cases it is extremely difficult to distinguish natural signals from anthropogenic impacts as well as to obtain reliable quantitative characteristics. In this study an attempt has been made to evaluate such contributions by reconstructing temporal and spatial variability of fluvial sediment budgets for small river basins with small reservoirs at their outlets and different land use histories. As important and efficient sediment interceptors, small reservoirs in agricultural areas are used as sources of valuable information on catchment scale erosion and sediment delivery (Verstraeten & Poesen 2001). In addition, reservoirs are also important as sinks of sediment-associated pollutants (Golosov et al. 2012) and organic carbon (Cole et al. 2007).

Small reservoirs located in dry valleys or on small streams and rivers are typical and important components of human-altered landscapes of the European Russia agricultural belt. Initially their widespread introduction in 19th Century was mainly aimed at interception of peak runoff for local water supply and (on perennial streams) providing continuous hydropower supply for water mills. At present such small reservoirs are utilized in diverse ways from local water supply for crops and livestock to local fishery and recreation.

Two small river basins located in the Seim River basin (Kursk Region, Western European Russia) have been investigated in this study. Selection of the case study sites considered in this study was based on representativeness of local topography, catchment morphometry (area, valley network structure), soil cover, land use and crop rotations. Several independent techniques including radioisotope tracers, soil surveys, geomorphic mapping, remote sensing data analysis and soil erosion modeling have been employed for evaluating sediment budgets for the studied basins. Detailed microstratigraphic separation of the reservoir infill sediment was based on sediment core descriptions and 137Cs depth distribution analysis Results of the investigations clearly show that the importance of catchment-scale land use changes for the reservoir sediment delivery and sedimentation depends largely upon the catchment area, sediment transfer distance, presence and efficiency of intermittent sediment sinks. Temporal dynamics of the latter also needs to be evaluated. The most general tendency observed is decreasing sediment delivery into reservoirs and increasing redisposition within small dry tributary valleys. Sediment budget and reservoir sedimentation variability of the two studied basins over the last decades have been largely controlled by natural tendency of decreasing spring snowmelt runoff and increasing frequency of heavy summer rainstorms accompanied with land use changes associated with collapse of the socialist-type economy, general economic disorder of 1990s – early 2000s and gradual recovery of industry and agriculture approximately since 2005.

REFERENCES

Cole, J.J., Prairie, Y.T., Caraco, N.F., McDowell, W.H., Tranvik, L.J., Striegl, R.G. Duarte, C.M., Kortelainen, P., Downing, J.A., Middelburg, J.J. & Melack, J. 2007. Plumbing the Global Carbon Cycle: Integrating inland waters into the terrestrial carbon budget. Ecosystems, 10, 171–184.

Golosov, V.N., Aseeva, E.N., Belyaev, V.R., Markelov, M.V. & Alyabieva, A.K. 2012. Redistribution of sediment and sediment-associated contaminants in the River Chern basin during the last 50 years. Erosion and Sediment Yields in the Changing Environment (Proceedings of a symposium held in Chengdu, China, October 2012). IAHS Publ. 356: 12–19.

Verstraeten, G. & Poesen, J. 2001. Factors controlling sediment yield from small intensively cultivated catchments in a temperate humid climate. Geomorphology, 40, 123–144.

Variability of total sediment supply of the Chao Phraya River, Thailand

B. Bidorn & S.A. Kish
Department of Earth, Ocean and Atmospheric Science, Florida State University, Florida, USA

J.F. Donoghue
Planetary Sciences Program, Department of Physics, University of Central Florida, Florida, USA

W. Huang
Department of Civil Engineering, Florida State University, Florida, USA

K. Bidorn
Department of Water Resources Engineering, Chulalongkorn University, Bangkok, Thailand

ABSTRACT

Fluvial sediment delivery to the ocean is an important factor in the environmental quality of coastal ecosystems and the evolution of deltas and coastal landforms. The sediment loads of the world's rivers are influenced by many factors, such as deforestation for agriculture, population pressure, water resources development, dam construction, and climate change (Walling et al. 2003, Walling 2006). A better understanding of river response to human activities and climate change is necessary in order to better manage coastal resources (Hoffman et al. 2010). The objectives of this study were to determine sediment transport characteristics of the Chao Phraya River, which is the major sediment supply source to the Upper Gulf of Thailand, and to evaluate the historical trends of total sediment supply for the Chao Phraya River basin. River surveys of the Chao Phraya River were carried out 9 times between 2011 and 2013. Survey data included river cross-sections, flow velocities, suspended sediment concentration, and bed load transport.

Analyses of these data indicate that suspended transport rates at the Chao Phraya River had a strong correlation with river discharge, and the relationship can be represented by a power equation with a goodness of fit of 0.81. It also appears that the bed-to suspended ratio varied in a narrow range between 0 and 5 percent. The suspended sediment load of the Chao Phraya River is strongly related to river discharge and accounts for more than 95 percent of the total sediment load. The relationship between total sediment load and river discharge, which has been daily observed by the Royal Irrigation Department since 1956, has been used to estimate mean annual total sediment yield for the river system. Based on trend analysis (Fig. 1), a significant long-term change of total sediment supply for the Chao Phraya River cannot be detected, even though two large dams (Bhumibol and Sirikit dams) have been constructed in the major tributaries, the Ping and Nan

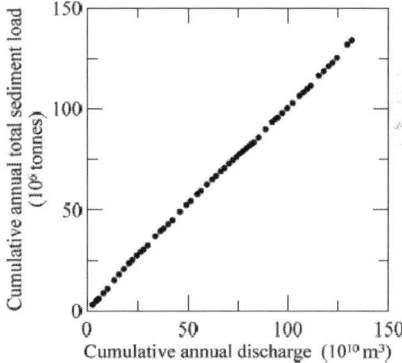

Figure 1. Variation of total sediment supply at the Chao Phraya River during the period 1956–2013.

rivers, in 1964 and 1972, respectively. The influence of damming and reservoir construction has apparently been outweighed by the increase of sediment transport caused by conversion of natural vegetation to agricultural use, which started in 1961 (Tangtham et al. 1998).

REFERENCES

Hoffmann, T., Thorndycraft, V.R., Brown, A.G., Coulthard, T. J., Damnati, B., Kale, V.S. & Walling, D.E. 2010. Human impact on fluvial regimes and sediment flux during the Holocene: Review and future research agenda. Global and Planetary Change, 72(3), 87–98.

Tangtham, N. & Boonyawat, S. 1998. Effects of land cover change and large reservoir operation on water balance of the Chao Phraya River basin. Kasetsart Journal (Nat.Sci.), 32, 511–519.

Walling, D.E. & Fang, D. 2003. Recent trends in the supended sediment loads of the world's rivers. Global and Planetary Change, 39, 111–126.

Walling, D.E. 2006. Human impact on land–ocean sediment transfer by the world's rivers, Geomorphology, 79(3), 192–216.

Responsible management of alpine rivers: The Arly Basin/Savoie, France

P. Ergenzinger
Weyersheim, France

C. de Jong
Institute of Imagery, City and Environment (LIVE), Faculty of Geography and Spatial Planning, University of Strasbourg, Strasbourg, France

ABSTRACT

On the 1st May 2015, the Arly Road between Albertville and Mégève (Upper Savoie) was destroyed by an estimated 30 year flood event caused by extreme precipitation and even the steel factory in Ugine was endangered by fluvial accumulation. Up to the Annexation of Savoy to France in 1860 the in-habitants of the Arly basin below Flumet never used the river bed for road travel but preferred to maintain two small roads on the safer mid-slope flanks on both sides of the valley, just above the zone of mass movements interacting with the river. A new road was constructed on the flood-prone river bed only after the Annexation of Savoy in order to reduce travel time and in demonstration of new political power. However, past flood events have induced frequent and expensive road maintenance. In order to protect the road, the Arly River was relocated to one side of the valley. During flood events, this causes asymmetric erosion with the river attacking only the slope opposing the road and thereby triggering intensive mass movements. These mass movements could mobilized entire valley slopes under extreme events and could become quite a substantial risk in the future. Due to active erosion of the lower river slopes causing significant multi-mass movements during the 2015 event, the sediment load substantially in-creased locally. In addition, substantial volumes of water were released as security measure from the hydropower dam during the event. As a consequence of the extreme event, amplified by the artificially in-creased discharge and sediment load, the river destroyed 400 m of road entailing 15 Million Euros of repair costs. Excessive erosion was not only caused by the geological setting but also due to the impacts of land-use and hydro-electricity. There has been intensive land-use change (for example an increase in forest area, a decrease in alpine meadows or intensive ski run development) as well as a reduction in coarse sediment below the hydropower dam of Flumet. Since the importance of snow cover and the quantity snowmelt runoff

Figure 1. Destruction of road along the River Arly, May 2015.

is decreasing, the importance of short term, intensive rainfall runoff is in-creasing. As a result, the river road will become even more costly in future (raising of the road, construction of multi-tunnel systems, diversion of river, reinforcement of river banks etc) and cannot be financed by the local administration alone.

Responsible management should go beyond the specific case by case approach at the short term but should encompass a more comprehensive, basin-scale, long term approach learning from lessons in the past as well as taking into account climate change and river impoundment and diversion. At present, planning is reduced only to the narrow river bed and the associated maintenance of road infra-structure related to the ski industry as well as heavy industry further downstream. Planning should take into account the local stakeholders, in particular the inhabitants. Extreme events in the past are forgotten too quickly and the same mistakes repeated too frequently. An innovative management approach should not only be based on the sediment transport capacity of the river but should also include the dynamics of sediment sources and the results of human interference.

Suspended sediment yield transportation by rivers of the Kamchatsky Krai into the Pacific Ocean

L.V. Kuksina & N.I. Alexeevsky
Moscow State University, Moscow, Russia

ABSTRACT

The amount of suspended sediments transported by rivers of the Kamchatsky Krai into the Pacific Ocean (including the Sea of Okhotsk and the Bering Sea) was estimated. The suspended sediment yield from non-studied territories was also assessed on the basis of geography-hydrology generalization and regression equations.

Using the monitoring data on suspended sediment yield in 63 gauges the total suspended sediment yield was estimated in the mouths of the major rivers in the Kamchatsky Krai. The basin area for them varies from 150 to 73500 km². The maximum amount of suspended sediments (~2.7 Mg tons per year in average) in the territory of the Kamchatsky Krai delivers the Kamchatka River which flows into the Pacific Ocean. The suspended sediment yield of largest river in the territory (the Pengina River flowing to the Sea of Okhotsk) is four times less. The specific suspended sediment yield (SSSY) in the Kamchatka River basin is 5 times more than in the Pengina River basin due to active volcanoes in its territory which are the sources of a huge amount of mineral particles in rivers. The specific suspended sediment yield of rivers of the Eastern coast of the Kamchatsky Krai significantly exceeds SSSY of rivers in the Western part of the territory due to distribution of the major volcanoes.

The suspended sediment yield increases with the growth of basin area, but dependency differs for river basins with area F less than 10000 km² and F > 10000 km² (Fig. 1).

Relying on the DEM for the Kamchatsky Krai relief such parameters as basin area, mean slope of basin, average river slope, basin altitude, and drainage density were estimated for a number of ungauged river basins, which were used for SSSY assessment. The results were compared with the map of SSSY distribution which was plotted on the basis of data from monitoring stations with taking into account the distribution of factors of suspended sediment yield formation. Table 1 demonstrates good correspondence of the values for the most part of analyzed rivers.

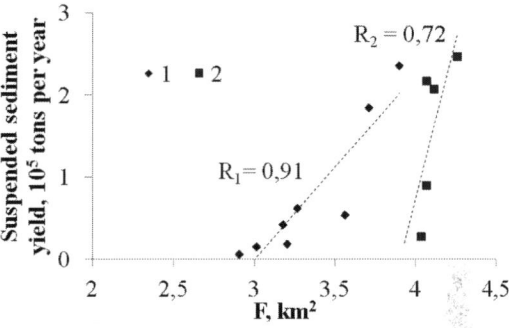

Figure 1. Dependence of suspended sediment yield on basin area. 1 – river basins with area F < 10000 km², 2 – river basins with area F > 10000 km².

Table 1. Specific suspended sediment yield of some ungauged river basins in the Kamchatsky Krai.

River	Basin area, km²	SSSY by equation, t/km²	SSSY by map, t/km²	Deviation, %
Bol'shaya Medvezhka	36.3	26.7	11–25	6.45
Khalaktyrka	49	61.6	51–100	–
Mitoga	149	2.26	<10	–
Utka	763	6.62	<10	–
Kihchick	1800	11.5	11–25	–
Unushka	223	3.63	<10	–
Oblukovina	2690	11.2	<10	10.7
Icha	4000	12.4	<10	19.4
Kovran	1120	14.0	10–25	–
Palana	2300	17.1	11–25	–

Divergence of values is observed for relatively small and large basins (deviation doesn't exceed 20%).

On the basis of these estimations of the total intensity of erosion in the territory of the Kamchatsky Krai and its variability in space and time will be evaluated.

Study on strategy of wide floodplain training in the Lower Yellow River

J. Li, E. Jiang & X. Zhang
Yellow River Institute of Hydraulic Research, Zheng Zhou, China

ABSTRACT

The Lower Yellow River Wide Floodplains are not only used for the flood detention and sediment desilting, but also for the living of a large amount of inhabitants. From the Yellow River management history, it has never stopped the debate about the wide flood-plain training in the Lower Yellow River between wide river strategy and narrow river strategy. Changes of water and sediment downstream after use of Xiaolangdi Reservoir, the wide floodplain of the fate of the issue has aroused great concern. This research focus on effect of the flood detention and sediment desilting and the disaster loss with different water and sediment situation. The situation contain three water and sediment allocations in 50 years including respectively 0.3 Bt, 0.6 Bt and 0.8 Bt sediment schemes, and two typical flood processes named the "58.7" and "77.8" flood. Three floodplain-used schemes are as follows: ① Without protecting levee scheme, that is to say, production dyke totally were broken away on the basis of present management and use of the Lower Yellow River Wide Floodplains. ② Protecting levee scheme, that means present production dyke will be consolidated and remolded as protecting levees against floods. ③ Division utilization scheme, different number of floodplain regions will be used for the flood detention and sediment desilting based on protecting levee scheme.

With methods of theoretical studies, historical data analysis, mathematical model computation and physical model tests, this issue explores water-sediment exchange mechanism between channel and floodplain. Interactive mechanisms between water-sediment movement and floodplain silting form when overbank flood happens have been found. The flood detention and sediment desilting function of the wide flood plain has been analyzed and the index system to be divided two parts and three layers for estimating effect of the flood detention and sediment desilting including 12 index and evaluation model have been constructed. The scheme of channel-floodplain optimal allocation and integrated disaster reduction measures have been raised. According to evaluation, the best operation scheme of wide floodplain considering flood prevention, silt alleviation and disaster reduction in future have come up.

The main achievement is as follows:

(1) The secondary suspended river is inevitable regardless of whether there exists the horizontal gradient ratio or not. In the natural state, although the development rate of secondary suspended river tends to slow down with the worsening of cross-sectional morphology, the secondary suspended river is irreversible. Only with manual adjustment means or engineering measures could we change the current situation of secondary suspended river.

(2) Based on the mathematical model computation and the wide flood plain flood detention model, two conclusions is acquired. First, the flood detention and sediment desilting function of wide flood plain is played the best under the scheme without protecting levee, but the disaster reduction benefit is weaken. Second, under the schemes with protecting levee and division utilization, the disaster reduction benefit is improved, but the flood detention and sediment desilting function is limited.

(3) In the wide flood plains of the Lower Yellow River, there exists great potential and ability of sediment allocation. Among all the sediment allocation modes, the best mode is the one of which the flood plain is warped with protecting levee.

In brief, we recommend the model of the wide floodplain training in the Yellow River. If flow rate is less than 6000 m³/s, keep the existing levees. It flow rate is more than 6000 m³/s, break the existing levees. For the flood loss, compensation should be made by the state. At the same time, we should actively encourage the people moved to the outside of the wide floodplain.

Study on sediment regulation approaches of Liu River

L. Lin & X. Guan
Songliao Water Resources Commission, Ministry of Water Resources, China

T. Yu
Key Laboratory of Groundwater Resources and Environment, Ministry of Education, Jilin University, Changchun, China

ABSTRACT

Liu river basin lies in semi-arid monsoon climate zone where characterized by sparse vegetation, infertile soil, strong wind and water erosion. In this basin, mountain areas accounts for 43%, followed by hills of 32%, sand dunes of 15%, and plains of 10% respectively. The Liu River is destined to be a sandy river because of its specific underlying surface characteristics (e.g geology, soil type, topography), also combined by human activities such as land reclamation, forests and grass destruction, overgrazing, and etc. Mean annual runoff of the Liu River is only 281 million m^3, but its multi-year average sediment concentration is quite high, up to 19.58 kg/m^3 at Xinmin station. According to monitoring data, more sediment discharge at Naodehai station located in upstream of Xinmin station, with annual average load of 13,859,700 t/a and sediment concentration of 51.59 kg/m^3 from 1954 to 2004. Therefore, the sediment in the Liu River mainly originates from the upstream region of Naodehai. Based on previous research of the Liu River, vast sediment accumulated in the riverbed from Naodehai to Xinmin, thereby forming a "suspend river". Sediment deposition has been seriously endangered flood control of the surrounding regions. In this paper, analysis was firstly conducted on the distribution and possible sources of sediments in the Liu River. Based on the analysis, sediment regulation advices and approaches were proposed from the aspects of controlling slope erosion, managing channel erosion and building reservoirs.

REFERENCES

Zhang Zhiling, 2008. The temporal and spatial variation of runoff and sediment discharge in Liu River Basin. Research of Soil and Water Conservation, 15, 1.

Zhang Guohui, 1999. Analysis of sediment variation law for Liu River. Journal of Liaoning Technical University, 177.

Table 1. Sediment characteristic of the Liu River.

| Station | period | Load (10^4 t) | | | | | | Concentration (kg/m^3) | | |
		Avg.	Max Value	Year	Min Value	Year	Max/Min	Avg.	Max Value	Date
Naodehai	1954–2004	1386	4904	1967	5.44	2004	901.95	51.59	1140	1972
	1954–1960	2884	4521	1954	2002	1960	2.12	67.6	556	1959
	1960–2004	1148	4904	1967	5.44	2004	901.95	48.92	1140	1972
Xinmin	1954–2004	551	1884	1969	39.17	2004	48.11	19.59	468	1968
	1954–1960	894	1265	1954	122.61	1958	10.32	22.4	116	1961
	1960–2004	516	1884	1969	39.17	2004	48.11	19.06	468	1968

The International Sediment Initiative (ISI) and its case studies

C. Liu
International Research and Training Center on Erosion and Sedimentation, Beijing, China

D.E. Walling
University of Exeter, Exeter, UK

M. Spreafico
University of Berne, Berne, Switzerland

J. Ramasamy, H. Dencker Thulstrup & A. Mishra
UNESCO

ABSTRACT

The management of sediment in river basins and waterways has been an important issue for water managers throughout history – from the ancient Egyptians managing sediment on floodplains to provide their crops with nutrients, to today's challenges of siltation in large reservoirs. The changing nature of sediment issues, due to increasing human populations (and the resulting changes in land use and increased water use), the increasing numbers of man-made structures such as dams, weirs and barrages and recognition of the important role of sediment in the transport and fate of contaminants within river systems, has meant that water managers today face many complex technical and environmental challenges in relation to sediment management.

UNESCO launched the International Sediment Initiative (ISI) in 2002. ISI aims to further advance sustainable sediment management on a global scale. This is achieved through the delivery of a decision support framework for sediment management that provides guidance on legislative and institutional solutions, applicable across a range of socio-economic and physiographic settings in the context of global change. ISI mobilizes international experience on sediment problems and their management through the compilation of a series of case studies representative of a broad range of physiographic and socio-economic conditions, which are made available as guidance for policy makers dealing with water and river basin management.

Case studies prepared to date as a key component of the ISI, include studies of the Nile River Basin, the Mississippi River Basin, the Rhine River Basin, the Volga River Basin, the Yellow River Basin, and the Haihe and Liaohe River Basins. Available in full from the ISI website at: http://www.irtces.org/isi/. The purpose of these case studies is to:

- Increase awareness of erosion and sedimentation issues;
- Increase understanding of erosion and sediment transport processes and associated sediment problems under different conditions;
- Improve the sustainable management of soil erosion and sediment transport by providing examples of monitoring and data processing techniques, methods for analysing their environmental, social and economic impacts; and effective management strategies and
- Ultimately assist in the provision of better advice for policy development and implementation and evaluation of management practices.

These existing ISI case studies are briefly introduced in the paper, which aims to provide an accessible overview of sediment problems and sediment management around the world for water managers and policy makers. Key issues relating to sediment management and recommendations for developing management strategies presented in this paper have been extracted from these case studies.

REFERENCES

Abdalla Abdelsalam Ahmed, 2008. Sediment in the Nile River system. UNESCO-ISI report, http://www.irtces.org/isi/.
Golosov, V. and Belyaev, V., 2010. The Volga River basin report. UNESCO-ISI report, http://www.irtces.org/isi/.
IRTCES, 2004. Sediment issues of the Haihe and Liaohe Rivers in China. UNESCO-ISI report, http://www.irtces.org/isi/.
IRTCES, 2005. Case study on the Yellow River sedimentation. UNESCO-ISI report, http://www.irtces.org/isi/.
Julien, P.Y. and Vensel, C.W., 2005. Review of sedimentation issues on the Mississippi River. UNESCO-ISI report, http://www.irtces.org/isi/.
Spreafico, M. and Lehmann, C., 2009. Erosion, transport and deposition of sediment – case study Rhine. UNESCO-ISI report, http://www.irtces.org/isi/.
UNESCO-ISI, 2011. Sediment issues and sediment management in large river basins – Interim Case Study Synthesis Report. UNESCO-ISI report, http://www.irtces.org/isi/.

Contribution in the study of sediment transport in northern Algeria

M. Meddi
Ecole Nationale Supérieure d'Hydraulique de Blida, LGEE, Algeria

ABSTRACT

There is currently an increasing erosion phenomenon in watersheds. Several authors presented blatantly the extent of the erosion phenomenon in Algeria (the most significant values in North Africa that can reach 2000 t/Km^2/year). Water erosion is the most common and severe form (Meddi et al. 1998; Meddi & Morsli 2001). It depends on the area resistance and climate aggressiveness. The intensity of erosion occurs mainly during flood periods, where a significant amount of sediments passes to impoundments.

Erosion, sediment transport and sedimentation are the cause of the degradation of agricultural soils, siltation of dams and a lot of damage whose costs are considerable. In Algeria, for instance, the hydro-mechanical development studies come up against the problem of the lack of data on sediment transport in order to assess its importance. This leads the engineer to use the available empirical methods for assessing the solid quantities transported annually by the river under examination. For this, we considered interesting to develop a model for calculating sediment transport, on an annual scale, based on two parameters available at the level of all the watersheds controlled by hydrometric stations, i.e., liquid flow and surface area. To give more meaning to the model, we used data from 71 Algerian stations and dams. After using the data of 55 measuring sites for calibrating the model, the validation of the model on 16 stations using the Nasch criterion was discussed. For the relationship to be linear, the variables were log-transformed. All Correlation coefficients are significant at the 95% threshold. The developed model is a power model with a multiple correlation coefficient of 0.87. The Nash criterion in the calibration and validation of the model is 78% and 83% respectively. This model must be operated with caution to prevent the errors that may be caused by the application thereof. The results show the existence of relationship between the specific degradation, the liquid discharge, the pluviometry represented by the rainfall erosivity factor and the area. The importance of the model lies in the availability of explanatory factors on practically all the Algerian watercourses. This forecasting model provides an estimation tool of the specific degradation for the planning and designing of dams and hydro technical structures as well as the development of catchment basins. The estimation of land losses is essential to preserve and manage the dams that are confronted with the silting up issue (Remini & Hallouche 2005). The model is also a tool for the study and dimensioning of flood and inundation control structures that northern Algeria is subject to on a regular basis. The estimation of land losses is also essential in the evaluation of the water erosion consequences on agriculture, which are represented by the widespread soils impoverishment and the decrease in the fertility of those soils (Merzouki et al. 1994).

REFERENCES

Meddi, M., Khaldi, A. & Meddi, H. 1998. Contribution à l'étude du transport solide en Algérie du Nord. Pub. AHS Pub. No. 249. Pp 393–398.

Meddi, M. & Morsli, B. 2001. Etude d'érosion et du ruissellement sur bassins versants expérimentaux dans les monts de Beni-Chougrane. Revue Z. Gemorph NF 2001; 45/4/443–452.

Merzouk, A., Rayan, J. & Kacemi, M. 1994. A perspective on soilerosion in Morroco's dry land semi-arid zone. Actes du colloque International des Sciences du Sol: " Sciences du sol au développement ", Rabat, Maroc, 6–8 Avril 1993, 12 p.

Remini, B. & Hallouche, W. 2005. Prévision de l'envasement dans les barrages du Maghreb. Larhyss Journal, ISSN 1112-3680, n° 04, Juin 2005, pp.69–80.

Application of airborne gamma-ray imagery to assist soil survey in the upper Pasak basin, Thailand

R. Moonjun
Department of Earth Systems Analysis, ITC Faculty, University of Twente, Enschede, The Netherlands
Land Development Department, Ministry of Agriculture & Cooperatives, Thailand

D.P. Shrestha
Department of Earth Systems Analysis, ITC Faculty, University of Twente, Enschede, The Netherlands

K. Duangkamol
Land Development Department, Ministry of Agriculture & Cooperatives, Thailand

ABSTRACT

Gamma-ray radioelements are a potential information source for soil mapping, since their abundance is related to soil geochemistry, specifically the chemical composition of parent materials and their weathering products resulting from geomorphic and pedogenic processes (IAEA 2003). Soils have been developed over sedimentation landscapes such as in alluvium and flood plain are uncertain due to the sedimentation process, which is difficult to characterize by field soil survey. Thailand has advantage of Airborne Gamma-ray Imagery (AGRI) covering whole country, for mineral survey but not yet for soil survey. Fuzzy classification method has been recommended as the potential techniuqe for digital soil mapping (McBratney et al. 2003).

The aim of this study was to evaluate the potential of AGRI for improving digital soil survey process in two phases: (1) a preliminary phase, where hypotheses of soil-geological unit relationships where soils are developed, and (2) a phase where soil map unit boundaries are generated using AGRI and DEM covariables.

The study was conducted in a well-characterized complex in flood plain and river terraces soil landscape: the 300 km^2 upper Pa Sak Valley in Petchaboon province, Thailand.

The relationship between AGRI data and geological units were examined by the distribution of radioelement response to selected soil sediment characteristics, based on a review of literature and supported by observation samples. The potential of AGRI to map sediment soils were examined by interpolate soil observation points applying fuzzy classification over the alluvium and flood plain. Where, soil map was produced from four covariables; three from gamma-ray radio element (%K, eTh and eU) and DEM using fuzzy classification.

To interpret the relation of gamma-ray radioelement with sediment geological units, the results show that, the Pleistocene sediments (Qt), consisting of mixed old alluvial materials on terraces, shows three different radioelement compositions. The Holocene sediments, indicated by Qa, area cover a relatively large area and they are high in all three radioelements.

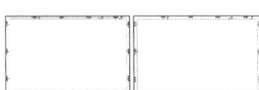

Figure 1. (left) Airborne Gamma-Ray Imager (AGRI), with geologic units and (right) predicted soil map from AGRI based on fuzzy classification with 92 observation points (black dots).

In the producing soil map, with a soil family level, the best result map was produced and four soil mapping units were interpolated as: 1) Fine silty, mixed, Aeric Vertic Endoqualfs, 2) Fine, mixed, Typic Endoaqualfs, 3) Loamy-skeleton, kaolinitic, Typic Paleustults and 4) Loamy-skeletal over fragmental, kalionitic, Petroferric kalionitic, Petroferric Haplustults. The result map was assessed the accuracy using 92 observation point data, the accuracy result are overall classification accuracy = 67.85% and 2 overall Kappa statistics is 0.6419.

In the hypotheses-generating stage, AGRI provided useful information in three forms (single signal, ratio, and lithology, material transport, and internal pedogenetic processes). In the mapping stage, AGRI showed deficiencies in the soil map production over alluvium terraces and flood plains, which provided a basis for future field sampling to correct these deficiencies.

AGRI suggested new boundaries, differentiating topsoil properties and the presence of plinthite. Further study, AGRI can be recommend for mapping some top soil properties together with other suitable covariables according to soil forming factors.

REFERENCES

IAEA. 2003. Guidelines for radioelement mapping using gamma ray spectrometry data In: International Atomic Energy Agency, Vienna.

McBratney et al. 2003. "On digital soil mapping." Geoderma 117(1–2): 3–52.

More bed load in rivers. Achieving a sediment balance close to the natural state in the canton of Bern

M. Pauli & L. Hunzinger
Flussbau AG SAH, Bern, Switzerland

O. Hitz
Civil Engineering Office Canton Bern, Section I, Bernese Oberland, Switzerland

ABSTRACT

The revised Water Protection Act has been in force in Switzerland since 2011. The act states that rivers are to be restored and that negative effects on the ecosystem caused by installations are to be overcome. Hydropower dams, river control structures, gravel extraction and other installations should not impair the bed load sediment balance of the waters. This means that the waters must transport enough bed load to fulfil the habitat requirements of animals and plants. Installations that affect the bed load sediment balance excessively must be remediated by 2030.

The authorities of the canton of Bern examined the bed load balance of the rivers on its territory and identified the installations that affect the bed load balance. In total, 600 km of river courses and 280 installations were examined.

Bed load yield in the catchment areas of the target water courses was estimated based on existing studies. In addition, the bed load transport associated with a reference state was estimated. Reference conditions are hypothetical. They exclude installations in the catchment area.

By comparing the bed load under current conditions with the bed load under reference conditions, that is to say with the required total bed load, it was possible to determine the degree of impairment of the bed load for individual river sections.

An approach proposed by Schälchli & Kirchhofer (2012) was used to define the required bed load. This approach postulates that the total bed load is sufficient if gravel bars that form in a river with natural morphology get their 30 cm thick top layer replenished each year. The average surface area occupied by gravel bars was determined using historic maps or aerial views of water courses close to their natural states.

Restoration of bed load transport only makes sense where benefits are expected. This is the case when a) the ecomorphology of a river is already natural or only slightly impaired or when b) the restoration of a channelized river will bring benefits for nature and landscape. In either case, the installations in the catchment area that contribute to the disturbance of the bed load transport were assigned the obligation to remediate.

In the canton of Bern installations affect the bed load transport on 56% of the length of the rivers of concern. Out of 280 examined installations, 56 have the obligation to remediate the continuity of bed load transport. However, one may observe regional differences according to the type of the installation:

Gravel traps reduce the bed load in rivers on the entire territory of the canton of Bern. So do dragging pits, although they have been decreasing in number and in the volume extracted for the last 20 years. In the lowlands large hydropower dams interrupt the bed load transport completely, whereas in the Jura Mountains small hydropower dams retain only part of the bed load. As bed load rates are already low in the Jura Mountains, the impact of the small dams is severe anyway.

The strategy to reactivate the bed load transport in the rivers of the canton of Bern was planned on a large scale. The strategy proposes constructive or management measures to remediate the sediment continuity of installations and sets a deadline for their implementation. The planning must be concretised in the coming years. An increase in annual bed load rate is most effective in rivers with large river beds that allow for morphologic patterns and habitats to form dynamically. Measure to increase the bead load rate will therefore be strongly interlinked with the restoration of rivers.

REFERENCE

Schälchli, U. & Kirchhofer, A. 2012: Sanierung Geschiebehaushalt. Strategische Planung. Ein Modul der Vollzugshilfe Renaturierung der Gewässer. Bundesamt für Umwelt, Bern. Umwelt-Vollzug Nr. 1226: 74 S.

Van Deemter's analysis of drainage to incised ditches in lowland areas

M.J.M. Römkens
USDA-ARS-Nat'l Sedimentation Lab., Oxford, Mississippi, USA
Department of Civil Engineering, University of Mississippi, Oxford, Mississippi, USA

ABSTRACT

Drainage of flat land areas in agricultural watersheds commonly takes place through embedded drains and/or incised open ditches. The effectiveness of these systems depends on a host of factors, including their spacing, depth, the wetted perimeter through which draining water accrues to open water bodies, the soil hydraulic conductivity, soil layering, etc. In this study the effect of the drainage system, drains vs. incised ditches, is determined using the analytical approach of conformal theory by van Deemter (1950) for a homogenous, infinitely deep aquifer under steady flow. Two ditch conditions were considered: (1) Figure 1 represents a situation of a rectangular ditch that does not contain free water and in which the vertical wall is a seepage zone over a finite depth. The groundwater table in the adjacent land mass is curvilinear and reaches the highest point midway between two similar parallel ditches. (2) Figure 2 represents a situation in which the vertical wall is impervious and in which the ground water level in the land mass at the ditch has a finite height, b, above the water level, h0 in the ditch. The solution approach taken by Van Deemter (1950) is similar to that for the case of drains

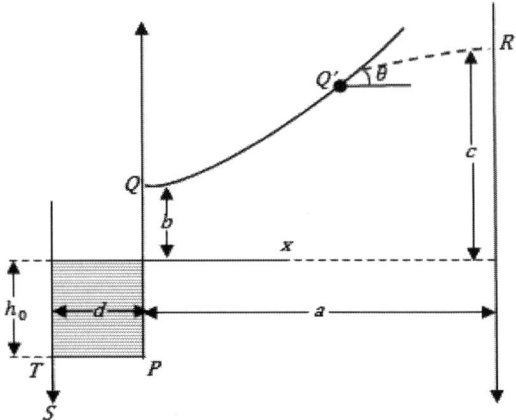

Figure 2. Diagram flow region for a rectangular ditch with an impervious vertical wall PQ and a water level depth, h0.

as discussed by Römkens (2013). Van Deemter (1950) presents a comparison of the water level in the land mass adjacent to the ditch relative to the water level midway between the ditches for various values of the precipitation/hydraulic conductivity regime. The data show appreciable differences between the flow regime near the drain and the open ditch, with the strongest effect for the drain system.

REFERENCES

Römkens, M.J.M. 2013. A theoretical treatise of drainage and seepage in bottomland areas adjacent to incised channels: The J.J. van Deemter analysis. Advances in River Sediment Research Fukuoka et al. (Eds). Taylor and Francis Group, London. ISBN 978-1-138-00062-9.

Van Deemter, J.J. 1950. Theoretical and numerical treatment of flow problems connected with drainage and irrigation. Verslagen Van Landbouwkundige Onderzoekingen. No 576. Dutch Ministery of Agriculture. Fisherie and Food supplies. The Hague, Netherlands. (In Dutch).

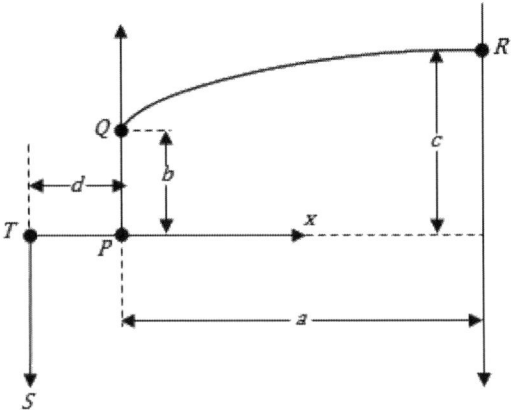

Figure 1. Diagram flow region for a rectangular ditch without free water.

An overview of hydro-sedimentological characteristics of intermittent rivers in Kabul region of Kabul river basin

N. Sadid, S. Haun & S. Wieprecht
Institute for Modelling Hydraulic and Environmental Systems, University of Stuttgart, Stuttgart, Germany

ABSTRACT

Intermittent rivers are classified as rivers with a high seasonal variability of flow. Such rivers may partially dry up during the year (usually from weeks to a few months), resulting in water as well as habitat discontinuity. However, due to the strong correlation to the hydrological characteristics of the catchment this occurs on a more or less predictable basis. The majority of Afghanistan's basins are drained by intermittent river systems and are the main water sources for irrigation and energy production. The river discharges and catchment precipitation comparison indicate that the high flow conditions originate from snowmelt and seasonal rain events, which result in slightly delayed discharge peaks in the rivers. The delayed discharge is considered very important for water management and especially for irrigation because the demand for water use increases when the temperature rises. The objective of this study is to give an overview on the intermittent rivers of Afghanistan and to out-line the major hydro-sedimentologic characteristics of these special river systems. Intermittent rivers undergo a strong seepage loss, where the downward component of flow velocity carries fine sediment in-to the coarse gravel bed and causes external as well as internal colmation of the river bed. This phenomenon can be seen from the grain size analysis of the armour layer of the river bed, which contains a considerable amount of fine sediments. As a result of river bed colmation, the groundwater recharge from the surface water is reduced until deposited sediments are remobilized by higher discharge events. In this study, the seepage loss through the river bed is analyzed by comparing the measured groundwater levels in wells located in the vicinity of rivers, together with the measured river discharge rates. Field observations of the river bed, surface and sub-surface's bed material through grain size analysis and suspended sediment data are used to analyze external colmation effects. This is shown on the example of Maidan River, a tributary of Kabul River in figure 1.

This intermittent river has a catchment area of 1,625 km² and remains dry from July to November

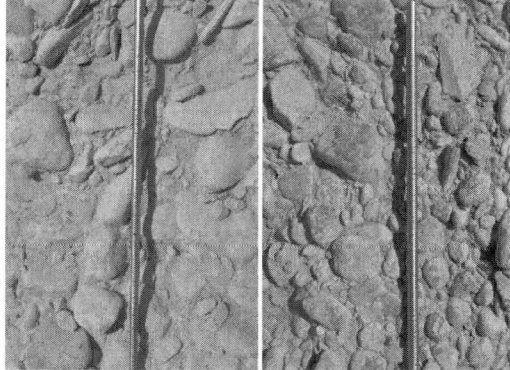

Figure 1. External colmation at Maidan River.

each year. The data concerning precipitation in the catchment, discharge rates, water levels in wells as well as sediment data are available. From investigated data can be seen that the monthly measured water levels in vicinity wells reflect the flow fluctuations in the river (discharge rates). This is a result of the strong seepage losses in intermittent rivers because the ground water levels are located far beneath the water level of the river during the dry period. The dry period causes also a reduction in vegetation cover of the catchment area, which may increase the erosion within the catchment during the wet season. This effect is observed from suspended load measurements, where the concentrations are much higher compared to a similar perennial river. Consequently, this heavy sediment load leads to reservoir sedimentation and so into a premature loss of storage volume. Amir Ghazi irrigation dam for instance, built on another intermittent river of similar characteristics in Kabul river basin, is not functional anymore due to excessive sedimentation. Intermittent rivers transport in addition to the suspended sediments also a considerable amount of bed load material.

Analyses on trends and reasons of runoff and sediment load of Yellow River stem

H.L. Shi & C.H. Hu
International Research and Training Center on Erosion and Sedimentation, Beijing, P.R. China

A.J. Deng
China Institute of Water Resources and Hydropower Research, Beijing, China

Q.Q. Tian
Ministry of Water Resources, Beijing, P.R. China

ABSTRACT

The Yellow River basin is vast in area with varying natural and geographical conditions, and the imbalance of sources of water and sediment throughout the basin is very prominent. However, as the mother river of the Chinese people, the Yellow River has suffered lots of impacts from human activities. Since the 1960s, large-scale artificial projects have been carried out, casting significant impacts on the water sources and sediment load of the Yellow River (showing in Table 1 and Figure 1).

The variation of runoff and sediment load of the Yellow River has suffered dramatic change under the impacts of the climate change and human activities. Based on the Mann–Kendal trend test for the data from 1950 to 2013 of the stem river, the sharp downward trends both proved in annual runoff process and annual sediment load process in the Yellow River, and their abrupt changes confirmed in 1986 and in 1980 respectively, according to the rank sum test. The factors of impacting on the changes of the runoff and sediment load are absolutely complicated. Through analysis on climate change, especially the precipitation, combined with the human activities, such as water diversion, reservoir constructions, water and soil conservation, etc., it is found that the increasing water consumption is the one of main contributions of the runoff decline after 1980, and the completion of Longyangxia Reservoir and its combined operation with Liujiaxia

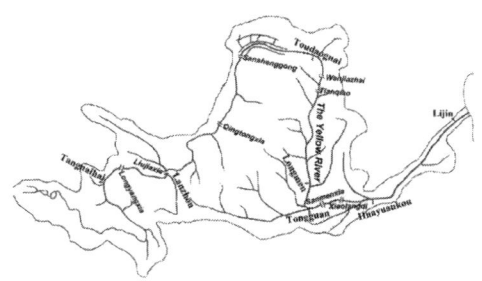

Figure 1. Sketch of Yellow River Basin.

Reservoir have exerted a direct bearing on the abrupt change of annual runoff; In addition to the reduction of annual runoff, which will inevitably lead to the annual sediment load decrease, the combined effect of the water-soil erosion control and the reservoir construction have played a leading role in the trend and abrupt change of sediment load decline.

REFERENCES

Gao et al. 2011, Changes in streamflow and sediment discharge and the response to human activities in the middle reaches of the Yellow River, Hydrol. Earth Syst. Sci., 15, 1–10

Hassan et al. 2008, Spatial and temporal variation of sediment yield in the landscape: example of Huanghe (Yellow River). Geophysical Research Letters, 35, L06401.

Liu et al. 2008, Changes in runoff and sediment yield along the Yellow River during the period from 1950 to 2006, Journal of Environmental Informatics, Vol. 12, No. 2, pp. 129–139.

Shi et al. 2002. Changes in sediment yield of the Yellow River basin of China during the Holocene. Geomorphology 46, 267–283.

Sui et al. 2008, Flow and high sediment yield from the Huangfuchuan watershed, International Journal of Environmental Science and Technology, Vol. 5, No. 2, pp. 149–160.

Sui et al. 2009, Changes in sediment transport in the Kuye River in the Loess Plateau in China, International Journal of Sediment Research, Vol. 24, No. 9, pp. 201–213.

Table 1. Basic information of main reservoirs on stem Yellow River.

Reservoir	Longyang-xia	Liujia-xia	Sanmen-xia	Xiaolang-di
Control area/10^3 km^2	131.4	181.8	688.4	694.5
Capacity/10^9 m^3	24.7	5.7	9.64	12.65
Date of commissioning	1986.10	1968.10	1960.10	1999.10

Identifying the sources of fine sediment to quantify the success of sustainable flood risk strategies

S. Twohig & I. Pattison
School of Civil and Building Engineering, Loughborough University, Loughborough, Leicestershire, UK

ABSTRACT

Enhanced levels of fine sediment entering our rivers have several negative impacts on the functioning of the system. The recent flooding of the Somerset Levels, UK in the Winter 2013/14 highlighted the potential impacts of reduced channel capacity as a result of sedimentation. Over 6500 hectares of agricultural land flooded for approximately 54 days. Fi-ne sediment accumulation in river channels has also been found to degrade fish spawning and invertebrate habitats. Nutrients and heavy metal contaminants are often associated with fine sediment, which reduces water quality. Increases in sediment delivery to river channels from changes in land use and cli-mate in the future are projected to increase the severity and occurrence of these issues.

Catchment managers need to put into practice measures which will reduce the quantity of fine sediment that is delivered to river channels. To do this they need to know where the sediment is coming from i.e. source and either prevent it being eroded or disconnect the pathway between source and channel.

The River Eye, Leicestershire is a typical UK lowland sand bed river with relatively homogenous geology. The 180 km^2 catchment is dominated by agricultural land use on heavy clay soils. The town of Melton Mowbray centered within the catchment has a long history of flooding. In 2002, the Melton Mowbray Flood Alleviation Scheme was implemented to reduce flood risk in light of the 1998 flood event. Online silt traps were installed to reduce fine sediment deposition within the town. Since installation, fine sediment is still a persistent problem for catchment managers within the River Eye, though its sources are unknown.

22 suspended sediment samplers were deployed throughout the catchment at strategic locations (upstream and downstream of the silt traps, confluences and Brentingby Dam) to identify which locations were delivering the most sediment. These devices were regularly emptied and analyzed.

The silt trap situated on the main channel was monitored to investigate its influence on local and

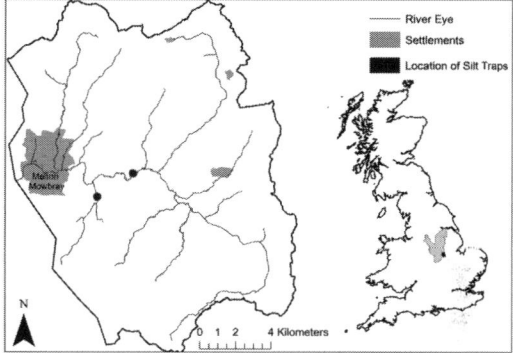

Figure 1. Location of Silt Traps in the River Eye Catchment.

reach scale flood risk. Four monitoring devices were installed to continuously measure river stage. These combined results provide an insight into the success of silt traps as natural flood defenses.

To determine the sources of fine sediment in this catchment, the approach of sediment fingerprinting was used. Originating in the 1990's, this technique has seen substantial recent attention, which start to question some of the basic assumptions behind the general approach, such as quantifying the uncertainty in the predictions, the non-conservativeness of some of the sediment properties, and the validity of correction factors and weightings.

The sampling strategy has been carefully de-signed to account for all the known land uses and at different levels of complexity. Standard laboratory tests were carried out on both source and channel samples, to distinguish between different land use categories in seemingly homogenous geology. This process provided an extensive database for the statistical assumptions of the sediment fingerprinting approach to be questioned and to identify sources of fine sediment. This knowledge can then be used by catchment stakeholders i.e. Environment Agency to inform flood risk management.

Stream flow modeling for a karst basin using coupled hydrological-hydrodynamic models: Case study of Lijiang River, China

Q.F. Wu, Y. Cai & S.G. Liu
Department of Hydraulic Engineering, College of Civil Engineering, Tongji University, Shanghai, China

Y.M. Jiang
Hydrology & Water Resources Bureau, Guilin, Guangxi Zhuang Autonomous Region, China

A.N. Makhinov & A.F. Makhinova
Institute of Water and Ecological Problems, Far East Branch, Russian Academy of Sciences, Khabarovsk, Russia

ABSTRACT

The karst basin is a highly non-linear complex system in the domain of hydrology, with a dual space structure of surface and subsurface. In some special zones like epikarst zone which are widely developed in Karst basins, hydrologic and hydrodynamic conditions usually are greatly different from those of normal areas. Therefore, it may be difficult to simulate the stream flow in Karst basins using some simple linear models.

This paper presents a comparison among three types of modified Xinanjiang models (XAJ) with different model structures for a sub-basin of the Lijiang River basin which is a typical karst basin. The complex model structure may result in the difficulty in model calibration for numerous model parameters. One-dimensional hydrodynamic models for Lijiang River were established by MIKE 11 HD coupled with Xinanjiang models.

Observed daily data from gauge stations during 1996–2000 and 2001–2002 were used for calibration and validation, respectively. The results show that these Xinanjiang models perform well with Nash-Sutcliffe efficiency (NSE) varying from 0.918 to 0.945, and most absolute relative error (RE) being less than 5%.

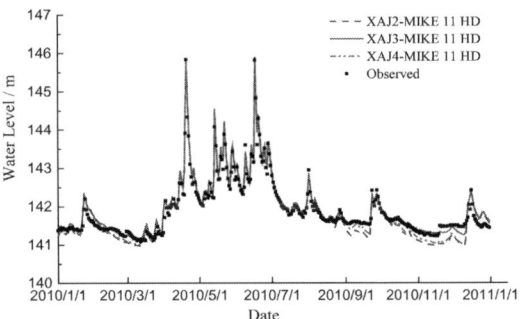

Figure 2. Hydrograph of coupled models at Guilin Station.

Furthermore, the Xinanjiang model with three components of runoff (XAJ3) performs better than the other two because of its appropriate model structure and number of parameters. It indicates that the coupled hydrological-hydrodynamic model is a suitable approach to model the stream flow in karst basins. For the coupled hydrological-hydrodynamic model, the different characteristics of simulated error during flood and recession period reveal the interaction between river water and groundwater in karst area of Lijiang River.

REFERENCES

Hartmann, A., Goldscheider, N., Wagener, T., Lange, J. & Weiler, M. 2014. Karst water resources in a changing world: Review of hydrological modeling approaches, *Reviews of Geophysics* 52(3): 218–242.

Jourde H., et al. 2006. Dynamics and contribution of karst groundwater to surface flow during Mediterranean flood, *Environmental Geology* 2006 (51): 725–730.

Li, R., Chen, Q., Tonina, D. & Cai, D. 2015. Effects of upstream reservoir regulation on the hydrological regime and fish habitats of the Lijiang River, China, *Ecological Engineering* 76: 75–83.

Zhao R.J. 1992. The Xinanjiang model applied in China, *Journal of Hydrology* 1992 (135): 371–381.

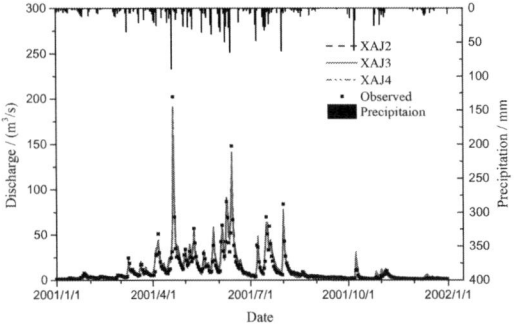

Figure 1. Hydrograph of XAJ models at Chaotian Station.

Mechanical effects of vegetation in soil conservation and soil erosion reduction

P.Q. Xiao, W.Y. Yao, Z.Z. Shen & C.X. Yang
Yellow River Institute of Hydraulic Research, Key Laboratory of Soil and Water Loss Process and Control on the Loess Plateau of the Ministry of Water Resources, Zhengzhou, Henan, China

ABSTRACT

Soil erosion is to-day one of the most serious economic problems in the world but especially on the Loess Plateau in China (Xiao *et al.* 2011). Improved knowledge of the effect of grass and shrub cover on resisting soil erosion can provide valuable information for soil and water conservation programs (Neave & Abrahams, 2002). Numerous studies have been conducted to document the effects of vegetation in reducing runoff and soil loss, but little research has been done to study soil erosion mechanical process on vegetation covered plots (Neris, *et al.* 2012, Hu *et al.* 2009). So, soil shear strength on different vegetation-covered slope was important to study the mechanical effects of vegetation on soil conservation and soil erosion reduction (LI, *et al.* 2013). The experiments were conducted at the hilly-gully region and water-wind erosion criss-cross region in of North part of shaanxi Prov., China, where is a highly erosion-prone region that is covered with thick, highly erodible loess material.

The results showed that soil shear strength increased linearly with the increase of vertical stress on different vegetation-covered slope. Shear strength was proportional to the vertical pressure of shear surface, and it was in accord with Coulomb's law. Soil shear strength, soil cohesion force and internal friction angle on grass and shrub slope were significantly greater than those of the bare slope due to the role of vegetation on reinforcing soil. Soil cohesion force on grass and shrub slope was 1.04–2.11 times larger than that of the bare slope. Internal friction angle on grass and shrub slope was 1.03–1.19 times larger than that of the bare slope. Cohesive force had a significant negative correlation with sediment yield. The sediment yield showed a decreasing trend with the increase of soil cohesive force. The sediment yield increased with the increase of flow shear stress, and they had a good linear relationship. The relationship between the critical shear stress of runoff and soil shear strength and cohesion was established, which was helpful to study erosion process from mechanical mechanism.

REFERENCES

Hu, X.S., Li, G.R., Hu, H.L., et al. 2009. Research on Interaction between Vegetation Root and Soil for Slope Protection and Its Mechanical Effect in cold and arid environments. Chinese Journal of Rock Mechanics and Engineering, 28(3): 613–620. (in Chinese)

Li, J.X., He, B.H., Chen, Y. et al. 2013. Root distribution features of typical herb plants for slope protection and their effects on soil shear strength. Transactions of the Chinese Society of Agricultural Engineering, 29(10): 144–152.

Neave, M. & Abrahams, A.D. 2002. Vegetation influences on water yields from grassland and shrubland ecosystems in the Chihuahuan Deser. Earth Surface Processes and Landforms, 27(9): 1011–1020.

Neris, J., Jiménez, C., Fuentes, J., et al. 2012. Vegetation and land-use effects on soil properties and water infiltration of Andisols in Tenerife (Canary Islands, Spain). 98: 55–62.

Xiao, P.Q, Yao, W.Y. & Römkens, M.J.M. 2011. Effects of grass and shrub on the critical unit stream power in overland flow. International Journal of Sediment Research, 26(3): 387–394.

Impact of sediment management on integrated water resources management

F. Yazdandoost & F. Farahani
Department of Civil Engineering, K N Toosi University of Technology, Tehran, Iran

ABSTRACT

Sediments carried into reservoirs have adverse effects both on the operation of the reservoir and the water management of the basin relying on anticipated volumes gained from the reservoir. Problems can exacerbate in cases where the basin faces water scarcity and/or excessive sediment loads from upstream rivers. It is therefore imperative to examine the role of sediment transport in the operation of the receiving reservoir within a greater perspective of the Integrated Water Resources Management (IWRM). A toolbox has been designed here to address the issue based on sustainability criteria. The toolbox comprises an estimating engine for the river sediment transport, an IWRM allocation model and a Multi Criteria Decision Making (MCDM) tool. The toolbox may be utilised to investigate the effect of various upstream sediment management scenarios on the role of reservoir in water resources management of the basin. Further assessments are made possible, in an integrated manner, based on sustainability criteria encompassing socio-economic and environmental characteristics of the basin.

Sefid-rud river basin in North West of Iran has been adopted here for the case study for its water resources dependence on the reservoir dam and its unique sediment characteristics. The basin is entirely located in the Guilan Province and as far as the water resources characteristics are concerned the supply is primarily regulated by the upstream reservoir dam and the demand is mainly affected by downstream land use dedicated to forests, unfed agriculture, pasture and urban areas. Figure 1 depicts the general location of the basin.

Application of the proposed toolbox has revealed projected unmet demands from the reservoir for the year 2041 based on values of reservoir capacity versus sediment transport initially calibrated for 2006. Three main demand sectors were considered as agriculture,

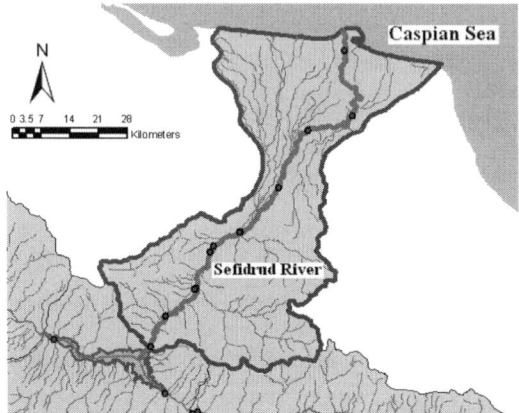

Figure 1. The general location of the basin.

industry and drinking water. Various demand scenarios were generated individually and in combination to assess the role of reservoir sedimentation on the water resources allocations. The DEFINITE 2 model was used to rank scenarios based on a multi criteria decision making approach. The toolbox may facilitate decision making based on integrated water resources management in cases directly affected by riverine sediment transport and reservoir sedimentation.

REFERENCES

Mehta, V., Aslam, O., Dale, L., Miller, N. & Purkey, D. 2013, Scenario-Based Water Resources Planning For Utilities in the Lake Victoria Region. Physics and Chemistry of the Earth, Vol. 61–62, pp 22–31.

Shi, Z., Ai, L., Fang, N. & Zhu, H. 2012, Modeling the Impacts of Integrated Small Watershed Mangement on Soil Erosion and Sediment Delivery. Journal of Hydrology, Vol. 438–439, pp156–167.

Integrated sediment transport modelling for rivers feeding lakes and wetlands

F. Yazdandoost & N. Khorami
Department of Civil Engineering, K N Toosi University of Technology, Tehran, Iran

ABSTRACT

Variation in normal sediments transported in rivers and watercourses, primarily due to excessive utilization and rapid land use changes in the basin, may threaten sensitive receiving wetlands and lakes. Loss of operational volumes as a result of reductions in supply and disruptions in the performance of wetlands in terms of quality and ecosystem are some of the resulting adverse impacts amongst many. A toolbox has been designed here to address the issue based on sustainability criteria. The toolbox comprises an estimating engine for the sediment transport, namely the SWAT model, a powerful GIS tool and a Multi Criteria Decision Making (MCDM) tool. Flow variations as well as nitrate concentrations would also be investigated alongside sediment quantities for various potential scenarios. Ranking of individual and combined scenarios may be obtained based on decision makers' priorities while the toolbox has the capability to incorporate various desired sustainability criteria.

The toolbox has been utilized for the case of Zarineh-Rud river basin feeding Lake Urmia. Lake Urmia, one of the largest saltwater lakes on earth and a highly endangered ecosystem, is on the brink of a major environmental disaster similar to the catastrophic death of the Aral Sea. Once with a surface area of approximately half a million hectares, Lake Urmia's shoreline has been receding severely with no sign of recovery, leading to a significant shrinkage in the lake's surface area currently decreased by around 88%. Lake Urmia is fed by a total of 60 (21 seasonal or permanent, and 39 periodic) rivers with the Zarrine-Rud being the main input, reckoned to provide around 40% of the total inflow into the lake. Evidently sediment transport conditions in Zarine-Rud will have grate impacts both on the morphology of the river course and the sensitive conditions of Lake Urmia.

Table 1. Results of various development scenarios.

Scenarios	amount of increase in flow rate from base model (m^3/s)	amount of change in sediment delivery from base model (ton/day)	amount of change in nitrate delivery from base model (kg/day)	Anticipated increase in the volume of the lake ($10^6 m^3$)	Anticipated increase in the height of the lake (mm)
Covering plants	5.5	-102	-47.7	173.44	34.69
Planting strips along contours	2.2	-74.8	-29.2	69.37	13.88
Terracing	1.7	-53.6	-20.3	53.61	10.72
Grassed waterway	15.5	-97	-33.5	488.80	97.76
Water transfer	6.2	+38.1	+13.1	195.52	39.10
Stopping water transfer from Zarineh-Rud to Tabriz	2.6	+14.3	+7.5	81.99	16.4
Limiting the operation of the Bookan dam	6	+22.4	+10.8	189.21	37.84

Results obtained using the toolbox indicates that impacts of erosion and sedimentation in river basins and receiving lakes and wetlands may be effectively investigated under various development scenarios thereby providing a decision support tool for operation and planning.

REFERENCES

Tripathi, M.P., Panda, R.K. & Raghuwanshi, N.S. 2003, Identification and Prioritization of Critical Sub-watersheds for Soil Conservation Management using the SWAT Model. Biosystems Engineering, Vol. 85 (3), pp. 365–379.

Slob, A.F.L., Ellen, G.J. & Gerrits, L. 2008, Sediment Management and Stakeholder Involvement. Sustainable Management of Sediment Resources: Sediment Management at the River Basin Scale, Elsevier B.V., pp.199–216.

Erosion risk mapping for Hulu Langat River basin

R. Zainal Abidin, N. Yusoff, M.S. Sulaiman & T. Mohamed Mustafa
Infrastructure University Kuala Lumpur, Kajang, Selangor, Malaysia

ABSTRACT

Hulu Langat river basin has become major area for water source in Klang Valley, Selangor, Malaysia. The topography of Hulu Langat basin which has major tributaries and a significant headwater features signifies the importance of Hulu Langat basin to surrounding communities. The erosion risk level along the Hulu Langat River was categorized by encapsulating the degree of soil erodibility and degree of erosiveness via the implementation of ROSE index and ROME scale. Rainfall erosivity is an erosive power of rainfall to cause soil loss. There are several factors that persuade the erosivity of rain such as the rain intensity, duration, the level of rain velocity and size of raindrops. Rainfall with high intensity and low frequency can produce more erosion compared to rainfall with high frequency and low intensity (Wei et al. 2009).

Table 1. Rose Index.

ROSE Index (MJ·mm/ha·hr)	Category
< 5000	Low
5000–10,000	Moderate
10,000–15,000	High
15,000–20,000	Very High
> 20,000	Critical

An alternative method to determine rainfall erosivity is "ROSE" Index (Roslan & Tew 1997) as shown in Table 1. In addition, weakness and physical properties of soils are inferred as soil erodibility. "ROM" scale (Roslan & Mazidah 2002) is a logical predictive calculation to indicate the degree of soil erosion tragedy or soil erodibility. The implementation of these scale and index successfully identifies the degree of erosion risk level along the Hulu Langat river basin. The erosion risk is further evaluated by mapping the monthly soil erosion risk maps, high erosion areas and pattern of net erosion and deposition for the high erosion rates in Hulu Langat river basin. The production of erosion risk map enables practitioners and policy makers to make decision for rehabilitation and other restoration work in future.

REFERENCES

Roslan, Z.A. & Tew, K.H. 1997. Rainfall Analysis in relations to Erosion Risk Frequency – Case Study (Cameron Highlands). Proceeding from 3rd International Conference on FRIEND, FRIEND '97, Postojna: Slovenia.

Roslan, Z.A. & Mazidah, M. 2002. Establishment of Soil Erosion Scale With Regards to Soil Grading Characteristic. 2nd World Engineering Congress, Sarawak, Malaysia.

Wei, W., Liding, C. & Bojie, F. 2009. Effects of Rainfall change on Water Erosion Processes in Terrestrial ecosystems: a review, Progress in Physical Geography 33(3):307–318.

The impact of sub-branch direction on sediment deposition in the condition of reciprocating mainstream flow

Mingcheng Zhu
Nanjing Hydraulic Research Institute, Nanjing, the People's Republic of China

ABSTRACT

Sediment deposition is an inevitable problem during the construction of hydraulic and navigation projects in the sub-branchs of tidal river reaches. For example, siltation in diversion canals of tidal river reaches, siltation in the downstream of tidal barriers, and siltation around power plant cooling water intake. Generally reciprocating flows were formed due to tidal effects in these areaes. As a result, sediment concentration of the sub-branch entrance reaches maximum value during high tide period. While ebb tide occurs, sediment starts silting and sediment concentration decreases. However, the sediment concentration inside the sub-branch varies according to whether there is a barraier upstresm.

For sub-branchs along Jiangsu coastal area, researchers have done a lot of works during recent decades. In 2002, Yan discovered that small tidal range and coastal salt marsh evolutionally forms wide and stable tidal creeks in northern region. In the south, however, a large number of narrow and dense grooves keep swinging and migrating obeying a law of siltation process. In the same year, Chen analyzed the formation and self adjustment function of tidal creeks at intersection area of two opposite direction tidal wave system. In 2005, Li put forward that regression tidal creek water from beach face is the main factors of sediment concentration asymmetry comparing flood and ebb tide. In 2010, Liu simulated development and deposition of coastal tidal flat using profile evolution model. In 2012, Wu presented the linear relation of tide ditch length and basin area which belongs to. When more intensive hydrodynamic conditions happens, multistage gully landform developes instead of more densive and deeper tidal creeks.

In addition to the effect of mainstream water and sediment conditions, channel length, direction, and bottom elevation of sub-branches also play a role on their sediment deposition. The corner angle at the entrance to the mainstream has influence on sub-branches's inflow sediment amounts and outflow velocity. This paper intends to use the method of a generalized mathematical model to calculate sediment deposition under several cases of different angles between sub-branches and mainstreams. The results will provide reference for relevant construction project in the future.

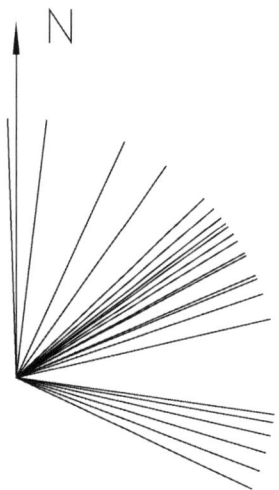

Figure 1. Main tidal creek direction towards to the sea in northern Jiangsu coastal area.

Development of sediment control structure for dam sedimentation counter measurement approach

J. Zulfan
UNESCO IHE Institute For Water Education, The Netherland
Research Center For Water Resource, Ministry of Public Works and Housing, Indonesia

N. Sasmito Slamet
Research Center for Water Resource, Ministry of Public Works and Housing, Indonesia

A. Prasetyo
Kyoto University, Japan
Research Center for Water Resource, Ministry of Public Works and Housing, Indonesia

ABSTRACT

Nowadays, many dams in Indonesia have sedimentation problem, one of which is Jatigede Dam in Sumedang, West Java Province of Indonesia. According to the Indonesian Ministry of Public Works, it is predicted that the rate of sedimentation of this dam reached approximately ± 7.7 million m³/year. If this condition occurs continuously, it would cause lots of damage such as reservoir shallowness and water shortage. There are many methods taken to reduce dam sedimentation, some of the frequent measures are dredging and flushing, which is very costly. Moreover, if it is not followed up with a comprehensive development in the upstream area then this methods will become less useful and ineffective. To overcome this challenge, the prototype of Sediment Control Structure was built and developed in Cikamiri River. This river is one of the Cimanuk river tributaries, which measured as the highest sediment rate supplier to Jatigede Dam. The structure was designed by integrating weir and sediment pond. Sediment pond structure is located off stream of the river to accommodate the maintenance activity.

The research was done by conducting continuous monitoring on water level, field measurement, and sediment sampling. The objectives is to detain the

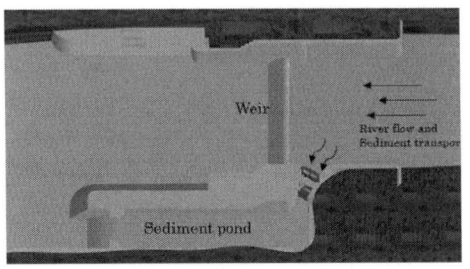

Figure 1. Sediment Control Structure.

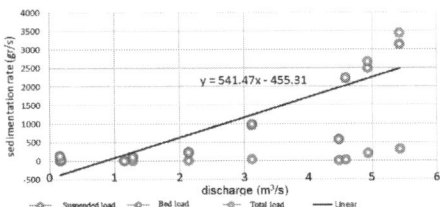

Figure 2. Sediment Transport rate in Cikamiri River.

sediment in the upstream river to reduce the sedimentation rate in Cimanuk River, which will eventually lowering the sediment supply into Jatigede Dam. Based on the study results, the sedimentation on Cikamiri river reach 22,228 m³/year.

Surprisingly, the Sediment Control Structure can detain the sediment around 4504 m³/year or 16% from the actual sedimentation on Cikamiri River. In a dry season the sediment flow about 1180 m³ and in a wet season it reach 3323 m³. This outcome shows us that the Sediment Control Structure is effective to detain the sediment in the upstream area before it flows to the reservoir.

REFERENCES

Garde, R.J. 2006. River Morphology, New Age International (P). Ltd., New Delhi.
Petersen, M. 1986. River Engineering. Prentice-Hall, Englewood Cliffs, New Jersey.
Research Center for water resources. 2012. Final Report of Monitoring Sediment Control Structure in Cibuah and Cikamiri River, RCWR, Bandung.
Research Center for water resources. 2013. Executive Summary of Sediment Control Structure. RCWR, Bandung.
Yang, C.T. 1996. Sediment Transport: Theory and Practice, The McGraw-Hill Companies, Inc, U.S.A.
Yiniarti. 2011. Sediment Transport. Master Programme on Water Resources. Departement of Civil Engineering, ITB, Bandung.

B. Sediment transport

Erosion and seepage failure around sheet-pile using two-phase WC-SPH model

A.M. Abdelrazek
Hydraulic Research Laboratory, Graduate School of Engineering, Hokkaido University, Hokkaido, Japan
Irrigation & Hydraulic Department, Faculty of Engineering, Alexandria University, Alexandria, Egypt

I. Kimura & Y. Shimizu
Hydraulic Research Laboratory, Graduate School of Engineering, Hokkaido University, Hokkaido, Japan

ABSTRACT

In this study, a Lagrangian formulation of the Navier–Stokes equations, based on the weakly compressible smoothed particle hydrodynamics (WC-SPH) method, was applied to simulate the seepage failure around sheet-pile.

Seepage around sheet-pile considers one of the most important factors that affect the stability and the performance of the cofferdams which commonly used in construction excavations.

SPH method is a mesh-free particle method and it is considered to be one of the most modern mesh-free particle techniques. It was originally invented for astrophysical applications; then it has been applied in a huge range of applications such as, free surface fluid flow (Monaghan 1994) and (Abdelrazek et al. 2014), multi-phase flow (Monaghan & Kocharyan 1995), snow avalanching (Abdelrazek et al. 2014), and gravity granular rapid flow (Abdelrazek et al. 2015). In SPH method, each particle in the domain carries all field variable information such as density, pressure, velocity and it moves with the material velocity. The governing equations in the form of partial differential equations are converted to the particle equations of motion, and then they are solved by a suitable numerical scheme.

In this simulation, the advantages of SPH will be exploited to simulate the soil–water interaction. Water is considered as a viscous fluid with week compressibility and soil is assumed to be an elastic–plastic material. The elastic–perfectly plastic model based on Mohr–Coulomb's failure criterion is implemented in SPH formulations to model the soil movement (Bui et al. 2007 & 2010). Interaction between soil and water is taken into account by means of seepage force and pore water pressure.

Numerical Simulation of the seepage around sheet-pile has been done; the numerical results are then compared with analytical solutions. The results have shown that the proposed model could be considered a powerful tool to simulate extremely large deformation and failure of soil.

REFERENCES

Abdelrazek. A.M., Kimura, I. & Shimizu, Y. 2014. Comparison between SPH and MPS Methods for Numerical Simulations of Free Surface Flow Problems, *Journal of Japan Society of Civil Engineers, Ser. B1 (Hydraulic Engineering)*, 70 (4), I_67–I_72.

Abdelrazek, A.M., Kimura, I. & Shimizu, Y. 2014. Numerical Simulation of Snow Avalanches as a Bingham Fluid Flow Using SPH method, *River Flow 2014*, 1581–1587.

Abdelrazek, A.M., Kimura, I. & Shimizu, Y. 2014. Numerical simulation of a small-scale snow avalanche tests using non-Newtonian SPH model, Journal of Japan Society of Civil Engineers, Ser. A2 (Applied Mechanics (AM)), 70 (2), I_681–I_690.

Abdelrazek, A.M., Kimura, I. & Shimizu, Y. 2015. Numerical Simulation of Granular Flow Past Simple Obstacles using the SPH Method, *Journal of Japan Society of Civil Engineers, Ser. B1 (Hydraulic Engineering)*, 70 (4), I_199–I_204.

Monaghan, J.J. 1994. Simulating free surface flows with SPH, *J. Comput. Phys*, 110, 399–406.

Monaghan, J.J. & Kocharyan, A. 1995. SPH simulation of multiphase flow, *Comput. Phys. Commune*, 87, 225–235.

Bui Ha. H., Sako K., and Fukagawa R. 2007. Numerical simulation of soil–water interaction using smoothed particle hydrodynamics (SPH) method. *Journal of Terramechanics*, 44, 339–346.

Bui Ha. H., Fukagawa R., Sako K. & Wells, J.C. 2010. Slope stability analysis and discontinuous slope failure simulation by elasto-plastic smoothed particle hydrodynamics (SPH). *Géotechnique*, 61 (7), 565–574.

Investigation on sandy riverbank failure eroded by water level rising

R. Arai, K. Ota, T. Sato & Y. Toyoda
Central Research Institute of Electric Power Industry, Abiko, Abiko-shi, Chiba, Japan

ABSTRACT

It is difficult to understand the mechanism of riverbank failure because of the combined actions of hydrological and geotechnical factors involved. To address problems (e.g., loss of land and agricultural land, and breaches in river levees) caused by such failures, their physical mechanism needs to be examined. A series of bank failure experiments was conducted in a soil tank in this study. Two types of sands were used, and two bank heights (50 cm and 25 cm) and a constant bank slope of 75 degree were considered to explore the major bank failure caused by rise in water level. In these experiments, the compacted sandy soil was eroded by loss of matric suction accompanying the rise in water level; thereafter, collapse occurred because of destabilization. The distinct largest failure event corresponding to rotational slide or cantilever toppling failure was captured using a handheld 3D scanner and particle image velocimetry (PIV) method. Causes of failures were discussed by comparing the relative differences in bank height and matric suction. Cantilever toppling failure was the most frequent failure type, stressing on the importance of modeling a cantilever toppling failure based on tensile strength, incorporating the effect of matric suction. The differences in the amount of failure sediment corresponding to rotational slide and cantilever toppling failure were also revealed through the failure velocity and behavior. These experimental data will be useful for modeling the failure sediment transportation corresponding to each bank failure type.

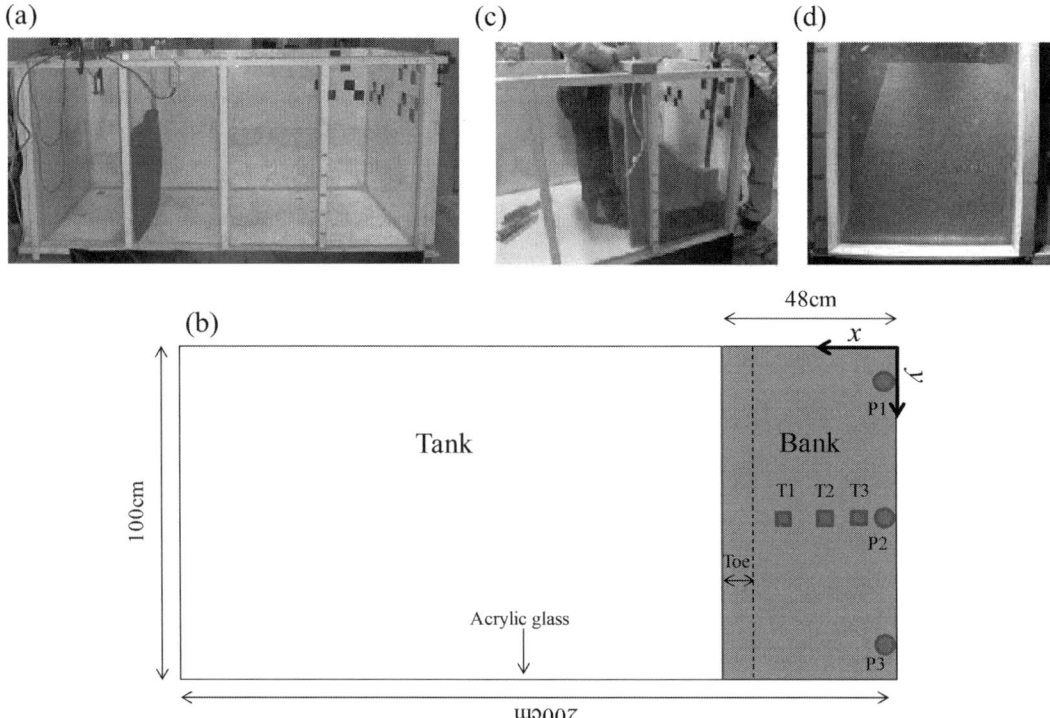

Figure 1. Experimental soil box. (a) lateral view; (b) plan view (P: soil moisture probe, T: tensiometer); (c) compaction by using a wooden rammer; and (d) bank cut to an angle of 75 degree.

Integrated investigation of space-time variability in bed load transport rates using remote sensing

M. Bakker & S.N. Lane
Institute of Earth Surface Dynamics, University of Lausanne, Switzerland

ABSTRACT

Quantifying sediment transport in general, and bed load transport in mountain streams in particular, remains a challenging task, despite the development of a wide range of in-stream measurement techniques and advances in numeric modelling. Uncertainties often result from temporal variations in sediment transport and related spatial variability in river bed sediment storage. 'Bed load pulses' occur on various scales and have been associated with variations in upstream sediment supply, the migration of sediment units (ranging from bedload sheets to bar complexes), changes in channel configuration and autogenic processes (Gomez et al. 1989). One approach to investigating these is to infer transport rates from morphological change, which makes bedload transport studies in general increasingly amenable to adoption of rapidly advancing remote sensing techniques. In this paper, we present an integrated, remote sensing based method for quantifying event-based bedload transport rates and sediment delivery rates in an actively braiding sedimentation reach of a mountain stream.

This study is set in the Borgne d'Arolla River, a tributary of the Rhône River that emerges from Pennine Alps in south-west Switzerland. Under normal flow conditions, all water is abstracted from the river for the use of hydropower in an adjacent valley. Glacially derived sediment that accumulates in sediment traps at water intakes is released down the river during intermittent purges. Purge frequency data from intake discharge records, combined with sediment trap dimensions and assumptions regarding sediment packing density and purging efficiency, provide reliable information on sediment delivery rates. We apply detailed remote sensing, both laser altimetry and drone based imaging, to map the bed topography of the sedimentation reach directly downstream of the intake (2 km × 0.2 km). These methods are used to quantify morphological change and to infer sediment transport rates based on the simple principle of mass (or volume) conservation (Lane et al. 1995). Sediment routing is performed in both 1D, the downstream direction, and 2D, using a steady state depth-averaged hydraulic simulation. Assuming no negative transport – there must be at least as much material in transport to meet the volume of deposited material measured at any one location – allows the reconstruction of minimum sediment delivery rates which we compare with those calculated from the purge record. These suggest a very promising level of agreement.

The results show strong spatial and temporal variation in sedimentation and erosion within the reach, leading to cyclical, wave-like sediment transport. Morphological or autogenic forcing, notably due to river bed width and bar/channel configuration, have crucial effects on the transfer of sediment within the reach and net sediment flux that emerges from the reach. On the contrary, the direct effect of hydrological forcing and upstream sediment supply on the transport through the reach is limited and quickly decreases in the downstream direction.

This study illustrates the potential of remote sensing techniques and their application in combination with conventional in-stream methods and numeric modelling to quantify and characterize sediment transport. Applications are not necessarily limited to coarse grained braided mountain streams but may also include flood events in ephemeral streams in (semi-)arid areas or (finer grained) morphological systems such as point bars or chute channels in meandering rivers.

REFERENCES

Gomez, B., Naff, R.L. & Hubbell, D.W. 1989. Temporal variations in bedload transport rates associated with the migration of bedforms. Earth Surface Processes and Landforms 14(2): 135–156.

Lane, S.N., Richards, K.S. & Chandler, J.H. 1995. Morphological Estimation of the Time-Integrated Bed Load Transport Rate. Water Resources Research 31(3): 761–772.

Application of Euler-Euler method in estimation of hydraulic structures scour

Sh. Basirat
Department of Engineering, Islamic Azad University, Science and Research Branch of Terhran

S.A.A. Salehi Neyshabouri
Department of Hydraulic Engineering, Faculty of Civil and Environmental Engineering, Tarbiat Modares University

ABSTRACT

Local scour around hydraulic structures have a significant impact on the design of them. Therefore, understanding the behavior of flow and changes of bed, influenced by the presence of structure on flow field is inevitable.

The aim of this work was to develop a numerical frame work like OpenFOAM to investigate modeling and simulation of fluid flows and sediment transport downstream of manmade structures. In essence, OpenFOAM is a high-level model that closely parallels the mathematical description of continuum mechanics.

In this paper, a three-dimensional Euler – Euler model is used for estimating the maximum scour depth downstream of hydraulic structures. With application of open source software, OpenFOAM and implementing proper functionality for scour simulation, model built and the effectiveness and weaknesses of Euler – Euler method were studied. Finite volume method is used for solving governing equations. To simulate the bed profiles observed during the experiments, the viscosity of the soil us is considered constant 1.0e–3, which have been determined by taking into account both physical and numerical constraints.

Results show that Euler-Euler model is appropriate for sedimentation and scour modeling and comparisons with Euler-Lagrange model shows that the present model have better behavior and needs less run time for calculating maximum scour due to jet flow. Sensitivity analysis suggests the importance of soil viscosity value and Drag and lift coefficients in predicting the scour depth in present modeling approach.

Reliability analysis of a 2D sediment transport model: An example of the lower river Salzach

F. Beckers, M. Noack & S. Wieprecht
*Department of Hydraulic Engineering and Water Resources Management,
University of Stuttgart, Stuttgart, Germany*

ABSTRACT

The first order scatter-analysis is applied to assess the reliability and uncertainties of a 2D morphodynamic model (Kopmann & Schmidt 2010). In comparison to other reliability methods (e.g. Monte-Carlo analysis), the scatter-analysis has benefits in terms of required simulation time since it uses considerably less simulation runs to determine a models' reliability. The consulted method assumes that investigative parameters are normally distributed and determines by means of the root mean square the confidence intervals. In case of twice the standard deviation (2σ) it leads to a confidence interval of 95%. This variance provides the input for each parameter that is investigated during the analysis.

The morphodynamic investigations are conducted with a calibrated 2D sediment transport model of the lower river Salzach. The applied software is Hy-dro_FT-2D (Nujic 2015). Main model characteristics are 12 km river length, 120 m mean river width and the usage of a representative hydrograph. Bed load transport is calculated using the formula of Meyer-Peter and Mueller. Six parameters are investigated including both model and river specific parameters, which are consecutively varied. These are the critical Shields parameter (Θ_{cr}), scaling factor of Meyer-Peter and Mueller equation (MPM), active layer thickness (p_{al}), acceleration factor (SCFG), roughness of river channel (k_{st}) and grain roughness (ks). Each simulation result is evaluated against a consistent reference simulation regarding bed elevation changes, bed load transport rates, grain size distribution and total riverbed evolution volume.

Figure 1 shows the results of the riverbed evolution volume of each conducted simulation against the reference simulation. It can be seen that the model parameters active layer thickness and the acceleration factor have a minor influence on the evolution volume compared to the scaling factor of the transport formula and the critical Shields parameter as well as to the investigated roughness coefficients.

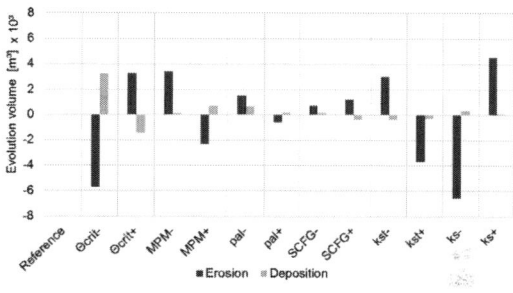

Figure 1. Effect on total riverbed evolution volume (erosion and sedimentation) of conducted simulations compared to reference simulation (modified parameters shown on x-axis).

Moreover, a spatial evaluation of the results indicates sensitive areas within the numerical model. In these areas, precise morphodynamic prediction is less reliable. As expected, sensitive areas are found in regions with poor descriptive data as well as in close vicinity to weirs, ramps and dams.

The applied scatter-analysis is found to be applicable to identify sensitive areas of the numerical model and to gain knowledge on the performance of both, model and river specific parameters. Due to the simplicity of implementation, the scatter-analysis is advisable for morphodynamic models to improve the understanding of the applied software and subse-quently of the simulated river system.

REFERENCES

Kopmann, R. & Schmidt, A. 2010. Comparison of different reliability analysis methods for a 2D morphodynamic numerical model of River Danube. River Flow 2010, Braunschweig, 1615–1620.

Nujic, M. 2015. Benutzerhandbuch Hydro_FT-2D, Erweiterung zu Hydro_AS-2D zur Simulation des Stofftransports, Ingenieurbüro Dr. Nujic, Hunziker, Zarn & Partner AG, Hydrotec Ingenieurgesellschaft für Wasser und Umwelt mbH (Manual in German).

Contraction rate of the flow, velocities, river bed stratification impact on the scour at the guide banks

B. Gjunsburgs & M. Bizane
Water Engineering and Technology Department, Riga Technical University, Riga, Latvia

ABSTRACT

Elliptical and straight guide banks usually are constructed to direct flow and sediments to the bridge opening, when part of the floodplain is blocked by approach embankment(s). Contraction of the flow leads to the flow modification; at the heads of the guide banks take place streamline concentration, sharp drop in water level, local increase in velocity, vortex structure, increased turbulence, local scour and as well general scour up and down of the bridge. The differential equation for the bed sediment movement under clear-water conditions is used, and a method for computing the scour development during multiple floods is presented. The method is confirmed by experimental data. Experiments in the flume with steady and unsteady flow, with different rate of contraction, derived method of calculation of scour development in time (Gjunsburgs et al. 2006, 2008) analysis, computer modelling of the scour with different contraction rate of the flow Q/Qb, different discharge distribution between channel and the floodplain confirmed considerable impact contraction rate of the flow on depth, width and volume of the scour hole at the guide banks. Contraction rate of the flow is not constant during the floods, contrary to the geometrical contraction of the flow. Local velocities, depth, width and volume of the scour hole are depending on the contraction rate: with increase of the contraction rate of the flow local velocity and depth of scour are increasing, as well with reduction of the critical velocity. River bed stratification leads to the increase depth of scour when fine sand layer is under the coarse sand layer or reduce it when coarse sand layer is under the fine sand layer. All results are confirmed in tests, theoretically and also presented in figures.

REFERENCES

Gjunsburgs. B., Neilands R.R., & Govsha E. 2007. Scour Development at Elliptical Guide Banks during Multiple Floods.
Proc. of 32nd Congress of IAHR, Corila, Venice, Italy, Vol.2, p.598 (on CD – 10 pages).
Lagasse, P.F., Richardson, E.V., & Zevenbergen, L.W. 1999. Design of Guide Banks for Bridge Abutment Protection. Stream Stability and Scour at Highway Bridges. Reston, VA: ASCE, 0-7844-0407-0.
Latishenkov, A.M., 1960. Questions of Artificially Contracted Flow. Moscow: Gosstroizdat (in Russian).
Levi, I.I. 1968. Hydrology Engineering. Moscow: Izdatelstvo Vishaja shkola (in Russian).
Melville, B.W. & Coleman, S.E. 2000. Bridge scour. Water resources publications, LCC, Highlands Ranch, Colorado, USA.
Neill, C.R. 1973. Guide to Bridge Hydraulics. Roads and Transportation Association of Canada, Press of University of Toronto, Toronto, Canada.

Study progress of bottom-block scour on Yellow River

Y. Cao, E. Jiang, J. Li & Q. Zhang
Yellow River Institute of Hydraulic Research, Zhengzhou, China

ABSTRACT

Bottom-block Scour(BBS)is a typical phenomenon occurred in the Little-North Reach of Yellow River and its tributary-Weihe River downstream during hyperconcentrated floods period. When it occurred, the river bed was scoured severely in a short time, the deposited fine sediment in the river bed (commonly known as "clay layer")were lift up into pieces or warps, some of them reached above the water surface (Fig. 1), and then were slumped, crushed, dispersed and taken away by flow. This strong erosion, within a few hours to dozens of hours, can make the riverbed degrade 1~2 m to almost 10 m. At the same time, it also caused swing and migration of mainstream, threaten the safety of river training works and the levees, take a high risk to flood control.

This paper review the research progress of our research team in the past decades.

Through soil mechanics test, the mechanics characteristics of soil in the Little-North Reach of Yellow River is studied in detail. The test results show that under a certain moisture content, the shear strength of clay layer most located above the critical damage line, and the coarse sediment layer most below the critical damage line, which is the reason why the river bed was scoured by pieces or wraps.

A detailed analysis for the adjustment rule of cross section during the Bottom-block Scour period was taken by digging deep into the original observation data. Normally, the cross section adjustment progress can be divided into four stages: pre-scour stage, channel bed elevation keeping relatively stable stage, rapid declining stage and continual scour followed by back-silting stage. The BBS occurred in the rapid declining stage.

Using the theory of transient flow, the critical discriminant formula of the Bottom-block Scour occurred condition is established, and based on the results of successful simulate the phenomenon of Bottom-block Scour, the key parameter of the formula is determined. The derived equation of critical discriminant formula for the Bottom-block Scour occurred condition is:

$$0.2 \frac{V^2 J}{g\delta} \geq \frac{\gamma_s - \gamma_m}{\gamma_m} \quad (1)$$

The prototype data and the flume data are used to verify the accuracy of the formula, and the results are shown that the formula is suitable. It can be used to forecast the BBS Occurrence.

The BBS damage mechanism of river training work is mainly due to the erosion resistance difference between the clay layer and the coarse sediment layer. On the basis of damage mechanism study and historic data analysis, the countermeasures were proposed for the Bottom-block Scour.

REFERENCES

GOU, Y-Y. 2004. Calculation and analysis of the bottom tearing scouring phenomenon in flood with hyperconcentration in the Yellow River[J]. Advances in Water Science, (15(2): 156–159. (in Chinese)

Jiang, E-H., Li, J-H. & Cao, Y-T. 2010 Research on mechanism of bottom tearing scour in Yellow River. Journal of Hydraulic Engineering, Apr. 41(2): 182–188. (in Chinese)

Liu, P., Dong, J. & Yu, C. 1998. Fluctuating uplift on rock blocks at the bottom of a scour pool by overfall jets. Science in China (Series E), 1998, 41(2): 130–139.

Figure 1. The Bottom-block Scour at the Xiaoshizui River Training Works on July 5, 2002.

Quartz silt deposition

S. Capapé & J.P. Martín-Vide
Technical University of Catalonia – BarcelonaTech, Barcelona, Catalonia

F. Colombo
SIMGEO UB-CSIC, Universitat de Barcelona, Barcelona, Catalonia

ABSTRACT

Common knowledge points to conditions with $u^* \gg W_s$ (with u^* the shear velocity and W_s the settling velocity) for a significant sediment suspension (Bagnold 1966). However, in tests with $^* \gg W_s$ multiple bed morphologies developed from the deposition of quartz silt (geometric mean size $Dg = 4.15\,\mu m$) over a plane, rigid, non-erodible surface in turbulent conditions and constant flow discharge for different initial sediment concentrations.

The experiments are performed in a rectangular tilting flume (SIMGEO UB-CSIC, Universitat de Barcelona) 15 m long, 0.37 m wide and 0.40 m high. The floor consists of rigid polished aluminum and the walls are transparent glass. The water and sediment mixture is constantly recirculated on the principle of mass conservation. Velocity, sediment concentration, water level and temperature are monitored. Direct town supply, drinking water is used.

Initial suspended silt concentration ranges between 1.28 g/l (E3) and 51.72 g/l (E12). Quartz silt particles

Table 1. Silt grain size.

	D_{10} (mm)	D_{50} (mm)	D_{90} (mm)	D_a (mm)	σ_a (–)
Initial quartz silt in suspension					
	1.25	2.37	12.48	4.15	2.38
Group 1					
	2.49	11.17	25.33	9.35	2.48
Group 2					
	1.96	9.78	23.24	7.97	2.78
Group 3					
	1.80	7.95	21.14	6.78	2.73

(Table 1) are non-cohesive by definition and flocculation is not relevant during the deposition. Furthermore, although this quartz silt contains ∼20% of clay size material ($D < 2\,\mu m$) the authors have not observed mass deposition characteristic of fine cohesive sediment deposition for concentrations much lower than 51.72 g/l (Mehta 1973).

The experiments usually last one day, with one experiment running for 72 hours (Fig. 1). At the end of the experiments, between 10% (for the shortest experiments) and 41% (for the longest) of the initial silt in suspension has deposited, irrespective of the initial sediment concentration.

The deposited silt is coarser than the initial silt in suspension. Measured $u^* - 0.016$ m/s is ∼1000 times greater than w_s for size D_σ. Suspended sediment concentration appears to decrease logarithmically and viscous effects are deemed relevant during the process of deposition.

REFERENCES

Bagnold, R.A. 1966. "An approach to the sediment transport problem from general physics." U.S. Geological Survey Professional Paper, 422-I, Washington, D.C.

Mehta, A.J. & Partheniades, E. 1973. Depositional Behavior of Cohesive Sediments. Coastal and Oceanographic Engineering Laboratory, University of Florida, Gainesville, Tech. Rept., 16.

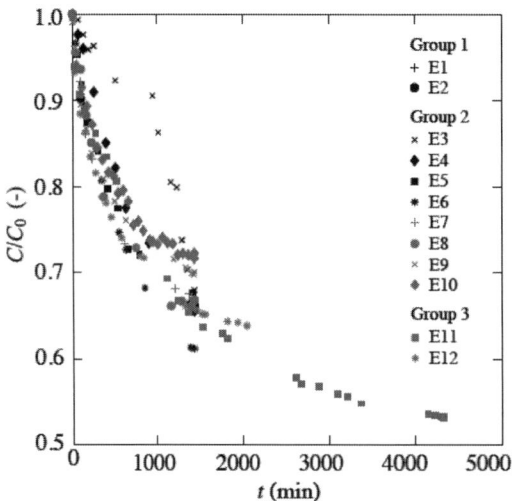

Figure 1. Sample of a figure caption.

Analysis of the sediment hydrography by modeling hyper-concentrated flow and sediment transport

Y.Y. Chiu & K.C. Yeh
Department of Civil Engineering, National Chia Tong University, Hsin-Chu, Taiwan

ABSTRACT

After rainfall intensity exceeds soil infiltration, water accumulates on the surface and the flows down the slope under gravity as overland flow. This rainfall-runoff process considered soil erosion or transportation of sediment had been simulated by numerical models which had been developed since 1960s. However, flows of water and sediment very over a large range of sediment concentration. The term hyper-concentrated flow is most often applied between water floods and debris flows, and could exist in both high and low discharges.

Usually, the hyper-concentrated sediment flows was considered to be a kind of Bingham flow. Liu & Mei (1989) studied the Bingham fluid on an incline plane and the experimental measurement is validated with the theoretical solution. According to the variety of concentration, Pierson (2005) pointed that a debris flow surge can dilute itself to a debris flood.

Hsu et al. (2012) used an integrated method, considering sediment supplies associated with soil erosion, shallow landslide and debris flow to estimate watershed sediment yield. However, the integrated method still has some limitations caused by the way of combined models in the study.

To investigate the transport behavior of hyper-concentrated flow, a two-dimensional numerical model, for hyper-concentrated flow and sediment transport, was developed in this study. The numerical scheme was finite difference; the model was identified for some experiments and numerical simulations; the accuracy of result was acceptable.

As a reference for basins planning, sensitivity analysis was performed to identify the model. Parameters of sediment, yield stress and viscosity were taken into account.

Farther more, for the purpose of investigating the characteristics of hyper-concentrated flow and sediment transport, a real case was simulated. A sub watershed of Dajia River Basin, Taiwan, was taken as a study site. The available data, such as hourly rainfall data and sediment records, have been used for model calibration and validation.

After combining with landslide yield, this numerical model could link sediment yield to the watershed area, and could also analyze of the sediment hydrography.

REFERENCES

Hsu, S.M., Wen, H.Y., Chen, N.C., Hsu, S.Y. & Chi, S.Y. 2012. Using an integrated method to estimate watershed sediment yield during heavy rain period: a case study in Hualien County, Taiwan, Nat. Hazards Earth Syst. Sci., 12, 1949–1960, doi: 10.5194/nhess-12-1949-2012.
Liu, K.F. & Mei, C.C. 1989. Slow spreading of a sheet of Bingham fluid on an inclined plane, J. Fluid Mech., Vol.207, 505–529.
Pierson, T.C. 2005. Hyperconcentrated flow transitional process between water flow and debris flow. Ch. 8 in Debris Flows and Related Phenomena, M. Jakob and O.Hungr, Eds., Springer, Heidelberg, pp. 159–196.

Flow and riverbed erosion-deposition simulation around submerged water intake

Z.F. Cui & D.C. Hu
River Research Department, Yangtze River Scientific Research Institute, Wuhan, China

ABSTRACT

The research on flow movement and the riverbed erosion-deposition around the water intake is essential for the safe operation of water intake engineering. And it is necessary to demonstrate the reasonableness of the position of the water intake from the terms of flow pattern, the river regime and reduce the impact of water intaking on the fairways and flood control etc.. As an example, water intakes of one power plant locate at downstream of Yangtze River were selected in this paper. And a 3D numerical model is used to simulate local flows(as shown in Fig. 1) and plane 2D flow and sediment model is used to simulate riverbed erosion-deposition, suspended sediment concentration around submerged water intakes, respectively. And recent riverbed evolutions around the water intakes were predicted and so on. The boundary conditions of 3D flow model provided by the calculation results of 2D model. The water-sediment condition of typical series and high flow year, moderate flow year and low flow year typical year were selected. And the reservoir operation impact of the Three Gorges Project and the upstream reservoir to the water-sediment process downstream were take into account. Then, the better location and appropriate elevation of water intake were recommended based on the calculations result

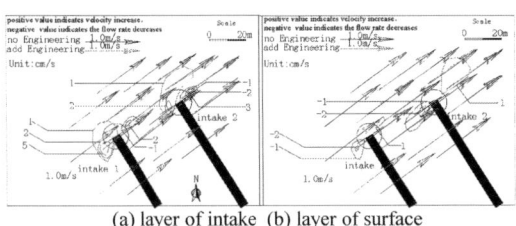

(a) layer of intake (b) layer of surface

Figure 1. Velocity variation around the intake of design 1.

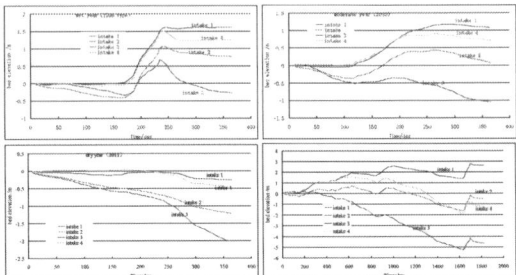

Figure 2. Variation process of bed elevation of four intakes gate.

of models through analyzing the change of riverbed erosion-deposition, bed elevation and suspended sediment concentration around water intakes(as shown in Fig. 2). The results show that the change of riverbed erosion and deposition around the water intake is not only affected by the conditions of incoming water and sediment, but also by the position of the local topography at the same river reach. Therefore, it is essential to select the appropriate location and elevation to the water intake engineering.

REFERENCES

Chen Huiquan, Xu Yulin, HE Yiying. Progress and experience of 50 years' study on cooling water circulation of thermal/nuclear power plants [J]. J. of China Institute of Water Resources and Hydropower Research, 2008, 6(4): 288–295.

Duan Zhike, LI huimei, Experimental Research on Sediment Problems in Intake Work in Thermal Power Plant[J]. J. of Water Resources and Hydropower Engineering, 1997, 7, 29–32.

Zhang Wei, Study on Regularity of Riverbed Evolution in Fluctuating Backwater Area of Xiujiang River, Guangxi Province[D]Nanjing: HoHai University, 2006.

Effect on bed load transport discharge of Chongqing reach by backwater of Three Gorges Reservoir in upper Yangtze River

Xu-hui Fu, Sheng-fa Yang & Yang Chen
National Engineering Research Center for Inland Waterway Regulation, Chongqing Jiaotong University, Chongqing, China

Jiang Hu, Sichen Tong & Yi Xiao
Key Laboratory of Hydraulic and Waterway Engineering of the Ministry of Education, Chongqing Jiaotong University, Chongqing, China

ABSTRACT

Since Three Gorges Reservoir runs at 175 m as normal level in 2008, the Chongqing reach have been the changeable back water zone. Gravels, rather than fine sediments are the key material which influence the river bed evolution according to more than 10 years field measurement along fluctuating backwater zone. And it very different from former researches. Then the bed load transportation near Chongqing reach is a hot point of researchers because of the influence on the inland navigation security and river bed evolution while Chongqing is the important inland navigation center of upper Yangtze river.

We collected some field measured bed load transport data and whole hydrology data of Cuntan hydrological station from 1964 to 2011 including flow discharge and water level data of average daily value. As Fig. 1 shown, the blue points of stage-discharge of Cuntan from 1954 to 2007 almost fits the curve before Three Gorges Project (TGP) backwater affected Chongqing reach. While the red points shows that water levels effected by TGP in low flow discharge less than 30000 m³s⁻¹ during the stage in front of dam run from 156 to 175 m. And the raising range is less than 15m almost in dry season from October to April of next year.

The data was analyzed according to flow power W_* as Yang (2011) mentioned with the parameters including energy slope J, average water depth H, average section flow velocity U and gravel diameter D_{50}.

The dimensionless flow power W_* change process of typical years at Cuntan hydrological station are shown in figure 2 while 2006 is selected as representation before backwater take effect and 2011 is the one after it. The figure indicates that the two process is less change from May to September during flood season except the flow discharge difference. And the gravel transportation are nearly same. The flow power W_* declined sharply during dry season from January to April because of the water surface raising obviously. But both W_* is very low

$$W_* = \frac{qJ}{\sqrt{gD_{50}^3}} = \frac{UHJ}{\sqrt{gD_{50}^3}}$$

whatever it changed and the gravel will hardly move in each year. The most important change of W_* happened in the end of flood season from September to October. There was mass gravel movement in 2006 for the flow power is very high. While the gravel motion decrease sharply in 2011 for the W_* is rather low. And it is the most important period to erosion the gravels silt in flood. The largest stream power period in 2011 except flood season is about on April and it is the new scour period for gravel after the backwater affect the area.

The change leads the gravel erosion delay to next April. The waterway in the zone is worse in March since the flow depth is rather low and silted gravel is still in navigation channel.

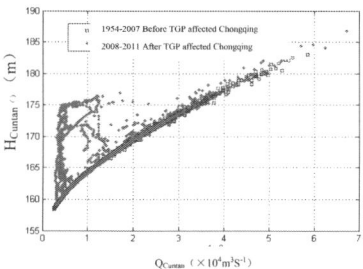

Figure 1.

REFERENCES

Coleman, S.E., J.J. Fedele, and M.H. Garcia, Closed-Conduit Bed-Form Initiation and Development. Journal of Hydraulic Engineering, 2003. 129: p. 956–965.

Coleman, S.E. and V.I. Nikora, Fluvial dunes: initiation, characterization, flow structure. Earth Surface Processes and Landforms, 2011. 36(1): p. 39–57.

Vries, J.S.M.v.T.d., et al., Analysis of dune erosion processes in large-scale flume experiments. Coastal Engineering, 2008. 55: p. 1028–1040.

Influence of high Paraná's River dunes variability in Itaipu's reservoir sedimentation

P.E. Gamaro, L.H. Maldonado & J.L. Castro
Itaipu Binacional Power Plant, Brazil

V.P. Bastolla
Atlantic Int'l University, Brazil

ABSTRACT

The estimation of sand transport in large rivers is an important parcel of the total load. Moreover, because of the measurement difficulties and the traditional sampling techniques are prone to large errors with limited spatiotemporal resolution and coverage and difficult and dangerous to employ during high discharges when most needed. Thus in large rivers many data of bed load is calculated as a percentage of the suspension sediment load. This insert an error not known which is not afforded to have in some situations. Paraná River is the 5th in the world in discharge and extension and his basin is well used to produce energy with 45 hydroelectric dam only in Brazilian territory. The last of them is the most important to Brazil and Paraguay. A bi national hydroelectric dam provides 20 % and 70% of Brazil and Paraguay's energy, being the larger producer of energy in the world (2.312×10^6 GW since 1984). From the last upstream dam until Itaipu reservoir, Paraná River runs freely and in his natural bed forming a 270 000 sq. km of basin not regulated, most of the sediment suspension load come from wash load.

Bed load is different, compound mostly from fine and medium sand, supplied from tributaries most from its right bank, find the appropriate conditions to dunes formation (Simons et al. 1965). This way bed load became the most important parcel of the total sediment load that flows into Itaipu's reservoir. Unfortunately, is not measured adequately and is only estimated by a percentage of suspended load without any analysis of the relation between them. Only in 2011 the technique of dunes displacement was then used to estimate bed load (Gauman et al. 2007). And in 2013 a modified method with the use of point technology that allow a better precision, and produced results showing that load changes not linearly dramatically with discharges, and not only from the main river, but also from floods of tributaries. Thus to obtain a more

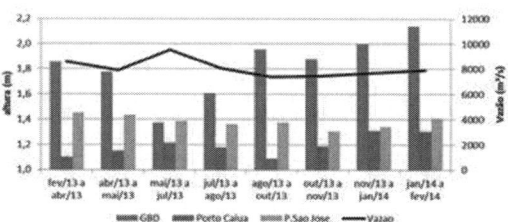

Figure 1. Bedload at three different gauge stations upstream reservoir over a year of monitoring.

realistic bed load flowing into the Itaipu's reservoir a continuous dunes tracking measurement program was established, with an advance of the original technique. Starting to monitored three gauge stations (entrance of the reservoir-GBD, 100 km- Caiua and 250 km P.S. Jose, upstream the reservoir). This procedure allowed an understanding that bed load estimated from suspended sediment load is under estimated (34% for higher discharges), and that only measuring continuously is possible to achieve a more precise bedload incoming into the reservoir. This paper presents results of a year monitoring every 30 to 45 days and compares to random bedload measurements using dunes tracking results, and the data estimated from percentage of sediment suspension.

REFERENCES

Fredsøe, J. 1981. Unsteady Flow in straight alluvial streams. Part 2 Transition from Dunes to Plane Bed. In Journal of Fluid Mechanics, v 102. 20–32

Gauman, D. & Jacobson, R.B. 2007. Field Assessment of Alternative bed load transport estimators. In Journal of Hydraulic Engineering –ASCE v.133 n.12 1319–1328.

Simons, D.B., Richardson, E.V. & Nordin, C.F. 1965. Bed Load Equations for Ripples and Dunes. In US Geological Survey Professional Paper 462-H. Washington D.C. 32p

Suspended load monitoring for sustainable hydropower development

M. Guerrero & A. Antonini
The Hydraulic Engineering Laboratory, DICAM, University of Bologna, Bologna, Italy

N. Rüther
Department of Hydraulic and Environmental Engineering, Norwegian University of Science and Technology, Trondheim, Norway

S. Stokseth
Statkraft AS, Oslo, Norway

ABSTRACT

Due to the increasing demand of CO_2 neutral energy not only in Europe but also in World, a relatively large amount of new hydro power plants (HPP) are built. In addition, will existing ones refurbished and renewed in order to run them more cost effective. A huge thread to HPPs is incoming sediments in suspension from the rivers upstream. The sediments settle in the reservoir and reduce the effective head and volume and reduce consequently the life time of the reservoir. In addition are the fine sediments causing severe damages to turbines and infra-structure of a HPP. For estimating the amount of in-coming sediments in suspension and therefore planning efficient counter measures, it is essential to monitor the rivers within the catchment of the HPP for suspended sediments. This work is considerably time consuming and requires highly educated personnel and is therefore expensive.

Consequently, will this study present a method to measure suspended sediment concentrations and their grain size distribution with a dual frequency acoustic Doppler current profiler (ADCP). This method is more cost effective and reliable in comparison to traditional measurement methods. The meth-od has been tested in large rivers against depth aver-aged suspended load measurements (Guerrero et al. 2013) as well as against detailed point measurements in the laboratory (Guerrero et al. 2014). Having more detailed information about the sediments being transported in a river, the hydro power plant can be planned, built, and operated much more efficiently and sustainable. The two horizontal ADCPs are in-stalled at a measurement cross section in the Devoll river in Albania. To verify the new method, the suspended load concentrations will be monitored also in the traditional ways at the same cross sections. Both, turbidity measurement devices included with an automatic sampling devices and an optical in situ measurement device (LISST SL by Sequoia Inc.) is installed to have detailed information of concentration and grain sizes over the depth.

Figure 1. Picture of the measurement cross section in the river Devoll in Albania.

Figure 1 shows the picture of the cross section where the ADCP measurements have taken place. The station delivers data automatically to a server. Therefore, it is ensured that also large flood events are covered. The current study will present data of the wet season 2015/2016.

This work is carried out as part of a larger research project on sustainable hydro power plants exposed to high sediment yield, SediPASS. SediPASS is funded by the Norwegian Research council and the Statkraft. Statkraft is supporting this project in the framework of a large R&D project on future handling strategies of sediments at hydro power plants.

REFERENCES

Guerrero, M., Szupiany, R.N. & Latosinski, F. 2013. Multi-frequency acoustics for suspended sediment studies: An application in the Parana River, Journal of Hydraulic Research, Volume 51, Issue 6, 1 December 2013, Pages 696–707.

Guerrero, M., Rüther, N. & Archetti, R. 2014. Comparison under controlled conditions between multi-frequency ADCPs and LISST-SL for investigating suspended sand in rivers, Flow Measurement and Instrumentation, Volume 37, June 2014, Pages 73–82.

Study on sedimentation velocity in transition zone

Y. Guo & L. Gao
Pearl River Hydraulic Research Institute, Ministry of Water Resources, Guangzhou, P.R. China

ABSTRACT

Sedimentation velocity is an important hydraulic characteristic and motion parameter of sediment. Researchers, in home and abroad, have done lots of theoretic and experimental works in this field (Shouyu 1989, Yuqing 1965, and Guoren 1963). Theoretically speaking, according to balance between effective gravity and flow resistance of sediment sinking in the water that the sedimentation velocity can be obtained as

$$W = \sqrt{\frac{4(\rho_s - \rho)gd}{3\rho C_d}} \quad (1)$$

W: sedimentation velocity [cm/s];
ρ_s: sediment density [g/km^3];
ρ: is water density [g/km^3];
d: diameter of particle or median diameter of particles in a mixture [cm];
C_d: rolling resistance coefficient.

According to experiment, C_d has relation with particles Reynolds number Re_d at laminar zone or Laminar flow to turbulent transition zone:

$$Re_d = Wd/\nu \quad (2)$$

Here ν is kinematic viscosity coefficient of water and change with the temperature of the water, its unit is cm^2/s.

Experimental studies show that, when $Re_d < 0.1$ it keeps laminar flow and $C_d = 24/Re_d$. Generally ρ_s doesn't change much and it often takes 2.65, then sedimentation velocity of laminar flow can be obtained as

$$W = 90d^2/\nu \quad (3)$$

Equation (3) is deforming of Stokes Equation. As in laminar zone that $Wd/\nu < 0.1$, then the sediment particle size d should satisfy the inequality (4) when it applies to equation 3:

$$d < (0.1\nu^2 / 90)^{1/3} \quad (4)$$

According to experiment, in turbulent zone that $Re_d > 500$. And resistance coefficient C_d can be got as $C_d = 1.2$, then we substitute it into equation (1) and get

$$W = 42.4d^{0.5} \quad (5)$$

As in turbulent zone that $Wd/\nu > 500$, so d that applying to equation (5) should satisfy the inequality

$$d > (500\nu / 42.4)^{1/1.5} \quad (6)$$

From equations (4) and (6) we can find that, when the water temperature is different and then the ν value is different, sediment particle size d is not a fixed interval, and shall change with water temperature and ν value. Many scholars at home and abroad have put forward some formula for the transition zone, which not only provides the fixed application range of different d, but also the formula structure is often very different, all this show that it is necessary to further study the formula for calculating the velocity of transition zone.

REFERENCES

Shouyu, C., 1989. Fuzzy, Optimization and Numerical Calculation of Hydrological and Water Resources System [M]. Dalian University of Technology Press.
Yuqing, S., 1965. Introduction of Sediment Dynamics [M]. China Industry Press.
Guoren, D., 1963. Sediment Movement Theory [M]. Nanjing Institute of Water Resources Science.

Estimation of maximum local scour depth around submerged spur-dike

S.Y. Hao
Nanjing Hydraulic Research Institute, Nanjing, China
College of Harbour, Coastal and Offshore Engineering, Hohai University, Nanjing, China

Y.F. Xia & H. Xu
Nanjing Hydraulic Research Institute, Nanjing, China
The State Key Laboratory of Hydrology-Water Resources & Hydraulic Engineering, Nanjing, China

ABSTRACT

The engineering goal of the first phase of the 12.5 m deepwater channel regulation project of Yangtze River downstream Nanjing is to extend 12.5m deepwater channel from Taicang to Nantong by means of regulating structures and dredging in the key location. The bed load of this river reach is fine sand which is easy to move so that the local scour protection around regulating structures such as submerged spur-dike used in the regulation project is important. Soft bottom protective mattress that can adapt the local scour is used to protect submerged spur-dike. There is few study on the estimation of maximum local scour depth around submerged spur-dike with bottom protection recently (Hao et al. 2013) and the estimation of local scour depth in engineering code is for emerged spur-dike. The study on the estimation of maximum local scour depth around submerged spur-dike with bottom protection is significant in engineering practice.

Based on the scale of submerged spur-dikes and flow-sediment condition in the regulation project (Xu et al. 2012), a normal model experiment with a similar scale of 1:60 is conducted to reveal the properties of maximum local scour depth around submerged spur-dike. The experimental flume is 40 m long and 8 m wide. In the middle of the flume is 30 m long movable-bed section paved with model sand which $d_{50} = 0.38$ mm. The bottom protective mattress model is made up of aluminium blocks pasted on the gauze. The submerged spur-dike is impermeable and its dimensions satisfy geometric similarity of the prototype submerged spur-dike used in the channel regulation project. Figure 1 shows the sketch of experimental flume layout.

The relationships of maximum local scour depth with the dimensions of submerged spur-dike, hydrodynamic condition and the width of bottom protective mattress are analyzed. An estimation formula of maximum local scour depth around submerged spur-dike with bottom protection is established through the experimental data by the principle of dimensional analysis:

$$\frac{D}{h} = 65 \left(\frac{v - v_c}{\sqrt{gh}}\right)^{1.2} \left(\frac{h}{H}\right)^{-1.6} e^{-0.038\frac{B}{H} - 0.14m} \quad (1)$$

where D = maximum local scour depth; v = approach velocity upstream; v_c = critical velocity for incipient motion of sediment; h = water depth before scour; H = height of spur-dike; B = width of bottom protective mattress; m = slope of spur-dike head in cross-section.

The formula yields well-fitting result with site measured values (Ji et al. 2000). The study result can provide technical reference for the design of local scour protection in channel regulation.

REFERENCES

Hao, S.Y., Xia, Y.F., Xu, H. & Wen, Y.C., 2013. Review of local scour around submerged spur-dike. Proceedings of the 16th China Ocean (Shore) Engineering Academic Congress, Dalian, China: 1301–1306.

Ji, Y.X., He, G.Q. & Lu, Y.J., 2000. Model selection and verification for a typical spur dike's scour depth in Yangtze River Estuary. Journal of Nanjing Hydraulic Research Institute, (4): 43–47.

Xu, Y., Fu, Z.M. & Zhang, H., 2012. Initial design report of the first phase project of the Yangtze River 12.5m deepwater channel regulation from the downstream of Nanjing (Taicang~Nantong reach). Shanghai: Shanghai Waterway Engineering Design and Consulting Company Limited.

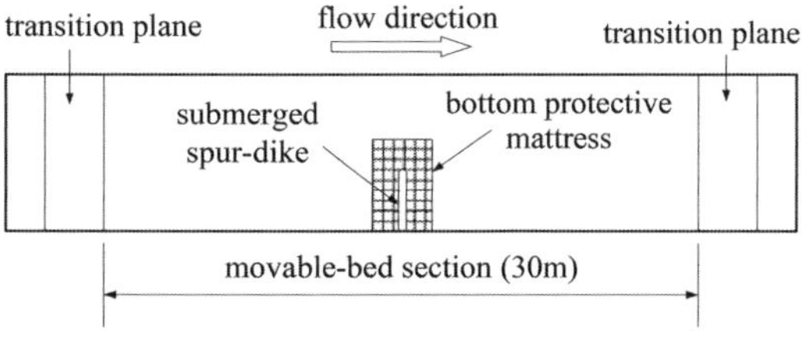

Figure 1. Sketch of the experimental flume layout.

Mathematical description of flow memory effects on graded bedload

K. Hassan & H. Haynes
Water Academy, School of Energy, Geoscience, Infrastructure and Society, Heriot-Watt University, Edinburgh, UK

ABSTRACT

Variability of antecedent flows has a fundamental control on the entrainment and transport of sediment in river systems. Specifically, the low flows between successive floods appear to have far greater influence on the stability of a river bed than previously assumed. Prolonged durations of low flows (termed "memory") increase bed stability, delay particle entrainment and reduce sediment transport.

Flume experiments were carried out using two poorly sorted sand-gravel mixtures of unimodal and bimodal distribution; five memory timescales ($T = 10, 30, 60, 120, 240$ minutes) of sub-threshold flow were employed and compared with baseline non-memoried data. Results show that increasing memory timescales up to 240 minutes raises entrainment thresholds up to 49%, while bedload is more sensitive, decreasing by up to 97%.

Memory effects were mathematically described using rising "exponents" (b) to quantify the degree of non-linearity of transport to shear stress. Also, changes in the structure of the bed due to memory were described by lumped "coefficients" (C). The proposed generic function: $q^* = C(\tau^*)^b$ uses non-dimensional values of sediment load (q^*) and shear stress (τ^*). Although this approach is similar to traditional bedload transport formulae (e.g. Paintal, 1971; Taylor and Vanoni, 1972; Parker, 1990), it is better able to describe the effects of longer memory timescales by using higher values of b and C. This systematic rise permits development of their recursive relations. For the unimodal bed (um) this yields Eq. 1a and 1b:

$$C_{um} = 3 \times 10^{22} (T)^{1.91} \quad \text{Eq. 1a}$$

$$b_{um} = 17.00 (T)^{0.012} \quad \text{Eq. 1b}$$

For the bimodal bed (bm) this yields Eq. 2a and 2b:

$$C_{bm} = 5 \times 10^{25} (T)^{4.50} \quad \text{Eq. 2a}$$

$$b_{bm} = 17.59 (T)^{0.079} \quad \text{Eq. 2b}$$

Table 1. Relationship between memory timescale T, coefficient C and exponent b.

Memory Timescale T (min)	Unimodal bed		Bimodal bed	
	C_{um}	b_{um}	C_{bm}	b_{um}
0	10^{19}	16.2	10^{16}	12
10	10^{24}	17.3	10^{33}	23.8
30	10^{25}	17.9	10^{29}	20.95
60	10^{27}	16.9	10^{33}	24.23
120	10^{26}	19	10^{35}	25.45
240	10^{30}	21.5	–	–

where T is the memory time scale in minutes. From the above relations, the predicted value of b and C against a selected value for T can be used in the generic transport function for predicting sediment load in memoried graded beds.

In Table 1, parameters C and b show highly non-linear relationships to T. Data for b increases to a maximum factor of 2.13 relative to the baseline data ($T = 0$). Similar comparison for C reveals far greater sensitivity with increases to the order of 10^{19}. Therefore, memory effects appear largely controlled by changes in bed structure, as associated with C.

The proposed mathematical functions validated over 80% of measured bedload data from the present flume experiments. Given the extreme variability of bedload in the near entrainment low flow regime (Bunte et al. 2004), this provides confidence in the findings of this study.

REFERENCES

Paintal, A. S., 1971. Concept of critical shear stress in loose boundary open channels, Journal of Hydraulic Research, 9(1), 91–11.

Parker, G., 1990. Surface-based bedload transport relation for gravel rivers, Journal of Hydraulic Research, 28(4), 417–436.

Taylor, B. D. and Vanoni, V. A., 1972. Temperature effect in low-transport, flat-bed flows, Journal of Hydraulic Engineering, 98(8), 1427–1445.

Evolution of Modaomen Bar at Pearl River Estuary

Y. He, C. Lu, J. Deng, Y. Yang & L. Yang
Key Laboratory of the Pearl River Estuarine Dynamics and Associated Process Regulation,
Pearl River Hydraulic Research Institute, Ministry of Water Resources, Guangzhou, China

ABSTRACT

Among the eight estuary outlets of the Pearl River Estuary, the Modaomen is the largest passageway for flood relief with a strong dynamic mechanism of runoff. It has complex tidal currents and wave dynamics with the influences of circulation and monsoon, resulting in various shapes and evolutionary characteristics of the mouth bar.

The Modaomen Outlet has a very strong dynamic mechanism of runoff. The mouth bar outside the Modaomen Outlet is of various shapes, due to the complicated tidal and wave dynamics under the influences of circulation and monsoon. Thus, the evolutionary patterns are complex. Therefore, the captioned research is especially important. As studied by J.B. Liul, the flow velocity tends to decline from surface to bottom during the spring tides in the Modaomen Outlet; however, the surface flow velocity is lower than the bottom one during the neap tides due to the larger vertical salt gradient (Liul, J.B. et al. 2011). Due to the influences of the Northeast Wind in winter, large amount of water moves in the southwest direction and the SW Alongshore Currents occur; however, due to the Southwest Wind in summer, the water currents move in the opposite direction to that in winter (Liu, C. et al 2010, Ou, S. et al. 2007).

This paper analyses the dynamic characteristics of the planar flow fields, vertical flow velocities and residual currents at the Modaomen Outlet, illustrates the causes of the Alongshore currents, and explored the recent evolutionary patterns of the mouth bar, based on the following data of the Modaomen Outlet from 2011 to 2012: the observation and measuring data of sea currents and winds in spring tides, moderate tides and neap tides; and the historical data of the upstream hydrology. The main conclusions are as follows:

(1) In the draught season, totally there was smaller flow velocity in the Outlet, ebb currents are weaker outside the Outlet, and the southwest ebb currents were weaker, due to the weak runoff. In the flood season, flow velocities of ebb currents increased obviously at the Outlet, the flood currents outside the Outlet moved mainly in the North-by-East direction and the ebb currents mainly moved southwestward. In the draught season, the Northeast Monsoon played an important role in deciding the sizes of surface currents; in the flood season, runoff plays an important role in deciding the sizes of surface currents.

(2) In the draught season, the sea area at the Modaoment Outlet has the characteristics of the stable Southwest Alongshore Currents, which are mainly caused by the South China Northeast Monsoon. In the draught season, the Alongshore Currents of different tidal stencils were endowed with various characteristics. Runoff and wind are the main factors, together with the influences of the landforms.

(3) The inner slope of the mouth bar shifted southwards, due to the erosion of the flood runoff on the bar top; the external slope shifted northwards, the bar top heightened and the south-north diameter shortened, due to the waves lifting the sediments and the alongshore currents transporting the sediments. Thus, the mouth bar evolved mainly in the SE direction, because the N-direction ebb currents gradually merged into the SW alongshore currents and they transported sediments.

REFERENCES

Liul, J.B. & Bao, Y. 2011, Spacial Distribution of Salinity and the Mechanism of Saltwater Intrusion in the Modaomen Water Channel of Pear River Estuary. American Institute of Physics. RECENT PROGRESSES IN FLUID DYNAMICS RESEAR- CH. AIP Conf. Proc. Vol 1376.

Liu, C., Xia, H. & Wang, D. 2010, The Observation and Analysis of Eastern Guangdong Coastal Down welling in the winter of 2006 (in Chinese). Acta Oceanologic Sinica 32(1).

Ou, S., Zhang, H. & Wang, D. 2007, Horizontal characteristics of buoyant plume off the Pearl River Estuary during Summer J Coast Res(50).

Modeling of non-capacity bed load transport in swash zone

Z. He, L. Tan & P. Hu
Ocean College, Zhejiang University, Hangzhou, China

ABSTRACT

The swash zone has the unique features, such as high unsteadiness, short duration, narrow space, which means bed-load sediment may not reach the equilibrium state instantly in reality. To better understand the physical phenomena, it is worthy to investigate whether bed load can adapt to the capacity regime within the swash zone.

Therefore, this study presents a non-capacity model for bed load transport and the swash flow. The governing equations are solved by the finite volume method along with a well-balanced centered Riemann solver for estimating the inter-cell numerical fluxes and bed slope source terms (Aureli et al. 2008; Hu et al. 2015). The present model is used to simulate the laboratory experiments in which the swash is produced by lock-release dam break flow propagating towards a mobile beach (Briganti et al. 2012).

Various scenarios with different sediment grain size, bed slope and initial still reservoir water levels are analyzed. The results show that the model can well predict the flow variables including swash depth, velocity, and sediment transport rate (Fig. 1). More importantly; the non-capacity model successfully captures the sudden slowdown of the water level increase and provides a more reasonable beach elevation (Fig. 2).

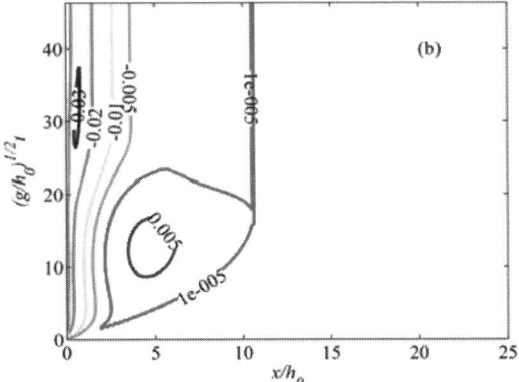

Figure 2. The computed beach deformation depth.

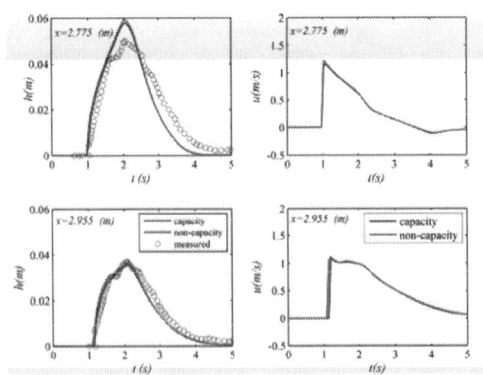

Figure 1. The swash depth and velocity.

REFERENCES

Aureli, F., Maranzoni, A., Mignosa, P., Ziveri, C., 2008. A weighted surface-depth gradient method for the numerical integration of the 2D shallow water equations with topography. Adv. Water Resour, 31, 962–974.

Briganti, R., Dodd, N., Kelly, D., and Pokrajac, D. 2012. An efficient and flexible solver for the simulation of the morphodynamics of fast evolving flows on coarse sediment beaches. International Journal for Numerical Methods in Fluids, 69(4), 859–877.

Hu, P., Li, W., He, Z., Pahtz, T., and Yue, Z. 2015. Well-balanced and flexible modelling of swash hydrodynamics and sediment transport. Coastal Engineering, 96, 27–37.

Influence of groundwater flow intensity on sediment motion

O. Herrera-Granados
Faculty of Civil Engineering, Wrocław University of Technology, Wrocław, Poland

ABSTRACT

The interaction between surface and groundwater flows in alluvial riverbeds is subject of numerous researches in recent years due to the necessity of better understanding the physical processes involving mass and momentum exchange between both flows. Seepage exists if the material that constitutes any hydraulic structure or river bed is permeable enough and if there is a difference between the upstream water level and the tail water level downstream. Due to the fact that groundwater flow rates are much smaller than open-channel flow rates; seepage is commonly neglected by river engineers in the analysis of civil and environmental engineering problems. Nonetheless, this small groundwater flow can represent an important factor in many physical processes.

This contribution presents the results of an experimental analysis concerning seepage's influence on sediment transport. This analysis is based on the results of several experiments that were carried out at the open-air laboratory of the Wrocław University of Technology in Poland. The output of the laboratory studies demonstrates that the artificially induced seepage affects the interaction between the flowing water along the channel and the motion of sediments from the bottom. Even though that the magnitude of the groundwater flow was significantly smaller in comparison with the magnitude of the open-channel flow (not bigger than 0.03% of the experimental shallow flows).

The groundwater flow was artificially induced by provoking hydraulic pressures at the lowest part of the bed, which was constituted by a fine-sandy soil. The bed elevation changes in time, were compared for different flow rates (with and without seepage). Additionally, another series of experiments were carried out for a constant flow rate but with different seepage intensities. This experimental research demonstrated that the flow through the porous medium does affect the bed degradation as a function of the hydraulic head, that provokes upward seepage. It is observable that the bed changes are dependent on seepages intensity, above all within the initial 2 meters of the flume, where secondary flows arose due to scouring. In general, the results are acceptable and consistent. Regardless the small magnitude of the groundwater flows; the impact of seepage becomes important because it is changing the behavior of the turbulent stresses. In consequence, the onset of sediment motion changes and sediment transport mechanics are affected. This fact can become a key issue for sediment management downstream hydraulic structures and future proposals for river training works.

Effects of bed-load on flow resistance and stability in step-pool systems

B. Hohermuth & V. Weitbrecht
Laboratory of Hydraulics, Hydrology and Glaciology (VAW), ETH Zurich, Switzerland

ABSTRACT

Mountain torrents are often obstructed with concrete or wooden check dams to limit channel incision. Besides the ecological deficits, these structures raise high construction and maintenance costs, and feature an abrupt failure mechanism. Up to date, no reliable engineering method exists to stabilize steep open channels with nature-oriented structures such as boulder step-pool systems.

Weichert (2006) proposed to stabilize steep open channels by adding large boulders to the existing grain size distribution of the channel bed. With increasing hydraulic load a stable and resilient step-pool system shall form. We investigated the effect of relative channel width, bed-load transport, grain size distribution and slope on the stability and formation of such self-formed step-pool systems. The initially flat bed was loaded with increasing discharge until a stable step-pools system was formed. Bed morphology scans and flow velocity measurements were performed once the bed had stabilized. For subsequent runs bed-load was added to the flume and the procedure was repeated for flows with bed-load transport.

For steep slopes of 5 to 11%, broad grain size distributions and the presence of distinct bedforms bed-load transport reduces flow resistance (Fig. 1). It leads to a filling of pools and smoothens the bed surface. A flow transition similar to the change from nappe to skimming flow was observed. This inhibits energy dissipation in the pools and at step-forming boulders. Non-dimensional hydraulic geometry equations for both clear-water and flow with bed-load are proposed (Fig. 1). A comparison of the new equations demonstrate a change in the main energy dissipation mechanism for flows with bed-load. Bed-load transport not only increases the mean flow velocity, but the data also indicates a change in the velocity profile. Both effects lead to an increased near-bed velocity and thus increase the drag force acting on step-forming boulders.

It was observed that the increased flow velocity for bed-load flow leads to the mobilization of step-

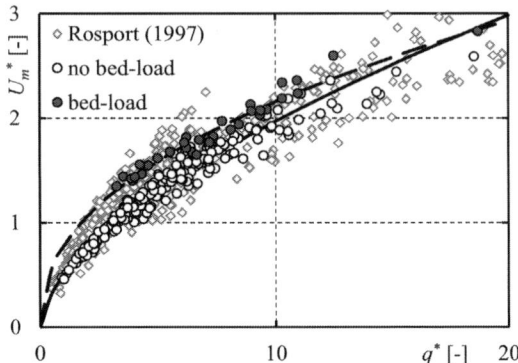

Figure 1. Dimensionless specific discharge $q^* = q/(gs^3)^{1/2}$ versus dimensionless mean flow velocity $U_m^* = U_m/(gs)^{1/2}$, with g = gravity acceleration, s = standard deviation of roughness heights. Rosport (1997) no bed-load data shown for comparison (—) fit for no bed-load, (— —) fit for bed-load data.

forming boulders. The total destruction of step-pool sequences always occurred at critical or near critical flow conditions with bed-load transport. Direct erosion of step-forming boulders was the dominant erosion mechanism for flows with bed-load. Existing stability approaches for steep channels with distinct bedforms are not able to capture the destabilizing effect of bed-load transport.

REFERENCES

Rosport, M. 1997. Fliesswiderstand und Sohlstabilität steiler Fliessgewässer unter Berücksichtigung gebirgsbachtypischer Strukturen (Flow resistance and bed stability of steep channels with consideration of typical bed-structures in mountain streams). *Mitteilung* 196. Institut für Wasserwirtschaft und Kulturtechnik, Universität Karlsruhe, Karlsruhe (in German).

Weichert, R., 2006. Bed morphology and stability of steep open channels. *Report* 192. Laboratory of Hydraulics, Hydrology and Glaciology VAW, ETH Zurich, Zürich.

State of the art on remote sensing methods for suspended sediment concentration in inland and coastal waters

A.E. Holdefer & K. Formiga
University of Goiás, Goiás, Brazil

ABSTRACT

Interest in remote sensing of suspended sediment concentration (SSC) is motivated by the environ-mental, economic and ecological importance of sediment transport in coastal and inland waters. The quantification of suspended sediment in rivers is important in studying the hydrologic, geomorphologic, and ecologic functioning of river flood plains and deltas. Applications include the optimization of dredging operations, assessing the environmental impact of construction activities, understanding geomorphologic change, evaluating fluxes of particulate organic carbon from rivers to the sea, etc. This study aims to review the state of the art for remote sensing methods of suspended sediments concentration (SSC) in inland and coastal waters.

METHODS

The SSC remote sensing retrieval methods are based on the characteristics of the radiation reflected by the water as a function of sediment concentration present in it. The first studies in remote sensing of SSC were mainly focused on the discovery and demonstration of the relationship between the con-centration of suspended sediments and the spectral reflectance, they are based on empirical methods developed by Ritchie et al. (1987); Ritchie et al (1990); Chen et al. (1992); Harrington et al. (1992). The methods can be divided according with the size of the water body, precision of the measurements and nature of the organic/in-organic mixture.

RESULTS

The SSC remote sensing retrieval methods are described as well as a compilation of empiric developed formulas relating sediment concentrations and spectral reflectance. These formulas are listed as the best fit for different water bodies, as well as different formulas for same water body in order to verify those with the best correlation.

DISCUSSION

SSC mapping using remote sensing is currently well established and routine, and the existing algorithms, although very different, show robust and consistent values for a big variety of water bodies, regardless the sediment concentration and the nature of the sediments present in it. The generation of products matured from the "basic" level of snapshots to include now time series for fixed locations and composite of multispectral images with better spatial/radiometric/temporal resolution. Challenges re-main, but new and exciting prospects have been identified, in particular the use of geostationary sensors and the exploitation of geostatistical correlations in space and time (Neukermans et al., 2008).

REFERENCES

Chen, Z., Curran, P. and Hansom, J.D., 1992. Derivative reflectance spectroscopy to estimate suspended sediment concentration., 77, pp. 67–77.

Harrington, J.A. Jr, Schiebe, F.R. and Nix, J.F. 1992. Remote sensing of Lake Chicot, Arkansas: Monitoring suspended sediments, turbidity, and Secchi depth with Landsat MSS data. *Remote Sensing of Environment*, 39(1):15–27.

Neukermans, G., Nechad, B. and Ruddick. K.2008. Optical remote sensing of coastal waters from geostationary platforms: a feasibility study. *Ocean Optics 2008 CDROM*.

Ritchie, J C., Cooper, M C. and Yongqing, J. 1987. Using landsat multispectral scanner data to estimate suspended sediments in Moon Lake, Mississippi. *Remote Sensing of Environment*, 23(1):65–81.

Ritchie, J.C., Cooper, M.C. and Schiebe, F.R. 1990. The relationship of MSS and TM digital data with suspended sediments, chlorophyll, and temperature in Moon Lake, Mississippi. *Remote Sensing of Environment*, 33(2): 137–148.

Study on the incipient velocity of biofilm-coated sediment

L. Huang, H.W. Fang, Q.Q. Shang, Y.S. Chen & G.J. He
Department of Hydraulic Engineering, State Key Laboratory of Hydroscience and Engineering, Tsinghua University, Beijing, China

ABSTRACT

Incipient motion of sediment, which is important in fluvial processes, can be affected by biofilm growth. In this paper, a series of flume experiments are carried out using non-uniform sediments, to investigate the entrainment of sediment with and without biofilm. Differences in entrainment and the velocity at incipient motion are measured over an eight week period, as well as the biomass dynamics. Sediment with biofilm is more stable, i.e., a greater incipient velocity, and the incipient velocity increases to a threshold level over time and then declines. Biofilm development is clearly an important control on the stability of sediments, especially in eutrophic water bodies.

In the critical condition of sliding, the drag force and the frictional force are under equilibrium. The forces acting on a single particle in fluid flow are shown in Figure 1: the effective gravity force F_G, the drag force F_D, the lift force F_L, the frictional force F_f, the cohesive force F_*, and the adhesive force due to biofilm F_A.

A theoretical expression of incipient velocity U_c that includes the cohesive and adhesive forces is derived, where the biomass dynamics $B(t)$ is directly incorporated. The incipient velocity formula in the sediment-water interface is:

$$\frac{U_c^2}{gd} = \frac{\gamma_s - \gamma}{\gamma}\left(6.25 + 41.6\frac{H}{H_a}\right) + \left(111 + 740\frac{H}{H_a}\right)\frac{H_a \delta}{d^2} + \frac{B(t)}{0.067\gamma d} \quad (1)$$

where: H is the water depth, H_a is the head of water equivalent to the atmospheric pressure, δ is the diameter of a water molecule ($=3 \times 10^{-8}$ cm), d is the particle diameter, $B(t)$ is obtained using a biomass dynamics model, see Eqs. (2) and (3).

$$\frac{dB}{dt} = \mu_{max} B \frac{1}{1+k_{inv,B}B}\frac{1}{1+k_d d} - c_{det}V(B-B_0) - c_{auto}B_b B \quad (2)$$

$$\frac{dB_b}{dt} = \left[\mu_{Bb}e^{\beta_{Bb}(T-T_{0Bb})} - c'_{det}B\right]B_b \quad (3)$$

where: μ_{max} is the maximum specific growth rate, $k_{inv,B}$ is the inverse half-saturation coefficient for biomass, k_d is the inverse half-saturation coefficient for sediment size, c_{det} is the detachment coefficient, V is the flow velocity, c_{auto} is the auto detachment coefficient, B_b is the active bacterial density, μ_{Bb} is the maximum specific growth rate for active bacteria, β_{Bb} is the coefficient of temperature, T_{0Bb} is the reference temperature, and c'_{det} is the bacterial detachment coefficient.

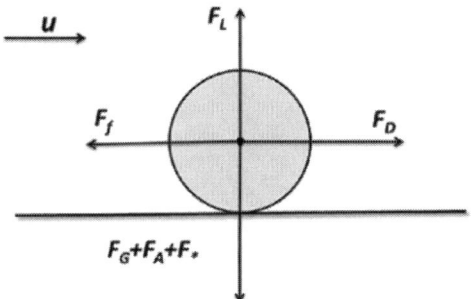

Figure 1. Forces acting on a particle of sliding in fluid flow.

REFERENCES

Fang, H. W., Chen, Y. S., Huang, L. & He, G. J. 2016. Biofilm growth on cohesive sediment deposits: An implication for the management of the Three Gorges Reservoir. Hydrobiologia, submitted.

Shang, Q. Q., Fang, H. W., Zhao, H. M., He, G. J. & Cui, Z. H. 2014. Biofilm effects on size gradation, drag coefficient and settling velocity of sediment particles. International Journal of Sediment Research 29: 471–480.

Riverbank erosion rates prediction incorporating soil erodibility and soil properties relationship: Bernam River, Malaysia case study

S.L. Ibrahim & J. Ariffin
Faculty of Civil Engineering, Universiti Teknologi MARA, Malaysia

A. Saadon
*Department of Civil Engineering, Faculty of Engineering and Technology Infrastructure,
The Infrastructure University Kuala Lumpur, Malaysia*

ABSTRACT

Riverbank erosion and lateral migration changes are very complex and complicated processes. Although it is obvious that all rivers will experience these phenomena, it may be less obvious that there are consistent underlying relationships between different riverbank erosion and lateral migration with parameters such as hydraulic and hydrodynamic characteristics, river bank geometry, soil properties, grain or flow resistance, channel characteristics and others such as vegetation condition. Relationship between soil erodibility and the riverbank erosion rates is considered as one of the primary processes concerned in the development of channel migration. This research investigates riverbank erosion in the Bernam River, Malaysia. Field monitoring data was collected to obtain sufficient amount of information for variables of riverbank erosion factors i.e. channel geometry, hydraulics characteristics and erodibility in order to explain the controls on temporal changes in riverbank erosion rates.

Field experimental data of soil erodibility measurement were gained using a novel, newly developed Jet Erosion Device (JEd) as shown in Figure 1. The JEd equipment is a fabrication of the previously developed Jet Erosion Test equipment. The purpose of this device is to measure the erodibility parameters of the different type of soils of the riverbanks. The working procedures are similar with some modification to adapt with the field data collection. Soil samples were extracted and tested in the laboratory to obtain the basic soil properties. Bank erosion data using erosion pins and bank profiling methods, channel geometry and hydraulics measurements were also obtained from series of field data collection.

Riverbank analysis using the variables obtained through dimensional analysis based on the field data collected were then conducted to identify the relationship of riverbank erosion with the hydraulics, channel and soil characteristics. The dimensionless form of the riverbank erosion relationship as follows:

Figure 1. Jet Erosion Device (JEd).

$$\frac{\varepsilon}{U_o} = f\left(\frac{\tau_o}{\tau_c}, \frac{gd}{U_o^2}, \frac{\rho_w U_o^2}{\tau_c}, \frac{\rho_s U_o^2}{\tau_c}, \frac{H}{d}, S, p, PI\right) \quad (1)$$

where ε is the riverbank erosion, U_o is the depth-averaged velocity for the reach (m/s), τ_o is the fluid shear stress (N/m² or kg/ms²), τ_c is the critical shear stress (N/m² or kg/ms²), d is the reach-averaged depth (m), g is the gravitational acceleration (m/s²), ρ_w is the density of water (kg/m³), ρ_s is the density of sediment particles (kg/m³), H is the bank height (m), S is the averaged longitudinal water-surface slope, P is the bed porosity and PI is the plastic index of the soil.

Statistical analyses using multiple linear and non-linear regression methods were performed to establish the empirical equation of riverbank erosion based on the selected dependent and independent variables with the incorporation of the soil erodibility parameters. Statistical parameters such as coefficient of determination, R^2, standard error of the estimate and discrepancy ratio percentages were used to aid in the selection of the most appropriate model. The predictive variables selected were based on their ability to explain the variation of riverbank erosion. Once a mathematical model in a form of erosion prediction equation was produced based on the regression of the field data, the results were interpreted and checked against any physical data. This validation process was done with field monitoring data and available secondary data.

Development of a bedload sensor for continuous measurement and its applicability

T. Itoh & T. Nagayama
Nippon Koei Co. Ltd. (NK), Tsukuba & Tokyo, Japan

R. Utsunomiya
JFE Advantech Co. Ltd. (JAC), Nishinomiya, Japan

M. Fujita, D. Tsutsumi & S. Miyata
Disaster Prevention Research Institute, Kyoto University, Kyoto, Japan

T. Mizuyama
National Graduate Institute for Policy Studies, Tokyo, Japan

ABSTRACT

There are two kinds of methods such as in-direct and direct measurements for bedload discharge. Bedload slot, which is called as Reid-type automatically recording bedload slot (e.g. Reid et al. 1980, Laronne et al. 1993) is one of direct method, and it cannot obtain total bedload discharge in the flood because of the volume limitation of the slot. There are mainly two kinds of bedload measurements in in-direct methods such as Japanese pipe hydrophone and geophone (Gray et al. 2010), and those usually need correlations between those signals and sediment volume measured by the other method. The measurement with loadcell can be useful for not only large particles but also sand particles due to development of the sensor. The submerged weight of sediment particles passing through the bed can measure using a plate, and it can measure continuously without disturbing bed surface condition. A small, automatically recording bedload sensor was developed with a plate whose plan size was designed as 1.0 m in long and 0.5 m in wide.

Herein, mass measured by loadcell in unit width is calculated by the time integration of the production of both sediment particle velocity and the averaged mass flux by two loadcells. The velocity can be estimated by the length between two loadcells and the time by the cross-correlation function of temporal changes of mass flux by loadcells. After initial tests (Itoh et al. 2014), flume tests were conducted in the water flow with bedloads to evaluate validity of measurement for moving particles, and were conducted using a straight channel in the steady flow with bedloads supplied steadily.

Figure 1 shows relations between inlet and estimated weight in the flume tests. Four kinds of sediment particles (M_1: 3.32, M_2: 11.8, M_3: 32.7 and M_3: 69.4 mm in a diameter, respectively) were used. There are large discrepancies in case of 3.32 mm in diameter, and those might depend on estimation methods using the cross-correlation function in the sediment laden flow. For example, the application of sediment

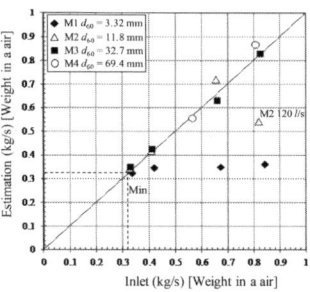

Figure 1. Relationships between inlet and estimated weight in laboratory flume tests (Weight is shown by measurement value in an air.)

moving layer (e.g., Egashira et al. 2005) can be effective to solve the problem. The system was installed in the flume of Hodaka Sedimentation Observatory in November of 2012 to obtain sediment runoff by floods, and data by flume tests using natural sediment particles are also shown.

REFERENCES

Itoh, T., Gotoh K., Utsunomiya, R., Nonaka, M., Nagayama, T., Tsutsumi, D., Fujita, M., Miyata, S. & Mizuyama, T. 2014. Experimental studies for monitoring of bedload using various sensors. Proceedings of the Interpraevent 2014 Pacific Rim (edited by Fujita, M. et al.), November 25–28, Nara, Japan: O–17.pdf in DVD.

Reid, I., Layman, J.T. & Frostick, L.E. 1980. The continuous measurement of bedload discharge. Journal of Hydraulic Research, 18(3): 243–249.

Laronne, J.B. & Reid, I. 1993. Very high rates of bedload sediment transport by ephemeral desert rivers. Nature, 336 11 November: 148–150.

Egashira, S. & Itoh, T. 2005. Paradoxical discussions on sediment transport formulas. River, Coastal and Estuarine Morphodynamics: RCEM 2005-Paker & Garcia (eds). 2006 Taylor & Francis Group, London: 33–38.

Gray, J.R., Laronne, J.B. & Marr, J.D.G. 2010. Bedload-surrogate monitoring technologies. U.S. Geological Survey Scientific Investigations Report 2010–5091.

Distributional characteristics of sediment concentration in lotus-root-shape compound channels

Z. Ji & C.H. Hu
State Key Laboratory of Simulation and Regulation of Water Cycle in River Basin, China Institute of Water Resources and Hydropower Research (IWHR), Beijing, China

F. He
Department of Sediment Research, China Institute of Water Resources and Hydropower Research, Beijing, China

ABSTRACT

In this paper, the distribution characteristics of sediment concentration in the lotus-root-shape compound channels are analyzed and compared by the data from flume experiments. The averaged sediment concentration of flood plain is smaller than that of main channel in the lotus-root-shape compound channels. The sediment concentration of plain is larger than a single channel because of the effects of momentum transfer between channel and plain. The ratio of the sediment concentration between plain and cross-section goes up with the sediment load. With the increment of water depth or the decrement of sediment load, the sediment concentration of channel and plain always debase, and the amplitude decreases alone the river. The mean sediment concentration in vertical direction reduces gradually from the center zone of channel to two sides of plain. The vertical distribution of sediment concentration doesn't obey Rouse-law. In general, the sediment concentration of channel is smaller, but that of plain is larger than these calculated data of the formula. The average annual sediment load in the Lower Yellow River was 1.6 billion tons before 1980s, ranked first in the world. The whole length of the lower reach is about 878 kilometers. The area of floodplain is about 3,544 km², accounting for 84% in the total area of the river. Thus, it is very important to study flow movement and sediment transport in the compound channel in the Lower Yellow River. A compound channel with the main channel and floodplains is a common phenomenon in fluvial rivers. When the discharge is smaller, flow runs in the main channel. Otherwise, flow would enter floodplains. Some effects of overbank flow must be considered in flood control, channel training and floodplain exploitation etc. Up to now, many results of the flow movement have been attained, but few research works on sediment transport in compound channels can be found.

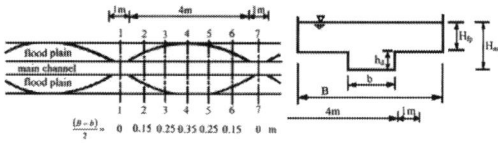

Figure 1. Sketches of lotus-root-shape compound channels. (left) Lotus-root-shape channel; (right) compound cross-section.

In this paper, the river reaches are generalized into the lotus-root-shape compound channels according to the plane character of the Lower Yellow River, and some experiments of sediment transport are conducted in the kind of channel. These research results benefits to the project harness in the Lower Reach of Yellow River. The generalized physical model of the lotus-rootshape model with 30 meters long. A self-circulating structure is adopted in the models and both sides and beds are fixed with cement. The symmetrical cross-sections consist of one channel and two floodplains, and both bed slopes, J, are 1.0×10^{-4}. For the lotus-root-shape channel, $b = 0.3$ m, but the (B-b)/2 changes from 0 to 0.35 m. The length of each lotus-root-shape reach is 4 m, and its side curvature is about 0.17. There is a one-meter transitional reach between every lotus-root-shape reach, shown in Figure 1. There are seven measured cross-sections for collecting data in a lotus-root-shape reach, but only the fourth cross-section has the same dimension as the straight channel.

Many experiments have been conducted in the both channels at different series of flow and sediment. The main experimental conditions include: Hmc = 0.030 m ~ 0.122 m, Q = 0.009 m^3/s ~ 0.022 m³/s, S = 4 kg/m³ ~ 25 kg/m³. The ash of coal power is adopted as the model sand, and its characteristic parameters are: median diameter, $d_{50} = 1.4 \times 10^{-5}$ m, specific gravity, 2100 kg/m³, uniformity coefficient, 1.73.

Discharge coefficients derived from sediment concentration to estimate discharge across a Sabo dam

K. Kawaike & H. Nakagawa
Disaster Prevention Research Institute, Kyoto University, Kyoto, Japan

N. Kim
National Forestry Cooperative Federation, Daejeon, South Korea

H. Zhang
Natural Sciences Cluster, Research and Education Faculty, Kochi University, Kochi, Japan

ABSTRACT

In numerical simulations of debris flow and evaluation of Sabo dam mitigation effects, one of the most important issues is estimation of discharge across a Sabo dam. In this study, under the assumption that the complete overflow equation and the free overfall equation are applicable to 1D simulation of debris flow, the coefficients of those equations are proposed and validated by comparing with experimental results.

Depending on the relative elevation of Sabo dam, river beds and flow surfaces on both upstream and downstream sides, the following two equations are applied to estimate discharge across the Sabo dam; complete overflow and free overfall equations.

$$M_0 = C_1 h' \sqrt{2gh'} \quad (1)$$

$$M_0 = C_2 h \sqrt{gh} \quad (2)$$

where M_0 is discharge per unit length, g is gravitational acceleration, C_1 and C_2 are coefficients. h' and h are overflow depth and fluid depth, respectively, shown in Figure 1.

From the experimental results, we propose the following equations for the coefficients of the above equations, C_1, C_2, derived from sediment concentration, C, on the upstream side of the Sabo dam.

$$C_1 = -0.38029 + 3.7331C \quad (3)$$

$$C_2 = 0.59751 + 0.3925C \quad (4)$$

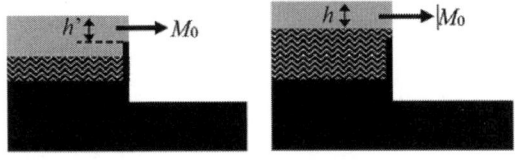

Figure 1. Complete overflow (left) and free overfall (right).

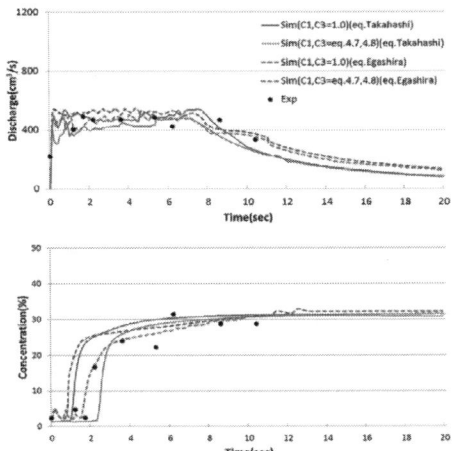

Figure 2. Comparison between experimental and simulation results of debris flow discharge (above) and sediment concentration (below).

Using those equations and coefficients, 1D debris flow simulations are carried out, where the Takahashi (1992) & Egashira (1993) equations are applied to bottom shear stress, erosion and deposition velocity estimation. Figure 2 shows the comparison between experimental results and simulation ones with different coefficient values. The proposed coefficient values improved the simulation results especially of sediment concentration in the both cases of Takahashi and Egashira equations.

REFERENCES

Egashira, S. 1993. Mechanism of Sediment Deposition from Debris Flow (part 1). J. of the Japan Society of Erosion Control Engineering, 46 (1), ser.186, 45–49. (in Japanese)

Takahashi, T., Nakagawa, H., Harada, T. & Yamashiki, Y. 1992. Routing Debris Flows with Particle Segregation. J. of Hydraulic Engineering, ASCE, 118 (11), 1490–1507.

Near-bed turbulence characteristics in unsteady hydrograph flows over mobile and immobile gravel beds

J. Kean, A. Cuthbertson & L. Beevers
Institute for Infrastructure & Environment, Heriot-Watt University, Edinburgh, UK

ABSTRACT

Improvements in sediment transport prediction require enhanced understanding of the turbulent flow conditions present within the wall region. Past studies investigating the effects of turbulence on sediment transport have suggested that turbulent burst and sweep events, in particular, exert a strong influence on the bed surface, and result directly in the transport of individual sediment grains (e.g. Sechet & Le Guennec 1999). Studies have also been conducted on the effects of flow unsteadiness on the near-bed turbulent structure (e.g. Nezu et al. 1997); however, these have not focused on the interactions with an erodible sediment bed. This paper looks specifically at the effects of differences between mobile and immobile beds on the near-bed turbulent burst and sweep events generated, as well as considering the influence of unsteady flow on these phenomena over both mobile and immobile gravel beds. Quad-rant analysis of the near-bed turbulent flow, using the method outlined by Lu & Willmarth (1973), indicates that the total number of overall Q2 + Q4 (burst/sweep) events detected above the immobile sediment bed is significantly greater than those detected above the mobile sediment bed (sample plot shown in Figure 1(a)). This is somewhat unexpected given that previous studies (e.g. Sechet & Le Guennec 1999) suggested a correlation exists between bed transport and the prevalence of burst/sweep events. Hole-size analysis of the quadrant plot provides a possible explanation for this apparent discrepancy, showing that when hole-size H' = 3 is applied to both data sets (Fig. 1(b)), the number of higher magnitude Q2 + Q4 (i.e. burst/sweep) events is greater in the mobile bed, compared to the immobile bed. A further finding was that the proportion of detected Q2 + Q4 events rose by up to 10% between the base and peak flow of the hydrograph rising limb, before reducing back to pre-hydrograph levels again during the falling limb. Further study to examine how the results of a hole-size analysis of quadrant plots may be correlated with observed sediment transport in unsteady hydrograph flows can now be carried out following these results, which have demonstrated that higher strength rather than higher volume of turbulent burst/sweep events are found where conditions are favourable for sediment transport.

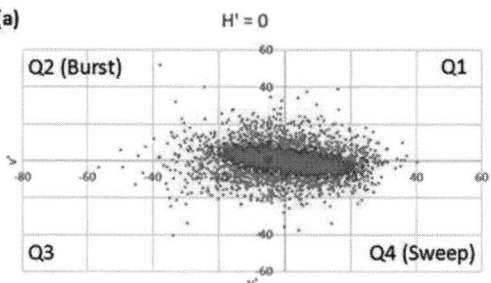

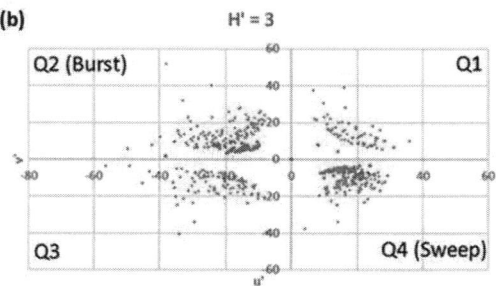

Figure 1. Quadrant plots at (a) H' = 0 and (b) H' = 3 at t = 0.5 h from start of hydrograph rising limb. Red signifies immobile results bed, and green signifies mobile bed results.

REFERENCES

Lu, J.J. & Willmarth W.W. 1973. "Measurement of the structure of the Reynolds stress in a turbulent boundary layer" Journal of Fluid Mechanics 60(3): 481–511.

Nezu, I., Kadota, A. & Nakagawa, H. 1997. "Turbulent structure in unsteady depth-varying open-channel flows" Journal of Hydraulic Engineering 123(9): 752–763.

Sechet, P. & Le Guennec, B. 1999. "Bursting phenomenon and incipient motion of solid particles in bed-load transport" Journal of Hydraulic Research 37(5): 683–696.

Numerical simulation of local scour around three circular cylinders in staggered array

H.S. Kim & M. Park
Korea Institute of Civil Engineering and Building Technology, Goyang, Korea

I. Kimura & Y. Shimizu
Hokkaido University, Sapporo, Japan

M. Nabi
Deltares, Delft, The Netherlands

ABSTRACT

Circular cylinder structures are easily observed in natural rivers and oceans. They significantly change flow patterns and then influence physical mechanisms around them. When the structures are placed on a movable bed, they cause an increase in local sediment transport and thus it results in scour in the vicinity of the structures. Local scour had been identified as one of the important mechanisms that induce damages or failures of the structures in rivers and oceans. Many experimental studies on local scour around cylinders have been conducted (Melville and Coleman, 2000). However, numerical studies about scour around circular cylinders are still rare.

There are several advantages of the numerical study on local scour around cylinders. Numerical simulation is free from scaling effect because it can be simulated in the prototype size. In addition, experiments of scour around cylinders are very expensive and difficult whereas numerical simulation is efficient and effective way to study flow and scour characteristics. Numerical studies have mainly used URANS models with a two-equation turbulence model. The scour process for a bed has been calculated using the sediment continuity equation with empirical bedload models. In particular, previous numerical investigations carried out the local scour around an isolated cylinder (Roulund et al, 2005).

There are few numerical studies on scour around cylinder groups, even though the cylinder groups have been widely used in river and marine engineering. Also, little has been done to quantify the effect of the cylinder array and interval on the scour depth and pattern. In this study, local scour around three circular cylinders in staggered array on a sandy bed is investigated using a large-eddy simulation coupled with sediment transport and morphodynamic models. The effect of the cylinder interval on the scour depth and process is examined.

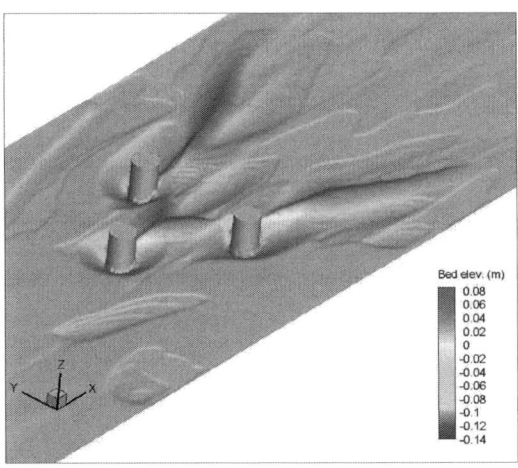

Figure 1. Bed elevation for 5D case at equilibrium state.

Fig. 1 shows the bed elevation with cylinder interval of 5D (D: cylinder diameter). Local scour is observed around individual cylinders and it is not overlapped. This is due to relatively large interval of cylinders. However, the scour depth and patterns at downstream cylinders are influenced by upstream cylinder. Scour depth at downstream cylinders increases about 1.3 times of that around upstream cylinder. This is attributed to fact that the turbulence vortices induced by upstream cylinder impinges on downstream cylinders.

REFERENCES

Roulund, A., Sumer, B. M., Frodsoe, J. and Michelsen, J., 2005. Numerical and experimental investigation of flow and scour around a circular pile. Journal of Fluid Mechanics, 534: 351–401.

Melville, B. W. and Coleman, S., 2000. Bridge scour. Water Resources Publication, Colorado.

Simulation of sediment hyper concentration in the lower Yellow River using variational data assimilation method

R. Lai, M. Wang & M. Wang
Yellow River Institute of Hydraulic Research, Yellow River Conservancy Commission, Zhengzhou, China

H. Wang
Press and Publication Center, Yellow River Conservancy Commission, Zhengzhou, China

ABSTRACT

The sediment hyper concentration in the Yellow River makes it difficult to simulate sediment concentration accurately when using traditional numerical models.

To improve the simulated accuracy of the numerical models, variational based data assimilation is applied in this paper. Variational based algorithm implements the minimum of the distance between observation and prediction, while, at the same time, taking an explicit dynamic system as a constraint condition. The obvious theoretical advantage of variational data assimilation is that it can provide exact consistency between the forecasts and the dynamics.

In the variational data assimilation method, a cost function is introduced first to describe difference between simulated sediment concentration and observations. One dimensional suspended sediment transport equation, taken as a constraint, is added into the cost function. Adjoint equation of the data assimilation system is used to solve the minimum problem.

The optimal procedure for the variational data assimilation of sediment concentration is as follows:

1. Define the assimilation window and the time step in the numerical model;
2. Calculate the values of discharge and water level using traditional numerical models;
3. Calculate the sediment concentration previously using sediment transport equations;
4. Once the observed sediment concentration is available, the Lagrange multiplier is estimated;
5. Update the cost function. ε, a number with small value, is used to determine whether the minimum cost function is reached. If not, the system will update the iterative step and back to calculate the sediment concentration again;
6. The assimilation system will continue until the cost function reaches the minimum value.

The usefulness and the capability of the variational data assimilation are demonstrated using historical flood data in the lower Yellow River, from Xiaolangdi to Gaocun. In the assimilation process, errors between the simulations and the observations are analyzed.

During the flood, the peak values of sediment concentration are about 82 kg/m^3 in Huayuankou station and 68 kg/m^3 in Jiahetan station, respectively. Although the direct simulation that simulated without data assimilation can simulates the peak values, the simulations are higher than the observations. However, after using data assimilation, the simulated accuracy improved.

It is believed that the flexible approach of the variational data assimilation can improve the accuracy of simulated sediment concentration. Finally, limitations of variational data assimilation and future work are discussed.

Experimental study on local scour protection of piers for Hangzhou Bay Bridge by using twisted duplex blocks

Z. Li, Y. Shi, R. Wang & J. Zhang
Zhejiang Institute of Hydraulics and Estuary, Zhejiang Surveying Institute of Estuary and Coast, Key Laboratory of Estuary and Coast of Zhejiang Province, Hangzhou, China

J. Wang
Ningbo City Bridge Development Co., Ltd., Ningbo, China

ABSTRACT

Over the past few decades, cross-sea bridges have played an important role in the social and economical development. For example, Hangzhou Bay Bridge, which locates in Hangzhou Bay in the East China Sea, is one of the longest cross-sea bridges in the world. Besides the great difficulties in the cross-sea bridge construction, local scour is an unavoidable and stubborn problem in the bridge management and potentially endangers the safety of the bridge. However, the mechanism of local scour is far from well-recognized and keeps as a hot topic in both academic and engineering fields. And the effective prevention from local scour concerns engineers and social managers.

In this paper, an experimental study was carried out by movable-bed physical experiments using Twisted Duplex Blocks (hereinafter referred to as Blocks), aiming to explore the mechanism of local scour of ramp piers that support and stead the sightseeing tower of the Hangzhou Bay Bridge. Considering the diameter of pier in the actual field is 2.2 m, natural bed sediment around the pier with median particle size (D50) of 0.049 mm and the limited space of experimental site, a normal physical model was chosen and the scale of the model was 1:80. The experiment was conducted in a laboratory flume with length of 40 m, width of 4 m and height of 0.5 m. Flow in the flume is re-circulated by using several variable frequency pumps. A false floor (16 m long × 4 m wide) with a ramp is installed at the both end of the flume. In the centre section of the flume, a recess (8 m long × 4 m wide × 0.3 m high), in which sediment materials and a pier with diameter of 2.75 cm are laid up. The median size of the sediment load of the physical model (Dw50) is caculated as 0.12 mm and wood flour was chosen as sedimental materials in the experiment. As Hangzhou Bay confronted with strong varying tides, the representative range of tidal height in the field is selected as 8 m to 8.4 m.

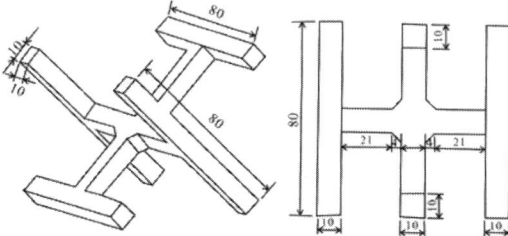

Figure 1. The sketch of structure of the Blocks (unit: cm).

The sketch of structure of Blocks shows as below (Fig. 1). The size of parts of the Blocks is 80 cm (1 cm in the physical model) long with a 10 cm × 10 cm (0.125 cm × 0.125 cm in the model) cross section. The layout density of Blocks ranges from 2.3 to 3.4 pieces per square meter.

Chiew (1992) adopted a method that prevents piers from scouring and changes the structure of flow stream around piers. Based on the experimental data obtained in the present study, the applicability and limitation of the method are discussed. In addition, several experimental groups, based on different densities of Blocks set around the piers and with/without filter layer installed under the Blocks by using rock and sand, were conducted.

The results indicate that different densities of layout of Blocks can reduce the depth of local scour and protect against the erosion of sediment around the piers. Furthermore, the results demonstrate that the effect of the protection is better in situation of Blocks installed with filter layers under themselves.

REFERENCE

Chiew, Y.M. 1992. Scour Protection at bridge piers. J.Hydr., Engrg., ASCE, 118(9):1260–1269.

Effect of proportion of wash load to suspended load on river erosion and deposition

C.T. Liao
Disaster Prevention & Water Environment Research Center, Hsinchu, Taiwan

K.C. Yeh, G.H. Liu & K.W. Wu
Department of Civil Engineering, Hsinchu, Taiwan

ABSTRACT

Wash load is the portion of sediment that is carried by a fluid flow in a river, such that it always remains close to the free surface. Generally, wash load grains tend to be very small and therefore have a small settling velocity, being kept in suspension by the flow turbulence. The composition of wash load is distinct because it is almost entirely made up of grains that are only found in small quantities in the bed. Characteristic motion of wash load is different from bed material load. In the past, lots of studies assume the wash load is 70~90% of suspend load in the river because of lacking fundamental data. This empirical method could not accurately definite about grain size and proportion of wash load; so sediment discharge could not be estimated precisely. The purpose of this study is developing a relationship of proportion of wash load with the influential factors in river, including the hydraulic, geometrical and sediment factors. Eq.(1) shows the relationship formula of wash load in suspended load that is established by using the multivariate regression analysis with data collected on lower Kao-Ping River, Taiwan during the typhoon flood events.

$$\frac{C_w}{C} = 5.58 \left(\frac{\bar{V}^2}{g.H}\right)(s)^{0.11}\left(\frac{D_{wash}}{D_{50}}\right)^{0.096} \quad (1)$$

where C_w/C is the proportion of wash load in suspended load; V is the mean velocity of flow; g is the gravity; H is the water depth; S is the channel slope; Dwash is the grain size of wash load. The effect of proportion of wash load on river deposition-erosion is carried out by adapting a one dimensional mobile-bed model called CCHE1D. The study compared different inflow sediment concentration with general empirical method, and the empirical formula of wash load. From the simulated results, it can be known that the different of bed variation simulated by the empirical formula from the observed values approximates 0.53 m which is superior to the result from general empirical method (about 0.62 m). In summary, the empirical formula can be available for lower Kao-Ping River, Taiwan. Therefore, the empirical formula of wash load provides not only an innovation method, but also some the ubsequent applications.

REFERENCES

Wu, W. & Vieira, D.A. 2002. One-dimensional channel network model CCHE1D 3.0-Technical Manual. Technical Report No. NCCHE-TR-2002-1, National Center for Computational Hydroscience and Engineering, University of Mississippi, USA.

Wu, W. 2007. Computational river dynamic. Taylor & Francis, London, U.K.

Experimental investigation on local shear stress and turbulence intensity over a rough bed with and without sediment using Laser Doppler Anemometry and Particle Image Velocimetry

P. Lichtneger, C. Sindelar & H. Habersack
University of Natural Resources and Life Sciences, Vienna, Austria

J. Kitzhofer
Dantec Dynamics GmbH, Ulm, Germany

E.A. Prager
Ing. Prager Elektronik HandelsGmbH, Wolkersdorf, Austria

ABSTRACT

Sediment transport presents an important process in river morphology. Incipient motion criteria, as well as formulas for bed-load discharge and maximum scour depth are important tools for hydraulic engineers. Typically, empirical approaches based on large-scale mean parameters like discharge, water depth, bed slope and a characteristic sediment grain size are employed (Meyer-Peter & Müller 1948, Van Rijn 1984). In recent years, the key role of turbulence on sediment motion is widely acknowledged.

Sediment motion is predicted based on small-scale parameters like sediment grain size and shape, local velocity gradients, local velocity turbulence and Reynolds' shear stress (Celik et al. 2013, Schmeeckle et al. 2007, Smart & Habersack 2007). The latter parameters represent a challenging request on the experimental technique and facilities for get-ting feasible data to derive new turbulence-based models and to allow the proper evaluation of large-scale based engineering approaches.

The present study focuses on capturing small-scale parameters by implementing Laser-Doppler Anemometry (LDA) and Particle-Imaging Velocimetry (PIV) in the presence and absence of sediment transport. An open-channel flow was simulated using a hydraulic flume with a built-in fixed bed roughness (a casting of a gravel bed). The measured velocity data provide information about the shear stress and turbulence intensity distribution in the symmetry plane and allow a unique direct comparison between the cases with and without moving sediment. Thus, a comparison of large- and small- scale based parameters related to sediment transport, especially the local boundary shear stress, are provided.

Further, the implementation of laser-optical measurement techniques in sediment-laden flows and related difficulties especially in connection with reflections are discussed. In the Figure 1 the setup of the measurement scheme is given:

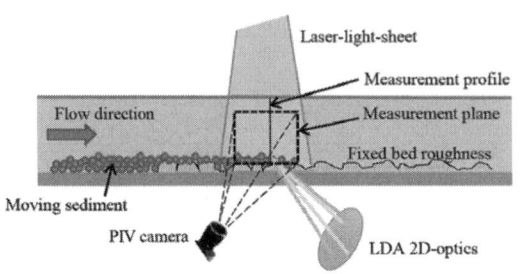

Figure 1. Experimental setup.

REFERENCES

Celik, A.O., Diplas, P. & Dancey, C.L. (2013). Instantaneous turbulent forces and impulse on a rough bed: Implications for initiation of bed material movement. Water Resources Research, 49(4), 2213–2227. doi:10.1002/wrcr.20210

Durst, F., Kikura, H., Lekakis, I., Jovanović, J. & Ye, Q. 1996. Wallshear stress determination from near-wall mean velocity data in turbulent pipe and channel flows. Experiments in Fluids 20. Pp 417–428.

Ferreira, R.M.L. 2011. Turbulent Flow Hydrodynamics and Sediment Transport: Laboratory Research with LDA and PIV. In. Rowinski, P. (ed.) Experimental Methods in Hydraulic Research. Geoplanet: Earth and Planetary Sciences.

Meyer-Peter, E. & Müller, R. 1948. Formulas for Bed-Load Transport. Proceedings 2nd International IAHR Congress, 1948 Stockholm, Sweden. 39–64.

Schmeeckle, M.W., Nelson, J.M. & Shreve, R.L. 2007. Forces on stationary particles in near-bed turbulent flows. Journal of Geophysical Research, 112(F2), F02003.

Smart G.M. & Habersack, H.M. 2007. Pressure fluctuations and gravel entrainment in rivers, J. Hydraul. Res., 45(5), 661–673.

Van Rijn, L.C. 1984. Sediment Transport .1. Bed-Load Transport. Journal of Hydraulic Engineering-Asce, 110(10), 1431–1456.

Analysis on transport characteristic of bimodal bed load & a new calculation model for non-uniform bed load

M.X. Liu & G.D. Li
State Key Laboratory Base of Eco-Hydraulic Engineering in Arid Area, Xi'an University of Technology, Xi'an, China

D.P. Sun & L.Q. Han
North China University of Water Resources and Electric Power, Zhengzhou, China

ABSTRACT

The transport of the non-uniform bed load is a complicated coherent physical process of the fluid and solid in river. Researches on the motion characteristics, transport laws as well as the action mechanism of the key factors are both helpful to develop the hydraulic theory and solve the engineering sediment problems. Based on the technologies of the image recognition and the instant monitoring for the bed load, series of flume experiments were carried out. Image recognition technology is used to recognize the number of the various size particles in the image taken by the high speed camera, which is meaningful to investigate the transport progress of the bed load.

By introducing the comprehensive flow intensity Ψ_b and the feature Froude number Fr_b, the transport characteristics of the bimodal bed load were researched under several discharge and sediment conditions, and the effects of particle non-uniformity on transport rate of bed load were especially investigated. Ψ_b is a comprehensive parameter of the flow intensity that reflects the sediment composition characteristics on the bed and the properties of the viscous sublayer. Fr_b is the feature Froude number that reflects the roughness of the bed.

By fitting the relationship curve between the transport intensity of bed load Φ' with the related sediment parameters separately, such as equalization diameter of non-uniform sediment, mixture ratio, grading differential, etc., it indicates that all of these parameters have influence on the bed load transport more or less. However, the transport intensity of the bed load presents a closer correlation with the mixture ratio. Mixture ratio is the key sediment factor affecting the transport intensity of the bed load. Other sediment parameters are also considered while establishing the transport rate formula of bed load.

By the means of dimension analysis among the transport rate, sediment parameters and flow parameters, the basic function form of the transport rate of the bimodal non-uniform bed load was proposed, which includes several factors such as the feature Froude Number, hydraulic slope, turbulence intensity, bed roughness and sediment composition, etc.

By means of fitting analysis on the experiment data, a formula for calculating the transport intensity of the bimodal bed load Φ' was established finally from the view of the comprehensive flow intensity Ψ_b. The two important coefficients in it are both associated with the sediment non-uniformity.

The effect mechanism of the sediment non-uniformity on the transport of bimodal bed load is analyzed preliminarily. The active particle interaction on the bed plays an important role in increasing the amount of the moving coarse particles and the total bed load, which caused the transport intensity of bimodal bed load presents a hump curve with the changing of sediment composition. The particle interaction mainly refers to collision effect and the collapse effect of the fine particles on the coarse particles. The collapse effect mainly refers to a progress that the surrounded coarse particle is exposed little by little as the gradual leaving of the surrounding fine particle on the bed, and may be easy to lose balance and move under a little flow fluctuation or particles collision.

REFERENCES

Almedeij, H., Diplas, P. & Fawjia, A. 2006. Approach to sediment from gravel for bed load transport calculations in streams with bimodal sediment. J. Hydraulic. Eng , 132 (11): 1176–1185.

Wilcock, P.R. & Crowe, J.C. 2003. Surface-based transport model for mixed size sediment[J]. Journal of Hydraulic Engineering, 129 (2): 120–128.

Patel, S.B., Patel, P.L. & Porey, P.D. 2015. Fractional bed load transport model for non-uniform unimodal and bimodal sediments. Journal of Hydro-environment Research, 9: 104–119.

Sediment transport along the Deepwater Navigational Channel of Changjiang Estuary, China

S. Lou, S.G. Liu & G.H. Zhong
Department of Hydraulic Engineering, Tongji University, Shanghai, China

G.F. Ma
Department of Civil and Environmental Engineering, Old Dominion University, Norfolk, VA, USA

ABSTRACT

Sediment transportation is an complicated process. It has great impacts on riverbed characteristics, morphology, aquatic ecosystems and water quality. High levels of sediment concentrations can reduce the amount of light reaching lower depths and inhibit the growth of submerged aquatic plants. Consequently it will affect species which are dependent on them, such as fish and shellfish. In addition, bed sediment is sometimes the main source of nutrients and contaminants. The resuspension and transportation of sediments participate in chemical and biological processes, and influence the nutrient or contaminants cycles in water column, which would have great impact on ecosystem and water environment.

Under the combined effects of river discharge, tide, wind, complex geometry and bathymetry, sediment transport in the Changjiang Estuary is rather complicated. In this paper, fine sediment transportation was studied in the Yangtze River Estuary through numerical simulation of FVCOM. FVCOM is an unstructured grid, finite volume, 3D primitive equation coastal ocean circulation model developed by UMASSD-WHOI joint efforts. It has been used widely and

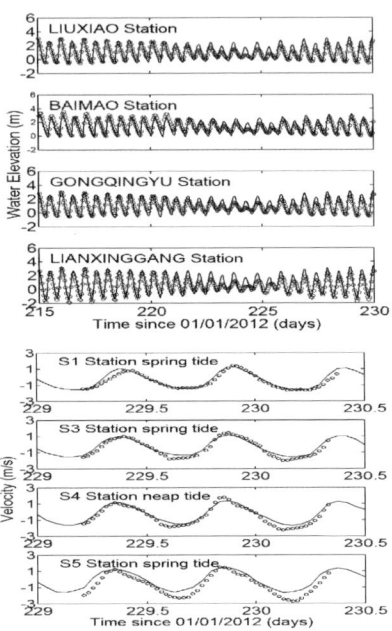

Figure 2. Parts of verification results (cycles: measurements; lines: simulation).

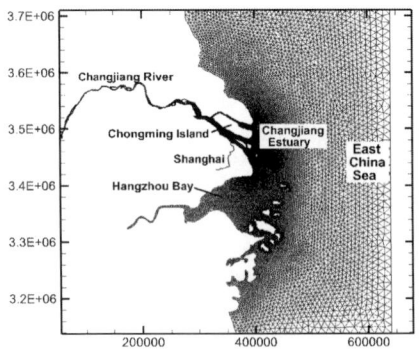

Figure 1. Study region and non-uniform grids.

successfully. Hydrodynamic module was established and verified. Figure 1 shows the study region and non-uniform grids. The agreements between simulated and measured water levels, depth-averaged current velocities and tidally averaged salinity distributions were excellent. Parts of results were shown in Figure 2.

After the verification of hydrodynamic module, fine sediment distribution and transportation were simulated. Based on the results, sediment flux was analyzed along the Deepwater Navigational Channel of Changjiang Estuary. Results in this paper could reveal the spatial and temporal distribution of sediment, and provide technological information for the Yangtze Estuary Deepwater Channel Regulation Project.

Study on the bed coarsening and limit scour depth of the lower reaches of the Three Gorges Reservoir

J.-X. Mao & X. Geng
China Institute of Water Resources and Hydropower Research, Beijing, China

ABSTRACT

The Three Gorges Reservoir has intercepted about 75.5 percent sediment of the upper reaches of Yangtze River and the relative equilibrium between the natural form and the runoff-sediment condition of the original river has been undermined, since the Three Gorges Project has been running for more than a decade. These changes resulted in strong scour and river bed roughness of the lower reaches of the Three Gorges Reservoir. For example, the median particle diameter of the bed material of the Chang 13 section (which was located at about 4.7 km downstream of Yichang Hydrological Station) was 0.24 mm in 2003 and it increased to 34.0 mm in 2010 (Fig. 1). With the armor layer being formed, the scour strength gradually reduced to zero and the riverbed reached its maximal scour depth. For instance, the annual aver-age scour depth was 0.26 m of the Yichang-Zhicheng reach and the annual scour depth was 0.45 m on aver-age of the thalweg longitudinal section during 2003-2010[1]. As a contrast, the annual average scour depth was 0.026 m of the reach and the annual scour depth was 0.1 m on average of the thalweg longitudinal section during 2011-2013. The data showed that this reach scoured weakly after the year 2010 and basically attained to the scour equilibrium. In this article, it used the theoretical formulas of non-uniform sediment incipient motion [2][3] (obtained by Han 1982) to calculate and study the river bed roughness and the limit scour depth of Yichang-Zhicheng reach (Fig. 2, the beginning of X-axis represents the Yi-chang Hydrological Station) with the measured data of the topography and the size distribution of bed material. The results showed that the calculated average limit scour depth was close to the measured average values during 2003-2010 of Yichang-Zhicheng reach if it calculated with the representative flow 45000 m³/s (the largest discharged flow currently of the Three Gorges Reservoir), and these data illustrated that the calculation method was reliable. If the representative flow was 55000 m³/s (the designed regulating flow of the Three Gorges Reservoir), Yichang-Zhicheng reach would continue

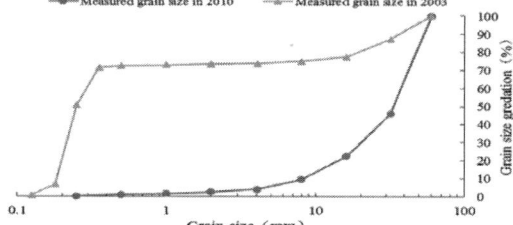

Figure 1. Rivebed coarsening of Chang 13 section.

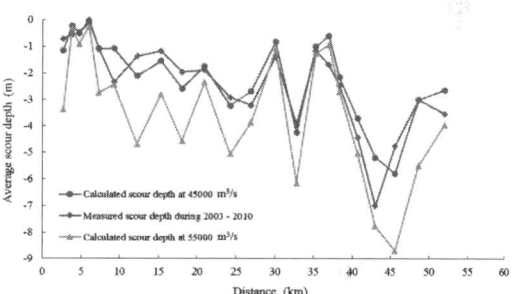

Figure 2. Average scour depth of some sections about Yichang-Zhicheng reach.

to scour and the average scour depth was about 1.30 m, and the scour depth of some sections would exceed 2.0 m.

REFERENCES

The Hydrological Office, Yangtze Water Resources. 2012. The flow and sediment features flowing into and out of the Three Gorges reser-voir, analysis of reservoir sediment and river scour downstream of the Three Gorges Dam.

Han, Q-W.1982. Laws of sediment incipient motion and sediment incipient velocity. Journal Of Sediment Research, 2, 11–26.

Han, Q-W. & He, M-M. 1996. Mechanism of non-uniform sediment incipient motion and incipient velocity. Yangtze River Scientific Research Institute, 3, 12–17.

Observation evaluating water and sediment runoff at Sabo dam in Kiso river basin

S. Matsuda, T. Nagayama, T. Ikeshima, K. Goto, Y. Nishi & T. Itoh
Nippon Koei Co., Ltd. (NK), Tsukuba, Japan

ABSTRACT

In Kiso river basin, the sediment movements such as debris flows are called "Jya-Nuke", in which "Jya" means debris flow movements seems to be snakes and "Nuke" means sediment runoff from torrents occurs suddenly. Sediment yielding is active and sediment is transported directly in the basin. Kamiyama-sawa No. 1 Sabo dam, in Kamiyama-sawa, a branch of Yokawa River, in Kiso River basin, is a narrow slit Sabo dam with transversal iron bars. The Sabo dam is 13 m in height, with 25 m in spillway width, a slit of 2 m in width and 7 m in height, and the ratio of slit for spillway width is 1/12.5. The iron bars in a slit consisted of six iron bars set with 700 mm clearance respectively. Around the Sabo dam, there are two types diameter of bed material. Mainly granite sand is around 2 mm, and the other one is granite boulder around 1 to 2 m.

In order to investigate a function of the Sabo dam, several sensors are installed such as flow depth (upstream and downstream), images from camera (slit part of Sabo dam from upstream and downstream, and storage area), turbidity for wash load and pulses form Japanese hydrophone for bed load, an electromagnetic velocity mater for bed shear stress. Monitoring started since June in 2013 concerning to sediment and water runoff.

After installation of sensors, there were two flood events on September 16th in 2013 (Accumulated rainfall depth: 95.5 mm, duration of rainfall: 14 hours, maximum rainfall intensity: 16.5 mm/h at Nagiso rainfall observation station) and on July 9th in 2014 (Accumulated rainfall depth: 93.5 mm, duration of rainfall: 6 hours, maximum rainfall intensity: 70.0 mm/h). Typical water and sediment runoff through the slit were observed for the first time on July 9th in 2014 (Fig. 1).

A peak of bed load and wash load runoff occurred at the same time or before a peak of flow depth at 1:30 and 18:40 on July 9th in 2014 (Fig. 1). After that time, in spite of the depth flow is decreasing stage of the flood, bed load runoff restart at the time(C) in Figure 1.

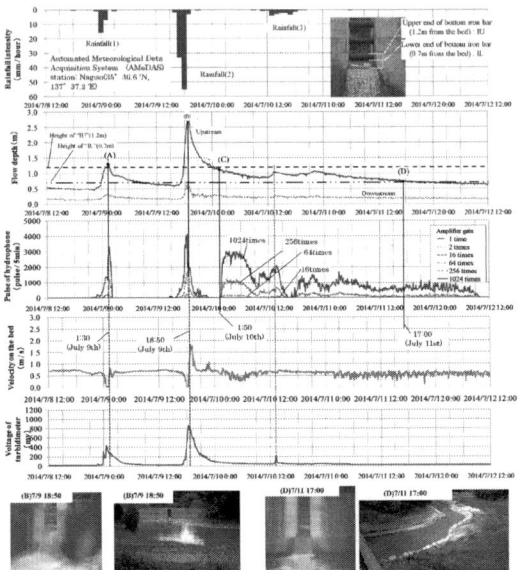

Figure 1. Temporal changes of water flow and sediment movements observed on July 9th in 2014

Sediment runoff was controlled by a bottom iron bar in the stage between the time (C) to time (D) due to the free surface in the slit is located at between top and bottom of height the iron bar. After the time (D), when the free surface is located at under the bottom of iron bar, bedload runoff is decrease smoothing.

The water runoff is controlled by narrow slit and horizontal bars results in delaying and smoothing of sediment runoff, especially in decreasing stage of the flood. Although there are only two floods events for discussions, we showed field observation about sediment control function by the slit type-Sabo dam. Monitoring data in another magnitude of floods needs continuously to clarify water and sediment control function by a slit-type with horizontal bars.

Turbulent flow and its characteristics over submerged obstacle marks

B.S. Mazumder
UGC Emeritus Fellow, Department of Aerospace Engineering and Applied Mechanics,
Indian Institute of Engineering Science and Technology (IIEST), Shibpur, Howrah, India
Department of Civil Engineering, Indian Institute of Technology (IIT), Mumbai, India

Haradhan Maity
Fluvial Mechanics Laboratory, Indian Statistical Institute (ISI), Kolkata, India

ABSTRACT

Turbulent flow has always been a challenge to the scientists, which is common in nature and has an important role in several geophysical processes related to a variety of phenomena such as river morphology, sediment transport, landscape modeling, atmospheric dynamics and ocean currents. As the turbulent flow is irregular, seemingly random (chaotic) and complex, till today no analytical solutions exist for turbulent flows. So their descriptions demand statistical approaches. We believe that even after 150 years, turbulence studies are still in their infancy. We are still discovering how turbulence behaves, in many respects. We do have a crude, practical, working understanding of many turbulent phenomena but certainly not a comprehensive theory, and nothing that will provide predictions of an accuracy demanded by designers.

One of the important problems in river morphology is scouring, and consequently sediment deposition on the river bed. Scour mark around any object placed on the sediment bed usually develops due to the interaction of the local flow with the sediment bed and the object. The study of scour marks around the objects on the sandy bed under the action of turbulent flows in natural environments has the potential to be useful to the researchers who undertake the dynamics of pipelines and short cylinders placed on river-, sea-beds and shallow water regions of coastal areas; and who examine the formations of crescentic scour marks in a recent stream for palaeo-current direction. Therefore, the use of submerged cylinders justifies the design of experiments for understanding of physics of fluids around the objects. This study is significantly important not only to estimate dynamics of coherent flow across scour holes to interpret similar scour holes available in the ancient sedimentary structures, but also in the context of basic problem of understanding the turbulence across the frontal scour marks of the obstacles, which induce strong turbulent eddies in their neighborhood.

The main purpose of the proposed work is to study experimentally the scour geometry generated around the static short circular horizontal cylinder placed on the sandy bed transverse to the flow; and to develop theoretical model for validation with the experimental data. More precisely, the emphasis is given to the following points:

1. Experimental set-up and data acquisition,
2. Experimental observations,
3. Statistical analysis of velocity data,
4. Variation of turbulence kinetic energy fluxes and conditional shear stress within and across the equilibrium scour,
5. Theoretical modeling and validation with experimental data.

Our knowledge on the evolution of scour width with time associated with statistical distributions is deficient. This investigation is aimed at studying the evolution of scour mark generated around the static short circular cylinders placed over the sediment bed transverse to the flow and to estimate the joint probability density function of fluctuating velocity components using Gram-Charlier type over the equilibrium scour geometry.

The quadrant threshold technique is used for direct estimation of observed data of conditional statistics of Reynolds shear stress for three crescentic scour depths and level surface; and subsequently, a theoretical model of probability density function of momentum flux variable is employed to verify the threshold technique. The theoretical model shows good agreement with the experimental estimates.

Highly seasonal suspended sediment and bed load transport dynamic in tropical mountain catchments

S.B. Morera
Instituto Geofísico del Perú & Universidad Nacional Mayor de San Marcos, Lima, Perú

A. Crave
Geosciences Rennes, Centre National of Research Scientific, Rennes, France

J.L. Guyot
Institut de Recherche pour le Développement, Lima, Perú

ABSTRACT

Hydrology and sedimentology development have been very limited in Peru in comparison to other sciences due to a lack of reliable suspended sediment data. A new national suspended sediment yield (SSY) dataset (1948–2012) has been collected and processed. Nevertheless, to understand erosion rates across the continent it is important to quantify the total sediment load (TSL) leaving the basin; also, a good knowledge of the transport processes for hydraulic design is required. Unfortunately, in Peru there is no current measurement of bed load, which could represent from 0% to 100% of the TSL. Field measurements of bed load transport are notoriously difficult and have large uncertainties, because it is both spatially and temporally highly variable in mountain catchments. This study aimed to quantify and characterize the TSL from the west-central Andes Mountains.

The Puyango-Tumbes (PT; 4708 km^2) and Zarumilla (ZA; 762 km^2) are two binational (Peru-Ecuador) adjacent catchments. These mountain catchments were monitored at El Tigre (45 m a.s.l.) and La Coja (22 m a.s.l.) stations respectively. Water discharge monitoring (1963–2015), suspended sediment samples (2004–2015) and bed load samples (2013–2014) were carried out in both stations. One monitoring strategy for total sediment load was established by combining: i) the most used bedload sampler called Helley-Smith; ii) the ADCP (Current Profiler Acoustic Doppler) device which was used to characterize changes in both local flow hydraulics and bedload transported across the vertical section; and iii) suspended sediment samples was taken at the surface of the river using a handle bottle. TSL was estimated from Figure 1 Results show marked seasonal variability at the PT and ZA; 98% and 99% of the whole total sediment load is transported during austral summer (Jan-May), respectively. The PT and the ZA transport are 340 t.km^{-2}.year^{-1} and 136 t.km^{-2}.year^{-1}, respectively; 98% and 75% are transported as suspended sediment, respectively, and the rest of the TSL is transported as bed load. High degree of spatial variability in TSL is observed in adjacent mountain catchments; also, the bed load fraction is highly variable

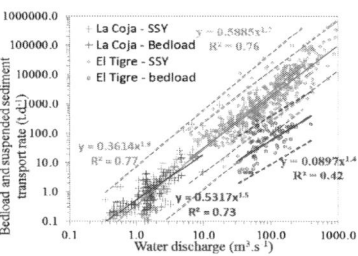

Figure 1. Bi-log. plot of daily suspended sediment (brown) and bed load (orange) rating curve, over 10 years (1182 data points).

in relation to TSL. Differences are attributed to special hydraulic characteristics (granulometry and the energy delivered for water discharge volume).

Estimating the bed load transport has been a central concern in mountain catchments during floods events, since most of them appear in a very short period of time. Highly accurate quantification of TSL supply requires continuous bed load monitoring. However, some limitations appear under these conditions, where strong turbulence surrounded by trees and pales frequently occurs. There is a clear need to complete the TSL rating curve for high water discharge. This abstract should be considered as being preliminary. Current monitoring (2014–2016) with advanced devices will improve understanding of total sediment load processes and mechanics associated with flood events.

REFERENCES

Morera, S.B., Crave, A., Guyot, J. & Condom, T. in process. Critical erosion rates and sediment yield in mountain catchments during extreme El Niño events: from the west central Andes to the Pacific Ocean (Peru). Earth and Planetary Science Letters.

Turowski, J.M., Rickenmann, D. & Dadson, S.J. 2010. The partitioning of the total sediment load of a river into suspended load and bedload: a review of empirical data. Sedimentology, 57, 1126–1146.

Prediction of bed load transport rate using Shiono and Knight model

N. Movahedi, A.R. Zahiri & A.A. Dehghani
Gorgan University of Agricultural Sciences and Natural Resources, Gorgan, Golestan, Iran

ABSTRACT

Natural rivers which have compound cross section are more complicated due to the interaction between main channel and flood plains. It is important for engineers to predict the lateral velocity distribution and boundary shear stress in order to accurate estimation of sediment transport rate. In this paper, a depth averaged form of streamwise Reynolds-averaged Navier-Stokes equation (RANSE) known as Shiono and Knight Model (SKM) (Eq. 1) is used to estimate lateral distribution of velocity.

$$\rho g H S_0 - \rho \frac{f}{8} U_d^2 \sqrt{1+\frac{1}{s^2}} + \frac{\partial}{\partial y}\left\{\rho \lambda H^2 \left(\frac{f}{8}\right)^{1/2} U_d \frac{\partial U_d}{\partial y}\right\} \\ = \frac{\partial H(\rho UV)_d}{\partial y} \quad (1)$$

Where ρ = fluid density; g = gravitational acceleration; H = depth; S_0 = river bed slope; f = Darcy-Weisbach friction factor; U_d = depth-averaged streamwise velocity; s = channel side slope (1:s, vertical: horizontal); λ = dimensionless eddy viscosity; and y = lateral coordinate the longitudinal river slope.

Many researchers tried to presented numerical and analytical solution for solving RANSE. Shiono and Knight (1991) by considering secondary flow term (Γ), proposed an analytical solution for estimation of lateral velocity distribution. The goal of this study is to solve this equation by finite difference method in gravel bed rivers and then computed distribution of the bed shear stress for prediction of bed load transport rate with various proposed formulas; such as Meyer-Peter and Muller, vanRijn, as well as Engelund and Fredsoe. An OTT velocity meter and a Helley-Smith sampler used to measure flow velocity and bed load data, respectively. The observed data were used to calibrate the main parameter of this model i.e. friction factor (f). As illustrated in Figure 1, the numerical solution can predict lateral velocity distribution accurately. Also the result showed Meyer-Peter and Muller formulae agrees well with the measured data.

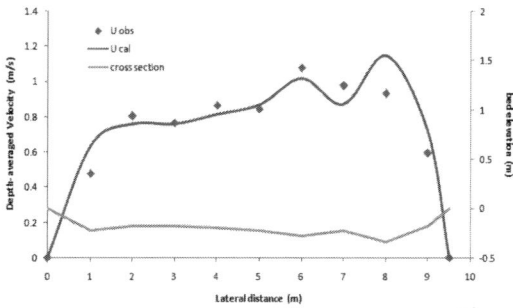

Figure 1. Lateral distribution of velocity.

REFERENCES

Meyer-Peter, E., Muller, R., 1948. Formulas for bed-load transport. In: Proceedings of 2nd IAHR Congress, IAHR, 39e64, Madrid, Spain.

Shiono, K. and Knight, D.W., 1988. Two-dimensional analytical solution for a compound channel. Proc. 3rd Int. Symp. on Refined Flow Modelling and Turbulence Measurements, Tokyo, 503–510.

Shiono, K. and Knight, D.W., 1991. Turbulent open-channel flows with variable depth across the channel. J. Fluid Mech., 222(1), 617–646.

van Rijn, L.C., 1984a. Sediment transport, part I: bed load transport. J. Hydraul. Eng., ASCE, 110 (10), 1431e1456.

River embankment failure and resultant flood and sediment inflow discharges due to overtopping river flow

H. Nakagawa, H. Mizutani, Y. Wang & K. Kawaike
Disaster Prevention Research Institute, Kyoto University, Japan

O. Kitaguchi
Central Nippon Expressway Company Limited, Japan

H. Zhang
Kochi University, Japan

ABSTRACT

The prediction of breach hydrograph is very important for the flood risk assessment. The peak outflow discharge from the river embankment breach can be calculated by using empirical relationships and physical based numerical models. Numerous empirical relationships for estimating the peak outflow caused by a gradual dam failure are presented by several researchers. However, the empirical relationships are derived from the limited number of data and mainly relates to depth and volume of water behind the embankment dam. Therefore, the peak outflow or breach hydrograph and sediment graph from embankment dam breaching can be calculated more approximately by using numerical models. Many researchers have developed physical based numerical model to predict the outflow hydrograph from embankment breach. The breached outflow discharge highly depends on rate of erosion of embankment. The behavior of unsaturated soil is different from saturated soil. However, most of the existing breach models do not consider infiltration of water inside the embankment. Therefore, it is important to study erosion rate of an unsaturated embankment for the prediction of outflow water and sediment discharges.

This study is focused on erosion of unsaturated river embankment due to flow overtopping and mainly on resultant flood and sediment inflow discharges into

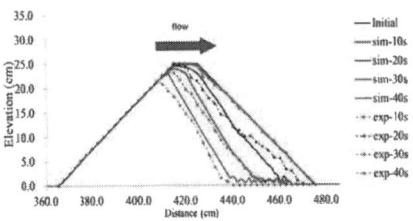

Figure 1. Comparison of simulated and experimental results of embankment shapes.

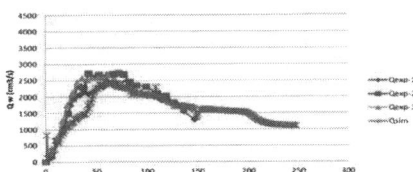

Figure 2. Comparison of simulated and experimental results of inflow hydrographs.

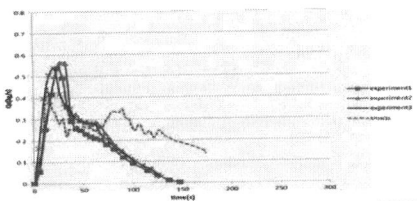

Figure 3. Comparison of simulated and experimental results of inflow sediment graphs.

land-side area. The experimental results of embankments made by non-cohesive fine sediments show that rate of erosion in small sized sediment is smaller comparing with bigger sized sediment. Shear strength due to suction plays vital role in erosion of unsaturated fine sediments. As the model which is considering the infiltration process inside the embankment body and resisting shear stress due to suction on surface erosion has already been developed by authors' research group, the main objectives of this study are to verify the developed model from the viewpoint of flood and sediment discharges through the experiments. The simulated results of embankment surface erosion, infiltration process, and resultant flood and sediment discharges are compared with some experimental results in this paper. Examples of simulated and experimental temporal eroded embankment shapes, outflow flood hydrographs and sediment graphs in case of sediment 7 are shown in Figure 1, 2, and 3, respectively.

Comparison of acoustic backscatter to turbidity for suspended sediment estimation in the Sacramento-San Joaquin Delta in California

M. Ozturk
Yildiz Technical University, Istanbul, Turkey

P.A. Work
U.S. Geological Survey, California Water Science Center, Sacramento, California, USA

ABSTRACT

Acoustic backscatter from devices designed to measure water velocities can similarly be used as a proxy for SSC, in some cases. Measurement quality depends on acoustic frequency, sediment size, size distribution, and time dependence.

In this study, acoustic backscatter profiles obtained from a 600 kHz, side-looking acoustic Doppler current profiler (ADCP) were used to estimate suspended sediment transport in Georgiana Slough in California's Sacramento-San Joaquin river delta (Figure 1). Here Topping et al. (2007) method were applied to a tidally forced river as a first time, which was improved by the researchers by taking into consideration several sites along the Colorado River in the Grand Canyon in the USA. By design, sound sent into the water column by the instrument reverberates back to the transducer. The phase shift in the return signal is range-gated to determine the velocity profile. The backscatter signal can also be range-gated to construct a backscatter profile.

The sound power is attenuated by both the medium through which it travels and the particles suspended in this medium. The sound that returns to the transducer has thus experienced a two-way travel loss, with some of this loss being dependent primarily on travel distance and the other component a function of suspended sediment characteristics and concentration.

The backscatter profiles were corrected for both fluid and sediment attenuation. The sediment attenuation coefficient was determined empirically from the slope of the fluid-corrected acoustic backscatter profiles. This process requires knowledge of only the frequency of the ADCP, water temperature and the salinity of the water.

Observed SSC, based on output from an optical turbidity sensor, varied between 10 to 50 mg/l for the most of the year round period (data are provisional and subject to review) but two flooding events of ∼40 days in early 2013 it reached 400 mg/l. After calibration, a linear fit between optical and acoustic estimates of SSC yielded a correlation coefficient R2 = 0.95 during these higher flow events. Improved predictions at

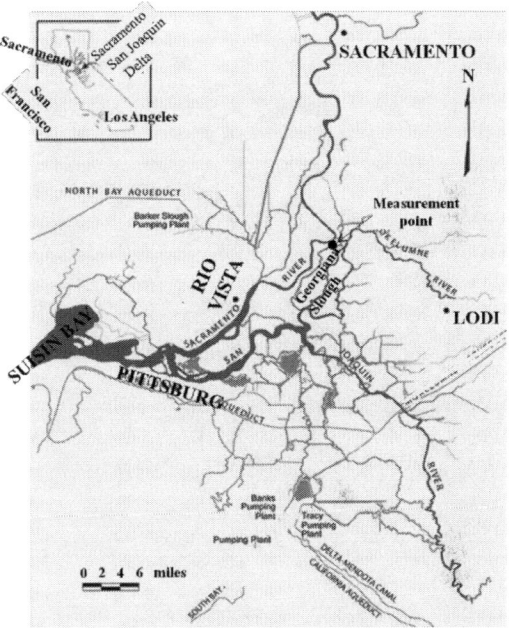

Figure 1. The location of Georgiana Slough in Sacramento and San Joachim Delta in Northern California (modified from California Department of Water Resources).

lower concentrations would likely require an acoustic sensor tuned to match the predominant size of the suspended sediment.

REFERENCE

Topping, D.J., Wright, S.A., Melis, T.S., and Rubin, D.M. 2007. High-resolution measurements of suspended-sediment concentration and grain size in the Colorado River in Grand Canyon using a multi-frequency acoustic system. In Proceedings of the 10th International Symposium on River Sedimentation, Moscow, Russia. August 1–4.

Proposing BEHI-NBS method for the estimation of river bank erosion on a river in Nepal

S. Pakuwal
Department of Environmental Science, Golden Gate International College, Tribhuvan University, Kathmandu, Nepal

S. Panthee
Department of Geology, Tribhuvan University, Kathmandu, Nepal

ABSTRACT

In this paper GIS based technique has been used to quantify the river bank erosion rate and test the applicability of BEHI-NBS method BANCS (Bank Assessment for Non-point Source Consequences of Sediment) model. BANCS model is based on the morphological characteristic of river, erodibility potential and erosivity of bank material used to predict the river bank erosion rate.

Rivers in high lands of Nepal are subjected to erosion forming mostly gully erosion. In Terai river flows through plain land forming meander and shifting, changing its pattern. The excessive scour of fine bed materials, meander development and shifting of channel or channel migration intensify bank erosion. Bank erosion has caused serious problem in Rivers in Nepal, engulfing the villages and towns, causing bridge and roads failure, loss of property and lives.

Process based model along with GIS as a tool was used to access the meandering migration, river bank erosion rate. Arc map 10 was used to analyze and find out the rate of shifting overlaying Google images of 2005, 2012 and 2014 and Landsat image 1984. Thirty survey location were selected on either bank (right and left bank) of river alternately within 5.49 kilometers long section of river on the basis of Google image and GPS in preliminary survey. To quantify the meander migration of channel, BANCS model (Bank Erosion Hazard Index and Near-bank Stress) proposed by (Rosgen, 1996) were reviewed and used and based on the model hazard map was prepared. The validity of the BANCS mod- el were tested using the erosion rate calculated from Arc GIS10 and field verification. For bank erosion potential, all the parameters used to quantify the bank erosion potential showed that, the study section has high bank erosion potential. Bank erosion hazard index and near-bank stress were found to be highly significant predictor of bank erosion rate.

REFERENCE

Rosgen, D. L. (1996). Applied River Morphology. Pagosa Springs. Wildland Hydrology Book, Fort collins, Co.

Turbulent hydrodynamics through cross-sections at upstream, interior and downstream of sparse vegetation patch in open channel flow

D. Pal, S. Maji & P.R. Hanmaiahgari
Department of Civil Engineering, Indian Institute of Technology Kharagpur, West Bengal, India

M.D. Bui & P. Rutschmann
Institute of Hydraulic and Water Resources Engineering, Technische Universitat München, München, Germany

ABSTRACT

Jute plantations are common in India and Bangladesh which are sparse and emergent. No comprehensive experimental investigation has been carried out so far to compare turbulent hydrodynamics in lateral direction at interior and exterior of a sparse and emergent vegetation patch. The proposed study investigates important turbulent features at different cross-sections located upstream, interior and downstream of sparsely vegetated open channel flow. The emergent vegetation patch is made by seventy uniform rigid acrylic cylindrical rods together with regular spacing between them. A Nortek ADV Vectrinoplus has been used to measure three-dimensional instantaneous velocities. It is found that time averaged streamwise velocities are decreasing interior of the vegetation along the streamiwse direction. The time averaged lateral velocities interior of the vegetation patch are directed towards the nearest sidewall. Vertical velocities are negative throughout the cross-section at upstream of the vegetation patch, however absolute magnitudes are decreasing in interior and downstream of the vegetation patch although vertical velocities are negative. Turbulent kinetic energy values are decreasing inside the vegetation patch and interestingly peak values are found to be occurring near the free surface in the interior of vegetation.

Experimental investigation of a cantilever failure for cohesive riverbanks

S. Patsinghasanee, I. Kimura & Y. Shimizu
Hydraulic Research Laboratory, Graduate School of Engineering, Hokkaido University, Hokkaido, Japan

ABSTRACT

This study elucidated the complex mechanisms of a cantilever failure with the slump block consideration by means of experimental works. Two types of cohesive materials were investigated under the similar hydraulic conditions using the acceleration sensors to clarify the cantilever failure processes. The main conclusions are as follows.

The experimental works were conducted in a straight rectangular flume. The water and sediments were recirculated using a constant-head tank of water placed at the upstream end of flume. Moreover, the water discharge remained constant at 6.45 l/s during the experiments. For the experimental design by considering the sidewall correction and geometrical scaling, the flume dimensions were 0.8 m in width, 8.0 m in length and 0.4 m in hight and the channel slope was set to 0.001. Additionally, the cohesive bank was 0.5 m wide, 2.0 m long, and 0.2 m high. The experiment conditions required to stop test were (1) when the failures proceeded throughout all of the banks or (2) when an equilibrium stage was reached (without a failure for a 3 h period). Furthermore, the sediment mixture composed of sand and silt with mean diameter of around 0.23 mm and 28.4 μm, respectively, was used. The sediment was initially wetted with water to achieve a water content between 32.2% and 39.6% for the silt-clay content of 20% and 30%.

The advantage of the experiment was the possibility to observe and record cantilever failure mechanism from the top and side views, and inside the bank during fluvial erosion, tension crack, cantilever failure and slump block. All mechanisms were recorded using five high-resolution video cameras and thirty six acceleration sensors. The acceleration sensors were installed inside the banks. Therefore, these sensors recorded the chronological failure mechanisms and identified the types of failure at points where the values varied greatly.

For the failure mechanism, the results showed that the fluvial erosion at the lower part generated an

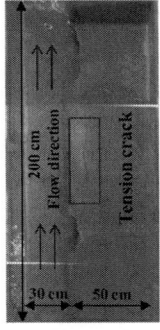 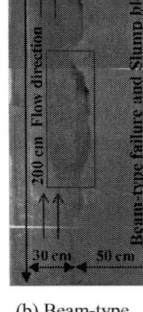 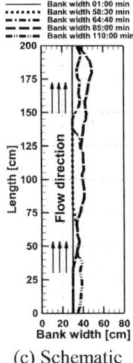

(a) Tension crack. (b) Beam-type failure and Slump block. (c) Schematic diagram.

Figure 1. The spatial bank width of the bank with 20% of silt-clay content ((a) and (b) show the tension crack and beam-type failure from 33.3 cm to 138.3 cm from the upstream end at 3,880 s of elapse time.)

overhanging block in the upper part of the banks. Tension cracks then developed at the upper surface of the banks (Fig. 1(a)). After that, the cantilever failure occurred along the tension crack line (Fig. 1(b)). The dominant failure mechanism was observed to be beam-type failure clarified using the acceleration signal. Moreover, the schematic diagram of spatial bank width is illustrated in Figure 1(c).

After the cantilever failure, slump blocks were observed on the riverbed in front of the bank, where they formed a sediment buffer that reinforced banks and reduced fluvial erosion. The slump block phenomena for the formation and deformation showed a significant effect on the silt-clay content and affected the bank geometry. Therefore, the reduction of the silt-clay content leads to smaller slump block dimensions as well as faster decomposition. Additionally, the random relationship between the slump block volumes and their decomposition times (the decay process of the slump block) has been reported in this study.

Bedload monitoring in a steep alpine stream: Results from the 2014 measurement campaign

R. Rainato & L. Picco
Department of Land, Environment, Agriculture and Forestry, University of Padova, Padova, Italy

L. Mao
Department of Ecosystems and Environment, Pontificia Universidad Catòlica de Chile, Santiago, Chile

ABSTRACT

In the mountain ranges, the bedload transport affects the downstream sediment delivery, the channel stability and thus the hazard assessment. Due to its impulsive nature, the investigation and assessment of the bedload is a challenging task and, consequently, field data are relatively scarce. The bedload predictive equations are frequently calibrated on laboratory experiments or specific field data, leading to generally low performances. In this sense, the employment of several direct and indirect monitoring methods can allow to obtain precious field data on the sediment dynamics. Collecting the material transported over a certain interval, bedload traps permit to obtain information on the rate and the grain size of bedload. On the other hand, single grain tracers enable to track the sediment displacements, transport distance and threshold conditions, allowing to integrate the information achievable by the traps (Ferguson & Wathen, 1998). Here we present the results obtained during the 2014 monitoring season performed in the Rio Cordon, a small instrumented basin located in eastern Italian Alps (Dolomites). The catchment drains a surface of 5 km^2, ranging among 1763 to 2763 m a.s.l, and exhibits a nivo-pluvial runoff regime. Since 1986, a permanent monitoring station records in continuous water discharge, bedload and suspended load of the Rio Cordon stream. Additionally, 320 m upstream to the monitoring station, along a cascade/step-pool segment of the main course, 250 Passive Integrated Trasponders (PIT) were seeded. In 2014, two moderate-magnitude flood events (both RI = 1.7 years) occurred, on June, 9 and November, 11, respectively. The flood occurred in June was a mixed snowmelt-rainfall event with peak of water discharge (Q$_{peak}$) and effective runoff (Re) equal to 2.06 m^3 s^{-1} and 16.6 10^3 m^3, respectively. The November flood was characterized by an alike Q$_{peak}$ (2.06 m^3 s^{-1}) while the volume of hydrograph that contributed to the bedload has been roughly two fold higher (Re = 33.3 10^3 m^3). Among the two events, the bedload significantly differs with 113 t of coarse material transported by the June event and barely 4.6 t mobilized by the November flood. While in June the

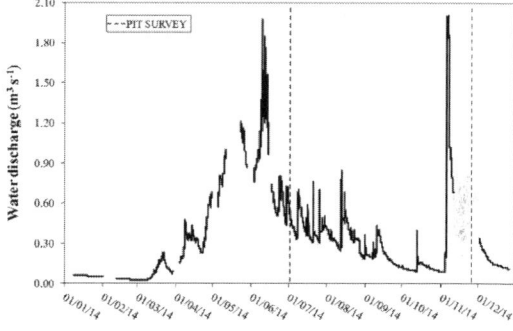

Figure 1. Discharge series in Rio Cordon during 2014.

source area was clearly identified as a debris flows channel located in the middle part of the basin, in November no field evidence was detected, suggesting that the coarse material may have been provided by minor bank erosion. The PITs monitoring performed after the floods recorded average travel distance (Li) equal to 115.91 and 95.18 m for the floods occurred in June and November, respectively. Overall, only few field dataset are available in literature on which both the bedload volumes and the transport distance of tracers were investigated in steep streams (e.g. Lenzi, 2004). Notwithstanding the short time scale, this work it is an attempt in this direction, focusing particularly in the comparison of two flood events characterized by apparently similar magnitude but that experienced clearly different sediment dynamics.

REFERENCES

Ferguson, R.I. & Wathen, S.J. 1998. Tracer-pebble movement along a concave river profile: Virtual velocity in relation to grain size and shear stress, *Water Resources Research* 34 (8), 2031–2038.

Lenzi, M.A. 2004. Displacement and transport of marked pebbles, cobbles and boulders during floods in a steep mountain stream. *Hydrological Processes* 18(10), 1899–1914.

Monitoring topography of laboratory fluvial dike models subjected to breaching based on a laser profilometry technique

I. Rifai[1,2,3], S. Erpicum[1], P. Archambeau[1], D. Violeau[2,3], M. Pirotton[1], K. El kadi Abderrezzak[2,3] & B. Dewals[1]
[1] *ArGEnCo Department, Research Group Hydraulics in Environmental and Civil Engineering (HECE), University of Liège, Liège, Belgium*
[2] *EDF R&D, National Laboratory for Hydraulics and Environment, Chatou, France*
[3] *Saint Venant Laboratory for Hydraulics, Chatou, France*

ABSTRACT

Fluvial dikes are commonly constructed for flow channelization, riverbed meandering prevention and as flood defense structures. The aging of dikes increases their vulnerability to extreme hydrological events that may cause their failure. This is even more critical as the presence of dikes leads usually to increased development in the protected floodplains.

Many laboratory experimental studies of dike failure due to overtopping were conducted, but most of them focused on normal configurations (i.e. dam break configuration), without accounting for the influence of a flow parallel to the dike. Results and recommendations cannot be simply transposed to fluvial dike cases, because the flow pattern in the near-field of the breach is strongly affected by the parallel flow in the main channel (Michelazzo et al. 2015), modifying significantly the breach erosional processes.

Several approaches exist for monitoring the breach formation, based on non-intrusive and distributed measurements. They include: (1) the photogrammetry method, which uses multiple viewpoint recording to reconstruct the 3D dike geometry (Frank & Hager 2015), (2) the side fringe projection (Pickert et al. 2011) and (3) the laser sheet sweeping *technique* (Spinewine et al. 2004). Here, we focus on the third method, which is less sensitive to artefacts that result from reflection on the water surface and turbidity.

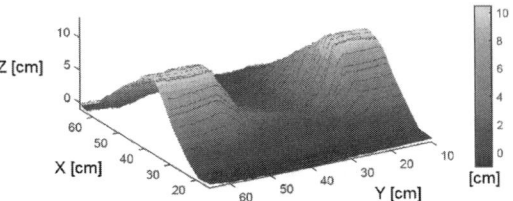

Figure 2. Interpolated DEM from multiple incident laser lines on the dike surface (red dots).

This method consists in determining the intersection point (x, y, z) of a view axis with the laser sheet plane (Figure 1). For each point of the deformed incident laser line, the view axis (Δ) is determined by the Direct Linear Transformation (DLT) method from the (u, v) coordinates of this point on the camera sensor.

Neither the location of the camera nor that of the laser are required for the calculation, which highlights the flexibility of such approach. In addition, this simple method allows an easy implementation and the integration of a refraction correction module.

Testing the laser profilometry on breached dikes yields satisfactory results (Figure 2). Though, further improvements are ongoing to correct bias due to lense distortions and water refraction in submerged areas.

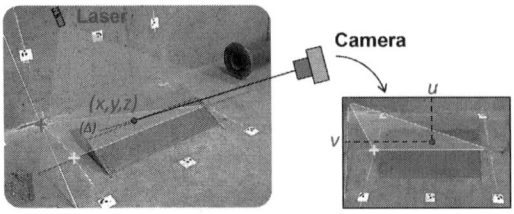

Figure 1. Schematization of the laser profilometry technique.

REFERENCES

Frank, P. & Hager, W.H. 2015. Spatial dike breach: sediment surface topography using photogrammetry. *IAHR World Congress 2015*. The Hague.

Michelazzo, G., Oumeraci, H. & Paris, E. 2015. Laboratory study on 3D flow structures induced by zero-height side Weir and implications for 1D modeling. *Journal of Hydraulic Engineering*, 141(6).

Pickert, G., Weitbrecht, V., & Bieberstein, A. 2011. Beaching of overtopped river embankments controlled by apparent cohesion. *Journal of Hydraulic Research*, 49(2).

Spinewine, B., Delobbe, A., Elslander, L. & Zech, Y. 2004. Experimental investigation of the breach growth process in sand dikes. *River Flow 2004*. Naples.

A data-driven fuzzy approach to simulate the critical shear stress of cohesive sediments

A. Schäfer Rodrigues Silva, M. Noack, D. Schlabing & S. Wieprecht
Institute for Modelling Hydraulic and Environmental Systems, University of Stuttgart, Stuttgart, Germany

ABSTRACT

The prediction of the critical shear stress for cohesive sediments is of great importance for sediment management but its measurement is labour-intensive and there is a lack of comparability due to different measurement systems. However, other parameters such as grain size distribution, bulk density and depth are easier to measure and more comparable.

For cohesive and mixed sediments a large number of physical, chemical and biological parameters influence the critical shear stress. Their interactions as well as their influence on the sediment's shear strength are not yet fully understood, hence, a certain degree of uncertainty regarding their incipient motion is involved. While analytical approaches are not able to consider this, fuzzy logic based models can handle uncertainty and imprecision. The aim of this work is to identify suitable parameters to estimate the critical shear stress using a multivariate fuzzy logic approach.

Because artificial sediments reflect the natural behavior insufficiently (e.g. Mitchener et al. 1996), this study is based on undisturbed sediment samples originating from two locations at the rivers Saale and Rhine. Due to the high complexity of the governing physical processes, a data-driven fuzzy model is chosen. A Takagi-Sugeno-type fuzzy inference system uses either linear or constant output membership functions and is thus well suited for optimisation and adaptive techniques (Tolossa 2012). The model structure is optimised using a neuro-adaptive learning method (ANFIS) provided by the MATLAB Fuzzy Logic Toolbox™.

At this stage of the study, the clay content and the sediment depth are the parameters, which give the best model performance. Figure 1 illustrates the model output compared to the measured data. The correlation coefficient between measured and com-puted data is $r = 0.84$.

The model gives reasonable results and shows higher correlation than analytical models found in scientific literature (e.g. $r = 0.54$, Ahmad et al. 2011), which are often derived from artificial sediment

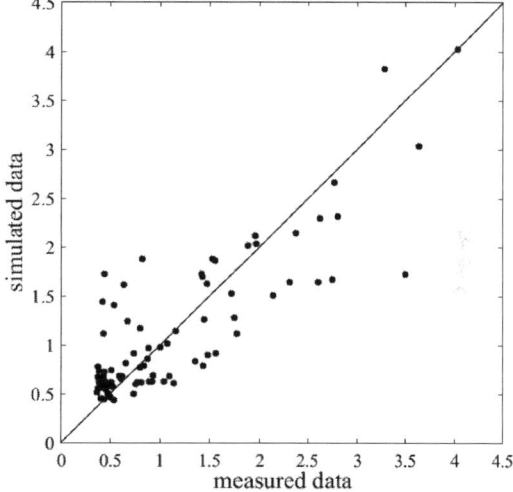

Figure 1. Measured versus simulated critical shear stress.

samples and thus focus on sub-processes only. The data-driven fuzzy model handles the lack of a complete understanding of the processes by describing the general behaviour with limited physical assumptions. Future work will include additional analyses of other sedimentary but also of biological and chemical parameters in order to improve the prediction capability of the fuzzy-model.

REFERENCES

Ahmad, M. F., Dong, P., Mamat, M., Wan Nik, W.B. & Mohd, M.H. 2011. The Critical Shear Stresses for Sand and Mud Mixture. Applied Mathematical Sciences, 5: 53–71.

Mitchener, H., Torfs, H. & Whitehouse, R. 1996. Erosion of mud/sand mixtures. Coastal Engineering, 30: 1–25.

Tolossa, H. G. 2012. Sediment transport computation using a data-driven adaptive neuro-fuzzy modelling approach, Institute for Modelling Hydraulic and Environmental System, University of Stuttgart, No. 211.

Discussion of the impact of pressure fluctuations on local scouring

W. Schanderl & M. Manhart
Chair of Hydromechanics, TU München, Germany

O. Link
Universidad de Concepción, Concepción, Chile

ABSTRACT

The wall shear stress is assumed to be the driving mechanism of both global and local erosion in a river bed. Considering the scouring process around a bridge pier placed in a sandy bed, this is plausible. The scour hole starts in the lateral front of the pier, where the maximum time-averaged wall shear stress is observed. Even though the wall shear stresses are basically the same, investigating the erosion process of cohesive bed material in vicinity of a bridge pier bears a major difference: Depending on the compaction of the cohesive material, the scouring process starts in the wake of the cylinder (Link et al. 2013).

Since there has to be an additional mechanism driving the erosion process of cohesive material, we aim at the discussion of the potential impact of pressure fluctuations on the scouring process by discussing both pressure coefficient C_p and friction coefficient C_f. Furthermore we estimate the force acting on the sediment layer due to local low pressure events.

To do so, we conduct highly resolved large-eddy simulation of the flow around a wall-mounted cylinder at three different Reynolds numbers of $20\,000 \leq Re \leq 78\,000$ based on the bulk velocity U_b and the diameter of the cylinder D (Schanderl & Manhart 2016). In the region of interest around the cylinder, the grid is refined by locally embedded grids (Manhart 2004).

Figure 1 shows an instantaneous sample of the friction coefficient around the cylinder. Visible is the footprint of the horseshoe vortex in front of the cylinder and a distinct von Karman vortex in the wake. Here, instantaneous wall shear stresses lead to a maximum absolute value of $C_f = 0.15$.

Horseshoe and von Karman vortex can also be observed in Figure 2, which shows the pressure coefficient at the bottom plate. We observe values of $-0.97 \leq C_p \leq 0.81$.

Regions of low pressure are located in close vicinity to regions of high pressure, leading to large pressure gradients. Flow inside the sediment layer due to these pressure gradients exerts an upwards directed force on the sediment, which might cause erosion.

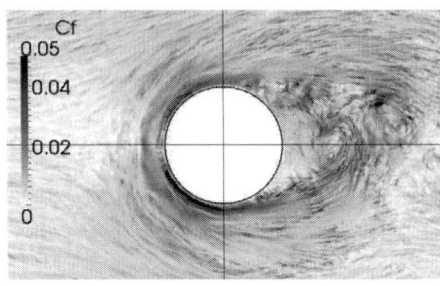

Figure 1. Instantaneous sample of C_f at $Re = 39\,000$.

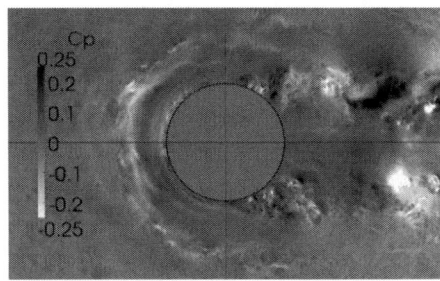

Figure 2. Instantaneous sample of C_p at $Re = 39\,000$. Note that this time step is not corresponding to the one of Figure 1.

We estimate the force acting on the sediment due to low pressure events and its time scale to establish. Furthermore we state this erosive potential to increase with the Reynolds number.

REFERENCES

Link, O. et al. 2013. Effects of Bed Compaction on Scour at Piers in Sand-Clay Mixtures. J. Hydraul. Eng., 139, 1013–1019.
Manhart, M. 2004. A zonal grid algorithm for DNS of turbulent boundary layers. Comput. Fluids, 33, 435–461.
Schanderl, W., Manhart, M. 2016. Reliability of wall shear stress estimations in front of a wall-mounted cylinder. Comput. Fluids, 33, 16–29.

Processes and effects of reversing currents on the erosion stability of wide-graded grain material

A. Schendel
Franzius-Institute for Hydraulic, Estuarine and Coastal Engineering, Leibniz University Hannover, Germany

N. Goseberg
Franzius-Institute for Hydraulic, Estuarine and Coastal Engineering, Leibniz University Hannover, Germany
Department of Civil Engineering, University of Ottawa, Canada

T. Schlurmann
Franzius-Institute for Hydraulic, Estuarine and Coastal Engineering, Leibniz University Hannover, Germany

ABSTRACT

Scour protection around hydraulic structures in fluvial and coastal waters is an essential component assuring a meaningful and durable design. Wide-graded grain materials, mainly composed of artificial quarry-stone material, are considered a flexible and efficient scour protection system. However, fundamental research studies, which contribute to verifying the stability of the material under fluvial, estuarine and coastal conditions as well as the applicability of wide-graded material as scour protection, are scarce. Therefore, large scale hydraulic model tests were carried out in order to investigate erosive potentials and the bed stability of wide-graded quarry-stone material. The present paper focuses on the underlying processes and the overall erosion stability of wide-graded material in estuarine conditions by simulating the influence of a tidal flow by reversing current.

The hydraulic model tests were carried out in the closed-circuit flume of the Franzius-Institute for Hydraulic, Estuarine and Coastal Engineering. The investigated material was wide-graded quarry-stone material with fractions ranging from 0.063–200 mm (Fig. 1). It was installed in a deep pit with a width of 1.0 m, a length of 2.7 m and in a thickness of 20 cm. The flow velocities were successively in-creased in seven load cases with mean velocities u_{mean} from 0.11 m/s to 0.77 m/s. In order to simulate tidal currents, the flow direction was reversed after each load case before exposing the bed with the next accelerated velocity. Flow velocities were determined by applying an acoustic Doppler velocimeter (ADV). For each load case and flow direction the amount as well as the grain size distribution of eroded bed load material was measured in sediment traps on both sides of the material bed.

During the experiment, a coarsening of the bed surface with increasing flow velocity was observed. The coarsening of the surface was seen in the grain size distributions of the eroded bed material (Fig. 1) and

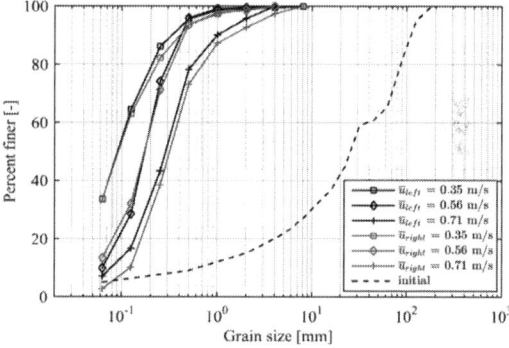

Figure 1. Grain size distributions of eroded bed load collected by sediment traps for three of the seven load cases.

indicated the development of a static armor layer at the end of each load case and a highly selective mobility of the available grain fractions. As a result of the reversing current and thus the renewed transport of previously protected grain fractions (hiding effect), additional material was eroded despite being exposed to the same flow velocity. However, the reversing current alone did not lead to a significant increase in the coarsening of the eroded bed load material between the initial flow direction (right) and the reverse direction (left) as shown in Figure 1.

In comparison to the previous experiments with unidirectional current (Schendel et al. 2015) less material was eroded for the equivalent load cases.

REFERENCES

Schendel, A., Goseberg, N. & Schlurmann, T. 2015. Erosion stability of wide-graded quarry-stone material under unidirectional current. J. Waterway, Port, Coastal, Ocean Eng. (accepted for publication).

Wilcock, P.R., Kenworthy, S.T. & Crowe, J.C. 2001. Experimental study of the transport of mixed sand and gravel. Water Resources Research, 37(12): 3349–3358.

Control of bridge abutment scour using triangular vanes

M. Shafai-Bejestan & N. Raee
Shahid Chamran University, Ahvaz, Iran

ABSTRACT

Bridge abutments redirect the flow and change the flow patterns; creating three dimensional flows around the abutment. Different vortices are developed around the abutment causing bed scour hole at bridge abutments. Over the past years great number of researches has been conducted to developed countermeasures for safety of the existing bridges. In general there are two methods of controlling scour in river engineering practice: (1) armor or covering the river bed material around abutments; and (2) modifying the approach flow patterns in the vicinity of abutment to reduce the strength of down flow vortex or by redirecting the approaching flow from the abutment. For existing bridges the covering methods, e.g. riprap material, have been extensively used because of its technical simplicity in construction and availability of the required materials. Constructing spurs upstream of the existing bridge abutment is also a flow altering measure which has been extensively studied (Li et al. 2005) with 100% protection. Constructing spurs however not only requires large amount of coarse gravel material; it should be covered with large size riprap material to be stabilized during flood. The triangular vanes, made by riprap material, also are new measures which have been studied recently for control of outer bank of river bends (Shafai-Bejestan and Bahrami-Yarahmadi, 2014). The vanes have a sloping crest that extends from the riverbank and forms a low angle (20–30 degrees) with the river bank. As it can be seen all the aforementioned techniques requires a large amount of riprap material. In many places, however, providing the required stone size for such purposes is too costly and therefore other materials has too be replaced. The six-leg concrete elements which are usually used in coastal engineering (Lebaron, 1999) also have been recommended for some river works such as river bank toe erosion. Each element consists of two concrete T-shaped pieces joined perpendicularly at the middle; forming six legs (see Figure 1). These elements can be installed either randomly or in a uniform pattern. In the present study experimental study is conducted to investigate the effect of series of triangular vanes constructed using sex-legs concrete elements on bed topography (Figure 2). Vanes were installed with different angle and different flow conditions. Tests also were conducted for different flow conditions and no vanes as base line tests. The results show that the vane can effectively redirect the main flow toward the river midway. The scour at bridge abutment is reduced. The new measure in some places provides a cost-effective solution compared to the riprap vanes. The new measures also can be easily and quickly constructed.

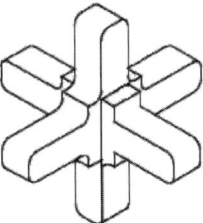

Figure 1. 3D view of a the six-legs element.

Figure 2. View of the experiment test.

REFERENCES

Lebaron J. W., 1999. Stability of A-jacks-armored rubble-mound break waters subjected to breaking and non-breaking waves with no Overtopping. M.Sc Thesis, Oregon State University, USA.

Li, H., Kuhnle R. and Barkdoll, B., 2005. Spur dikes as an abutment scour countermeasure. Impact of global climate change, ASCE proc., 1–12.

Shafai-Bejestan, M., Bahrami-Yarahmadi, M., 2014. Study of the effect of triangular vane on bed topography in a 90° mild bend, River Flow 2014, 2157–2161.

Estimation for the riverbank collapse volume with sandy-riverbank in the desert reach of the upper Yellow River

A. Shu, X. Zhou, G. Duan, F. Li & S. Wang
School of Environment, Key Laboratory of Water and Sediment Sciences of Ministry of Education, Beijing, China

ABSTRACT

Yellow River, the second longest river in China, exhibits the characteristics of wandering channel, more river branches and higher sediment concentration. Sandy-riverbank collapse in desert reach is very prominent in the fluvial processes of the upper Yellow River, especially during flood period. The study area has been selected as the lower part of the upper Yellow River from Xiaheyan (Ningxia Province) to Toudaoguai (Inner Mongolia Province), which is called briefly as NM channel and located at the desert region with a total length of 1080 km. The riverbank collapse can be described schematically as in Figure 1.

Field data obtained from sandy riverbanks shows that the vertical riverbank profiles before and after collapse are similar. Both the slope angles are equal to the internal friction angle of the riverbank material, which is in agreement with the results of experiments on the riverbank scour process conducted by Nagata et al. (2000). Based on the field observations and theoretical analysis for the NM channel, calculation of the riverbank collapse volume can be simplified into the erosion volume by water flowing. In this paper, a method calculates the riverbank toe erosion volume is introduced. Only the interaction between flow and riverbank is considered in the process, and the riverbank height is not involved, which changes constantly with sand movement.

The amount of material eroded can be calculated using the residual shear force method, in which the erosion rate ε is related to the shear stresst and shear strength τ_c by

$$\varepsilon = k(\tau - \tau_c) \qquad (1)$$

where k is the soil erodibility and is determined experimentally.

The amount of erosion is

$$E = \int_Y^H k(\tau - \tau_c)dy \qquad (2)$$

where H is the water level and Y is the minimum water depth at which riverbank erosion occurs.

At a sandy riverbank, it is a difficult and indeed dangerous task to survey the flow velocity vertically. Experiments were performed to obtain a relationship between the velocity at a given depth and the surface velocity, which can be expressed by

$$\frac{u}{v} = 0.56\left(\frac{y}{H}\right)^2 + 0.76\frac{y}{H} + 0.81 \qquad (3)$$

On the basis of residual shear stress and using an experimentally determined vertical velocity distribution, the amount of erosion is given by

$$E = \int_Y^H k\left\{\frac{\gamma \alpha^2 v^2 n^2}{H^{1/3}}\left[0.56\left(\frac{y}{H}\right)^2 + 0.76\frac{y}{H} + 0.81\right] - \tau_c\right\}dy \qquad (4)$$

After all, the calculation was carried out using Matlab software.

The method above has been preliminarily presented for riverbank collapse volume C and verified by using measured date from Wuhai observation site, located at the Yellow River passes through the Ulan Buh Desert. The results compared favorably with the measured values.

REFERENCE

Nagata, N., Hosoda, T. & Muramoto, Y. 2000. "Numerical analysis of river channel processes with bank erosion." Journal of Hydraulic Engineering, ASCE, 126(4), 243–252.

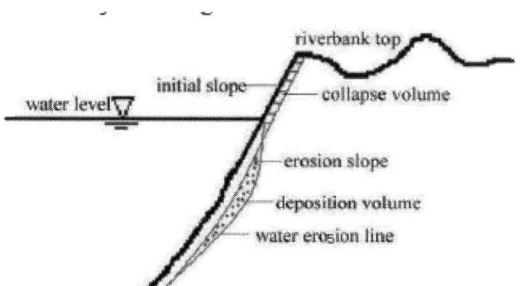

Figure 1. Sketch of the riverbank collapse process.

Incipient motion for gravel particles in cohesive mixture of clay-silt-gravel

U.K. Singh & Z. Ahmad
Department of Civil Engineering, Indian Institute of Technology Roorkee, India

A. Kumar
Department of Civil Engineering, Jaypee University of Information Technology, Waknaghat, India

ABSTRACT

Incipient motion condition for the cohesive sediment is significantly different from the cohesionless sediment due to dominancy of physio-chemical properties of the cohesive sediment. On mixing the cohesionless sediment with cohesive sediment, the resulting mixture possesses certain amount of cohesive property; therefore, it is treated as cohesive sediment mixture. In past, several experimental studies have been conducted on incipient motion for different cohesive sediment mixtures like clay-sand, clay-silt-sand, clay-gravel, clay-sand-gravel, etc. However, study has not been yet conducted for clay-silt-gravel mixture to the best knowledge of the authors. The unequal mobility of sediment particles occurs in the case of incipient motion of non-uniform sediment especially in the mixture having fine to coarse particles. In the present study, the occurrence of incipient motion is considered once the gravel starts moving in the cohesive sediment mixture of the clay-silt-gravel.

Experiments were carried out in a tilting flume having a test section of 6.0 m length, 0.75 m width and 0.18 m depth. Clay-silt-gravel mixture was placed on the bed in which the clay percentage was varying from 10% to 50% while silt and gravel were taken in equal proportions. The laid clay-silt-gravel mixture was compacted under dynamic compaction condition with the help of cylindrical roller and hand rammer. After preparing the bed, mixture samples were taken out from the bed for the determination of their bulk density, unconfined compressive strength, and moisture content.

The appearance of the erosion pattern on the bed surface has been observed for varying percentage of clay in the sediment mixture. Various factors affecting the incipient motion for gravel particles in the cohesive mixture have also been studied.

The incipient motion condition was visually observed and the flow condition at which the gravel particles start moving as bed load is treated as incipient motion condition. At the incipient motion, flow parameters were measured for the computation of critical shear stress. After the end of each run, the channel bed was visually inspected and its profile was measured. It was found that the gravel particles dominate on the bed surface for 10% clay content in the mixture. However, gravel particles dominancy reduces as the clay percentage increases. The erosion was appeared in the form of line on the bed surface for 30% clay content in the sediment mixture. As the clay percentage increases to 40% and above the erosion appeared in the form of bunches (i.e. like mass erosion). The observed critical shear stress for the mixture was compared against the cohesionless sediment having the mean size same as that of cohesive sediment mixture. It was found that critical shear stress for cohesive mixture is significantly higher, when clay percentage varied from 30% to 50%, compared to the cohesionless sediment. Parameters like clay content, bulk density, water content and unconfined compressive strength have been taken into consideration to investigate their effect on the critical shear stress of cohesive sediment mixture. It was found that clay percentage, bulk density and unconfined compressive strength has the increasing trend with the dimensionless critical shear stress. The clay percentage was found as dominant factor that govern the incipient motion condition for gravel particles in the present study of cohesive sediment mixture.

Assessment of sediment rating curves for mountain rivers in Malaysia

S.K. Sinnakaudan, M.R. Shukor & S.I.H. Ismail
Faculty of Civil Engineering, Universiti Teknologi MARA Pulau Pinang, Permatang Pauh, Pulau Pinang, Malaysia

A.Z. Abdul Razad & M.N. Ahmad
TNB Research Sdn. Bhd., Lorong Air Hitam, Kawasan Institusi Penyelidikan, Kajang, Selangor, Malaysia

ABSTRACT

Sediment transport measurements at the inflowing rivers into reservoirs which are located at the mountainous catchment area has been always a daunting task. However, to meet the continues requirement on reservoir sedimentation monitoring and management initiatives, Total Sediment Load (TL) rating curves (RC) are developed mostly based on low flow data and TL equations area adopted to give reliable prediction for medium to high flow data. However, with the present state of knowledge, the selection of best TL equation for natural rivers remains a major doubt among engineers and practitioners. Thus, the present study attempts to evaluate suitability application of the 10 existing TL equations to predict the RCs specifically for 2 inflowing rivers namely Sungai Bertam and Sungai Kenyir into Kenyir Reservoir which is the largest made reservoir in Peninsular Malaysia.

A total of 48 sets sediment and hydraulic data which covers low to high flow ranges were obtained from Sinnakaudan & Abdul Razad (2014) and adopted in the analysis. The study streams data encompass wide range of flow discharge (0.034 m³/s–23.96 m³/s) with

Table 1. Discrepancy ratio analysis for the selected TL equations.

Approach	Percentage (%) of data falls within r, (0.5–2.0)
Laursen (1958)	0
Bagnlod (1966)	0
Engelund & Hansen (1967)	0
Graf (1968)	0
Yang for Sands (1968)	0
Shen & Hung (1972)	31.25
Ackers-White (1972)	0
Yang for Gravel (1984)	3.25
Sinnakaudan et al (2006)	0
Sinnakaudan et al (2010)	31.25

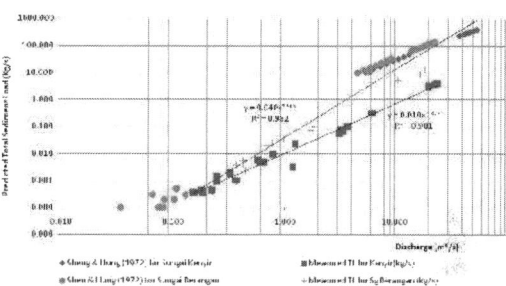

Figure 1. Comparison between measured and predicted Rating Curves.

diverse particle size ranging from fine sand to boulder size of bed material (d50).

The accuracy of existing equations was assessed by discrepancy ratio (r) analysis whereby predicted and measured TLs (medium to high flow data) were compared and accepted if the ratio is within the acceptable range from 0.5–2.0. It was found that Shen & Hung (1972) and Sinnakaudan et al. (2010) equations can predict less than 32% accuracy (Table 1). However, Shen & Hung (1972) shows consistency in prediction for low to high flow data. As such, it was adopted for TL RC development for the said rivers. The actual and predicted rating curves are given in Figure 1.

REFERENCES

Jarrett, R.D. 1984. Hydraulic of high gradient streams. Journal of Hydraulic Engineering, 110(11), 1519–1539.

Sinnakaudan, S.K., Sulaiman, M.S. & Teoh, S.H. 2010. Total bed material load equation for high gradient rivers. Journal of Hydro-environment Research, Vol. 4, 243–251.

Sinnakaudan, S.K & Abdul Razad, A.Z. 2014. Development of Total sediment load of High Gradient Rivers of Sg. Lemoi, Sg Bertam and Sg. Telom for Ulu Jelai Hydroelectric Scheme in Pahang. TNB Reaserch Sdn. Bhd (unpublished).

Super Deltaflex – Advanced development of transit time acoustic flow measurement

W. Stedtnitz
Former Stedtnitz Maritime Technology Ltd. (SMT) Eganville, Ont. Canada

T. Schott
Berner Fachhochschule BFH, Burgdorf CH

ABSTRACT

Observation of flooding events over years and years have shown that all kinds of ultrasonic flow measuring devices – Transit time and Doppler systems alike – fail, when data is most needed: in case of peak flooding! This happens because with higher velocity the sediment content is rising dramatically thus leading to catastrophic acoustic attenuation.

Events in Europe, even more so in situ measurements in sediment-loaded rivers near the Himalayan Mountains have shown that the acoustic attenuation rises with sediment load, but never blocks completely. That means it should be possible to obtain reliable data, if the signal to noise ratio can be significantly improved.

This can be done by new hardware and software designs. Freely Programmable Silicon-chips allow compact and cost effective solutions.

We report about improvements in the 30 to 50 dB order by applying mathematical data processing routines like pulse compression, cross-correlation and clever signal analysis.

REFERENCES

BfG-JAP-Nr.: 1895, 71 Seiten, Bundesanstalt für Gewässerkunde, BfG-0976.
Dixon, R.C. 1994. Spread Spectrum Systems, With Commercial Applications.
ISO 6416 International Standard, Hydrometer, Measurement of discharge by the (acoustic) method, Ref. No. ISO 6416–2004 (E).
ISO 6418 International Standard, Liquid flow measurement in open channels, Ultrasonic velocity meters, First edition, 1985-02-15, UDC 532:534–8, Ref. No. ISO 6418–1985.
IOSR Journal of Electral and Electronics Engineering (IOSR-JEEE), e-ISSN:2278-16, p-ISSN: 2320–3331, Volume 8, Issue 2 (Nov. – Dec.2013), 33–41.
John Wiley & Sons, Inc. ISBN 0-471-59342-7.
Kliche, W. & Siedschlag, S., Testbericht 1995. Test Ultraschalldurchflußmeßanlage AFFRA.
Morgenschweiss, G., Rudolph, A. & Heitefuss, C. 1992. Die Durchflußmeßstelle Mühlheim / Ruhr. Sonderdruck aus Jahresbericht Ruhrwassermenge 1991. Ruhrverband Essen, S. 32–44.
Naga Jyothi, A. & Raja Rajeswari, K. 2013. Generation and Implementation of Barker and Nested Binary Codes.
Stedtnitz, W. 1989. Abflußmessung mit Ultraschall, Ein Erfahrungsbericht aus Nord-amerika, Deutsche Gewässerkundliche Mittei-lungen, S. 120–127 DK556.535.3.08/.088:534.321.9-4:534.2(71–73).
Urban, H.G. 2002. Handbook of Underwater Acoustic Engineering, Atlas Elektronik GmbH, Naval Systems Division ISBN 3-936799-04-0.

Transport of sediment in the presence of stable clast

M.S. Sulaiman & R. Zainal Abidin
Infrastructure University Kuala Lumpur, Kajang, Selangor, Malaysia

S.K. Sinnakaudan
Universiti Teknologi MARA, Penang, Malaysia

ABSTRACT

Presence of surface bedform in the form of stable clast is believed to be exist at micro-scale. Stable clast is believed to modify the flow field and provide sink and source for incoming sediment particles. The dynamics of flow pattern and bed load rate in the presence of stable clast are always puzzling due to wide range of particle size and presence of roughness elements. Thus, this issue can be treated using two methodological approaches namely local scale and reach scale solutions. Incorporating the flow fields and turbulence statistics at a very local scale leads to the prediction of occurrence of bed load transport using probability approach. Bed load and turbulence data were physically measured at mountainous region having diverge surface bedform in its presence. Understanding transport occurrence at finer scale is vital for instream rehabilitation, river restoration and installation of sediment sampler on river beds. However, treating the dynamics of flow and bed load at reach scale is significantly differed from local scale approach. Transport of sediment in the presence of stable clast at reach scale is best equipped using continuous transport prediction or reach-averaged bed load model. Wide ranges of data from King et al. (2004) are used for further investigation at reach scale solution. The presence of stable is verified from previous report and morphodynamic analysis. Transport of sediment using existing prediction model postulate the less suitability and a need for modification. Thus, a similarity approach is used to develop the bed load model which can predict the bed load transport at reach scale. This new equation is successfully predicting the transport of sediment in the presence of stable clast. This equation alleviates the dependence on wide range of particle size as previously demonstrated by past model.

REFERENCE

King, J.G., Emmett, W.W., Whiting, P.J., Kenworthy, R.P. & Barry, J.J. 2004, Sediment transport data and related information for selected coarse-bed streams and rivers in Idaho, Gen. Tech. Rep. RMRS-GTR 131, 26 pp., Rocky Mt. Res. Stn., U.S. Dep. of Agric. For. Serv., Fort Collins, Colo.

Research of bed load distribution density based on image recognition technology

D.P. Sun, H. Chen, Y. Sun, A. Gao, M.J. Dong & L.Q. Han
North China University of Water Resources and Electric Power, Zhengzhou

M.X. Liu
Xi an University of Technology, Xi an, China

ABSTRACT

With some imperfection on traditional method, it is necessary to explore a new real-time bed load transport detecting system which can collect the transport rate and components of it. Based on the image recognition technology, the author developed the real-time bed load transport detecting system, so that makes dynamic bed load transport process detecting come true. The detecting system is composed of image recognition system and software recognition processing system, which can ascertain characteristic value (bed load transport rate, the lateral distribution of transport rate, the grain composition of bed load and so on) and variable processes of them by the method of river dynamics. Through the real-time image recognition, the system is able to attain the number of bed load particles(quality), speed, trace and horizontally-projected area and some other basic parameters, thus distinct three different function modules, dynamic detecting bed load distribution, sediment discharge dynamic calculation and dynamic kinetic energy calculation, are explored based on these basic parameters. In order to detect its reliability, this system is applied to the non-uniform bed load movement and statistical analyzed in four aspects, which is mainly contained: First, the dynamic variation rule of the non-uniform bed load size distribution; Second, probability distribution, turbulent characteristics and traces of particle instantaneous velocity; Third, the change rule of the transport rate of fractional grain and points belt; Fourth, bed load kinetic energy change rule and transformational relation with sediment discharge. According to experimental results, this kind of non-contact test method can not only make dynamic automatic testing come true, but also make the results accurate and reliable. This method enriches the measuring points of bed load, thus is beneficial to the in-depth study of the rule of bed load; what's more, it can also provide more accurate and reliable technological, which means for the research of the interaction of water and sediment in their transport process.

Combination of permeable and impermeable spur dikes to reduce local scour and to create diverse river bed

A. Tominaga
Nagoya Institute of Technology, Nagoya, Japan

S.H. Sadat
Kabul University, Afghanistan

ABSTRACT

Spur dikes are used primarily for protecting levee from erosion during flood in alluvial rivers by means of velocity reduction and flow deflection mechanisms. Recently, their environmental functions are taken into account since they can create diverse riverbed configurations, such as deposition and local scour. However, an impermeable spur dike is exposed to severe local scour and sometimes they are broken down. Meanwhile, it is known that effects of a permeable spur dike, such as pile-group, on flow structure and bed deformation are reasonably mild (Zhang et al. 2005, Sadat & Tominaga 2013). Therefore, the combination of permeable and impermeable spur dikes is useful to provide suitable conditions for ecological aspects.

In this study, flow structure and bed deformation around a single spur dike downstream of a pile-group in an open channel are imvestigated experimentally and numericaly. The purpose was to find out an optimum distance between pile-group and spur dike for reducing the local scour. For comparison, a single impermeable spur dike (conventional type) without the pile-group was also evaluated. A series of experiments under emerged flow and clear-water scour conditions were conducted. In the cases with an upstream pile-group, the local scour was faded with a noticeable decrease in the depth and amount of bed erosion. The scour depth and volume were decreased as the distance increased from zero to four times the spur dike length. When distance is 4 L, the maximum scour depth reduces to a minimum 44% smaller than that of case C0. The maximum scour-depth and volume of scouring were higher in the case with 6 L distance than 4 L. Thereby, a 4 L distance in this study is a favorable distance in regard to the maximum scour depth and volume of scouring.

Flow velocity near the nose of spur dike was higher in the conventional spur dike. In the cases with a pile-group, flow near the spur dike becames lower relative to the conventional case. These results are confirmed by a numerical simulation. We applied a two-dimensional numerical model with the depth-averaged k-ε model and conducted reproduction verification of the flow around the combination of permeable and impermeable spur dikes. These flow structures around the combination of pile-group and spur dike were well simulated by a 2D numerical model and the reduction of bed shear stress near the spur dike by the pile-group was demonstrated.

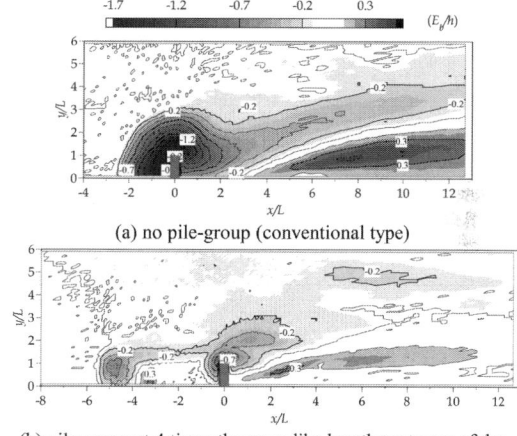

(a) no pile-group (conventional type)

(b) pile-group at 4 times the spur-dike length upstream of the spur dike

Figure 1. Example of the bed deformation contours.

REFERENCES

Sadat, S. H. and Tominaga, A, 2014, Influence of pile group density on minimizing local scour of a double spur dike group, *Journal of Japan Society of Civil Engineers, Ser. B1*, 70(4), 1_85–1_90.

Zhang, H., Nakagawa, H., Ishigaki, T., Muto, Y. and Khaleduzzaman, A. T. M., 2015, Flow & bed deformation around a series of impermeable & permeable spur dikes, *Proc. Int. Conf. on Monitoring, Prediction and Mitigation of Water-related Disasters*, Kyoto, pp 197–202.

Estimated response of Nieuwe Waterweg Rotterdam to deepening of the navigation channel

A.P. Tuijnder, L.M. Perk, R.C. Steijn, B.T. Grasmeijer, J. Adema, N. Geleynse & J. Cleveringa
Arcadis NL, Advisory group River, Coast and Sea, The Netherlands

L.C. van Rijn
Department of Physical Geography, University of Utrecht, Utrecht, The Netherlands

ABSTRACT

In the context of an environmental impact assessment the response of the Nieuwe Waterweg (New Waterway) to deepening of the navigation channel was evaluated. The New Waterway is the primary channel that discharges the water from the River Rhine to sea. It's median discharge is (Q50) is approximately 1350 m³/s, only during peak discharge the Haringvliet floodgates open and the discharge to the Haringvliet becomes larger. The New Waterway is the primary access to one of the busiest ports in the World. In order to make the inland parts of this port accessible to larger ships, dredging of the navigation channel is planned. The current guaranteed bed level, the guaranteed nautical depth (NGD), is 15 m –NAP, or approximately 15 m below the local ordinance level (NAP) that approximates the aver- age sea level. The channel will be deepened to 16.3 m –NAP. This level increases stepwise towards the city of Rotterdam.

The research questions in this study are: What are the effects of the deepening on the hydro- morphological system, locally and upstream? How does the deepening affect the yearly amount of dredging maintenance in the navigation channel and the harbor basins? What effects does the increased maintenance dredging have on the turbidity of the water?

The approach followed in this study is a combination of literature study, 2D and 3D hydrodynamical modelling at a fine and a coarse resolution using Waqua in Simona, a 1D sediment transport study using the TSAND model by van Rijn (2015), a 3D morphological model in Delft3D and empirical relations to estimate the discharge cross section.

The hydrodynamics have been modelled and studied for the 5, 25, 50, 75 and 95 percentile discharges of the Rhine at Lobith (at the Dutch-German border) for a spring tidal cycle, a neap tidal cycle and an average tidal cycle. Salinity effects were included in all cases as these proved crucial for capturing the natural flow patterns.

The TSAND model has been extended for the purpose of the study. TSAND is a 1DV model that incorporates the effects of tidal velocities including

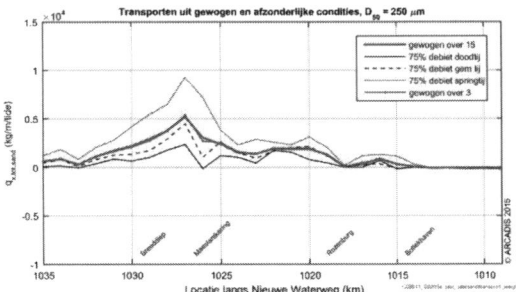

Figure 1. Average TSAND transport rate from 15 conditions (blue) and 3 conditions: spring neap and mean tide (red).

asymmetries, the rivers flow velocity profile, density currents and their distribution over depth and the sediment composition (sand and mud). This 1DV model was fed with data from the detailed 3D hydrodynamic simulation. The net transport was calculated as the weighted average over the 15 transport conditions. From these results we were able to confirm that the use of bed shaping discharge gives reliable results with much less effort (See Fig. 1). The bed shaping discharge was used to calculate the development in the reference situation and the deepened situation. The overall picture of the area is that it is sediment importing from both the fluvial and marine side. The sedimentation of the coarser fraction primarily takes place in the navigation channel while the harbor basins act as efficient sediment traps for the fine suspended sediment (silt and mud). The sedimentation in the harbor basins is linked to the presence of the turbidity maximum where suspended sediment concentrations reach several hundred milligrams per liter. Further effects regarding the deepening will be presented, the analysis will be finished in the coming weeks.

REFERENCE

Van Rijn, L.C. (2015, subm) Description of TSAND model. Full reference will be provided in extended abstract.

Variation in total sediment rate prediction using different fall velocity methods

S.I. Waikhom
Civil Engineering Department, Government Engeenering College, Surat, India

S.M. Yadav
Sardar Vallabhai National Institute of Technology, Icchhanath, Surat, Gujarat State, India

ABSTRACT

It has been evident from the study of many researchers that sediment terminal velocity is a key parameter responsible for quantity of sediment sustained by the flow in open channels for transportation. Even minor changes in the value of fall velocity, when calculated by different methods, lead to significant variation in estimated total sediment transport/flow rate. Researchers like Shu- Qing Yang (2003), Shen & Hung (1983) & Chih Ted Yang (1979) have implied the use of fall velocity to compute total load transport rate in their equations. Owing to the critical contribution in sediment transport in rivers, many attempts have been made since long to estimate fall velocity precisely. In the present study equation of Cheng (1997) followed by equation of Concharov (1962), Julien (1995), Soulsbey (1999), Van Rijn (1989), Zhang (1989) & Zhu & Cheng (1993) have been used to calculate terminal velocity for wide range of data sets & the values of fall velocity predicted is utilised in estimating total load transport rate by Shu-Qing Yang (2003) formula. Predictive capability of total load transport rate by using different fall velocity formula is checked with the aid of statistical parameters (Discrepancy Ratio, Mean Absolute Percentage Error, Thiel's Coefficient of Inequality & Square of Pearson product-moment correlation coefficient). Evaluation reveals that, Shu-Qing Yang (2003) predicts well for R. A. Stein (1965) data set using Cheng (1997) fall velocity formula with minimum discrepancy. Few data points are found to be scattered from the line of equality, with mean absolute percentage error ranging from −53 to +5. Also, for all the other data used, Cheng (1997) fall velocity formula gives better results in predicting total load transport function of S. Q. Yang (2003) with percentage error ranging from −30 to +30 followed by Zhu & Cheng (1993) & Zhang (1989) formula ranging from −54 to −6.86 & −60.58 to −9.85. Thus the anomaly in result needs contextual understanding for application of sediment transport rate prediction.

REFERENCES

Brown, P.P. & Lawler, D.F. 2003. Sphere drag and settling velocity revisited. J. Environ. Eng., Celik, I., and Rodi, W. (1991). "Suspended Sediment Transport Capacity for open Channel Flow." J. H.E ASCE, Vol. 117,
Cheng, N.S. 1997. Simplified settling velocity formula for sediment particle. J. H. Eng., ASCE, 123(8),
Dietrich, W.E. 1982. Settling velocity of natural particles. Water Resource. Res., 18(6), 1615–1626.
Ferguson, R.I. & Church, M. 2004. A simple universal equation for grain settling velocity Journal of Sedimentary Geology, 74, 933937.
Garde, R.J. & Ranga Raju, K.G. 1977. "Mechanics of Sediment Transportation and Alluvial Streams Problems," Wiley Eastern Ltd., p. 171.
Hallermeier, R.J. 1981. Terminal settling velocity of commonly occurring sand grains. Sedimentology, 28(6),
Ibad-zade, Y.A. 1992. Movement of sediment in open channels. S. P. Ghosh, translator, Russian translations series, Vol. 49, A. A. Balkema, Rotterdam, The Netherlands.
Rouse, H. 1937. "Modern Conceptions of mechanics of fluid turbulence"Trans. ASCE, VVVol. 102, pp. 463–543.
Rubey, W. 1933. Settling velocities of gravel, sand andsilt particles. Am. J. Sci., 225, 325–338.
Soulsby, R.L. 1997. Dynamics of marine sands, Thomas Telford, London.
Yalin, M.S. 1977. Mechanics of Sediment Transport, 2nd Ed., Pergamon Press, New York, N.Y.
Yang, S-Q. & Lim, S-Y. 2003. Total load transport formula for flow in Alluvial channels, Journal of hydraulic engineering, Vol. 129, pg 68–72, ASCE, ISSN 0733-9429/2003
Zanke, U. 1977. "Berechnung der Sinkgeschwindigkeiten von Sedimenten." Mitt. Des Franzius-Instituts fur Wasserbau, Technical Univ., Hannover, Germany.

Characteristics of sediment movement and river-bed morphology at mountainous stream confluence region

X.K. Wang, E. Huang & X.N. Liu
State Key Laboratory of Hydraulics and Mountain River Engineering, Sichuan University, China

X.F. Yan & H.F. Duan
Department of Civil and Environmental Engineering, The Hong Kong Polytechnic University, Hong Kong

ABSTRACT

Confluences are commonly observed in the natural river network of China, in particular in the southwestern mountainous regions. It is important to study and understand the evolution of the stream morphology for better design and management of channel engineering. This study selected the confluence zone of Shenxigou and Baisha stream, which is located in Dujiangyan City, Sichuan, as the prototype to build up the down-scaling physical model in the laboratory. Experiments were carefully designed and tested to investigate the characteristics of sediment movement and channel bed morphology at this confluence zone. Meanwhile, the numerical simulations are also conducted and calibrated by the experimental tests. The details in the confluence region were simulated to understand interaction between the sediment and water motions.

To study the sediment characterization at the stream confluence, an artificial deposition zone (point bar) was set in the first experiment at the downstream of the confluence on the tributary side. Besides the designed deposition zone, the fixed bed is exposed without bed materials. The evolution of the point bar was studied for examining the scour effect of different discharge ratios of the two streams. It was found that the scour effect becomes more profound for the case of one main stream flow only than that of the confluence flows from two streams. Meanwhile, for both cases, a narrow and long sediment ridge was observed at the downstream of the deposition zone. This result confirms the difference of the sediment movement for the different flow configurations (i.e., single-stream flow or confluence flow). To explain, the flow structures and hydrodynamics at the confluence region were explored for different flow conditions.

As the result of sediment movement, the evolution of channel bed morphology can be easily caused, which was investigated by the second experiment with movable bed configuration. The results showed clearly the sorting effect on the bed surface at the downstream confluence zone. Basically, three regions including

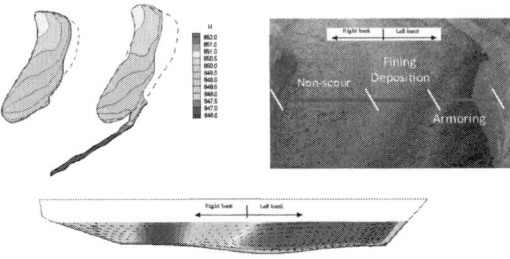

Figure 1. The characteristics of (a) deposition zone evolution, (b) bed surface sorting effect and (c) simulated flow field in the confluence zone.

non-scoured region, fining deposition region and greatly armoring region were observed and laterally distributed from the right bank to the left bank in the studied confluence region. Meanwhile, the morphology of the pool-riffle-pool bed type is formed from the upstream to downstream in the mainstream of this confluence zone.

To examine the details of the confluence effect, the numerical simulation was conducted and compared to the experiments. After validation calibration by the experimental cases, the numerical results were used for analyzing the flow structures and hydrodynamic processes in order to explain the observations and results in the experimental tests. Finally, the results and findings of this study are discussed for the theoretical development of river dynamics and practical implications of river engineering.

ACKNOWLEDGEMENT

The work was supported by Sichuan Province Science and technology support program (No. 2014SZ0163) the National Natural Science Foundation of China (No. 51579163), and the Open Foundation of State Key Laboratory of Hydraulics and Mountain River Engineering Sichuan University.

Experiment for bed erosion focusing on combination of horizontal distance and overlapping height between main and counter Sabo dam

H. Watabe, K. Kaitsuka, M. Sugiyama & T. Itoh
Research and development center, Nippon Koei Co., Ltd., Tsukuba, Japan

H. Muramatsu, T. Nagayama, H. Ogawa, T. Miike & A. Miyamoto
Sabo Department, Nippon Koei Co., Ltd., Tokyo, Japan

Y. Yamada
Sabo Frontier Foundation, Tokyo, Japan

T. Mizuyama
National Graduate Institute for Policy Studies, Tokyo, Japan

ABSTRACT

Erosion take place in downstream area of main dam and counter dam in flood, and vertical erosion reach to bottom of Sabo dam, it lead loose stable structure of Sabo dam.

There are 2 patterns for erosion in downstream area of Sabo dam, one is the local scouring by overflow in between main and counter dam, the other is bed erosion by imbalance of sediment transportation along downstream reach of counter dam. Apron and counter dam for countermeasure of erosion are in downstream area of Sabo dam. Horizontal distance and overlapping height of Sabo dam is set by empirical formula, and it is not always countermeasure for bed erosion against the design flow. More effective countermeasure for erosion in downstream area of Sabo dam and between main and counter dam should be conducted.

In present study, erosion in downstream area of Sabo dam and between main and counter dam against the combination of overlapping height and horizontal distance between dams is examined experimentally. Large scale of water discharge is assumed in this experiment.

It is not easy to eroded in case that counter dam is set as high against main dam. However, erosion depth in downstream of counter dam is noticeable in comparison with depth of embedment because the height of over flow from counter dam is big. It is also eroded three-dimensionally if the counter dam is set in long distance from main dam. If counter dam is set in condition that over flow from main dam hit to the counter dam, it is difficult to erode between dams and downstream area of counter dam due to the decreasing of energy of overflow.

In case of small scale of water discharge, although over flow from main Sabo dam does not hit to top of counter dam, impact of over flow is not noticeable

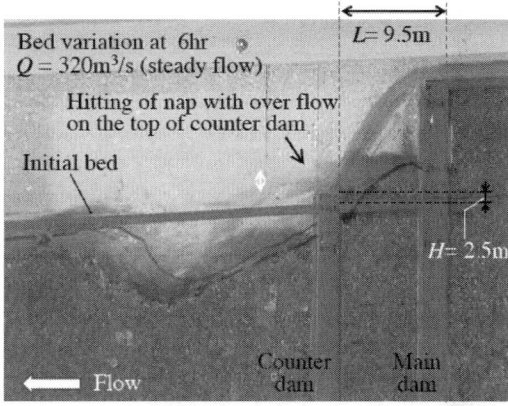

Figure 1. Erosion in downstream area of Sabo dam.

against the erosion. Bed erosion decrease if counter dam is set in appropriate position that overflow from main dam hit and overlapping height is big.

We are going to obtain some data under the condition of various magnitude of water discharge.

REFERENCES

Hayashi, S. 1983. Hydraulic Studies on the Phenomenon of Scour at the Base caused by free falling Nappe over Sediment Control Dams, The bulletin of the Faculty of Agriculture, Mie University, 66: 101–189.

Mizuyama, T., Yamada, T. & Yazima, S. 1990, Design of slit Sabo dam apron, Report of PWRI, No. 2835.

Homogeneous two-dimensional Poissonian model applied to the suspended movement of pollutant and non uniform fine sediment in open channel flow

G. Wilson Júnior
Federal University of Rio de Janeiro, COPPE/UFRJ, Rio de Janeiro, Brazil

C.S.G. Monteiro
Federal University of Rio de Janeiro, LAMEMO/UFRJ, Rio de Janeiro, Brazil

ABSTRACT

The Random Theory proposes a kinematic analysis of the suspension of pollutant and/or solid particles in open channel flow. It avoids problems relating to: (i) the non-linearity of the equations, (ii) the complexity of the liquid and solid interactions, (iii) the non-uniformity of the sediment properties, and (iv) the unawareness of liquid and solid movement's changes. For a sediment particle, the suspended $2-D$ trajectories $\omega(x, z, t)$ result from the combination of two $1-D$ chronologic and interdependent displacements series in the $i = 1, 3$ senses, intercalated with periods of time when the grain does not move in these senses. After restate some successful $1-D$ works carried out in natural streams, creeks and rivers, using chemical, radioactive and fluorescents tracers, this paper has the following objectives: (i) to show that the Random Processes also describe the $2-D$ movements – longitudinal and vertical – of sediment and pollutants in open channel flows; (ii) to present the $2-D$ Poissonian models of sediments and contaminants in longitudinal and vertical suspension movements; (iii) highlight the importance of temporal and spatial Intensity Mobility Functions in the definition, calibration and validation of random models; and (iv) to show how to consider the grains sizes, to describe their behavior in suspension.

Random models are obtained from analytical expressions that define the mobility functions $\lambda_{1i}(t, n)$ and $\lambda_{2i}(x_i, k)$, $i = 1, 3$, which consider the sediment and pollutant particle characteristics, as well the hydrodynamic properties. Their analytical expressions are obtained considering the mobility of the particles as a function of time, of the distance traveled in one direction and of its past performance in time (n) and distance (k), in each direction (i). For the $2-D$ Homogeneous Lagrangean Random Poissonian Processes, the grain mobility functions assume constant values. Each pair of values of the mobility functions describes the sediment grain movements in one direction. Examples of the evolution in time of the concentration of

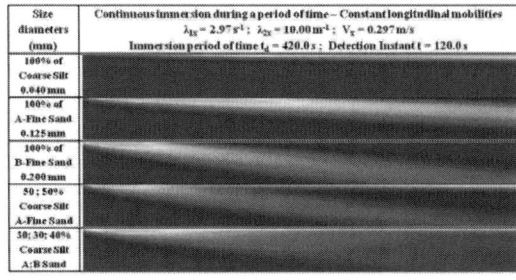

Figure 1. Plumes of fine uniform and non-uniform sediments in suspension. Laboratory Channel: L = 20.00 m; H = 2.00 m. $\lambda_{2z} = 10.00\,\text{m}^{-1}$. Coarse silt ($\lambda_{1z} = 0.0014\,\text{s}^{-1}$; $V_z = 0.0014\,\text{m/s}$). A-Fine sand ($\lambda_{1z} = 0.0141\,\text{s}^{-1}$; $V_z = 0.0014\,\text{m/s}$). B-Fine sand ($\lambda_{1z} = 0.360\,\text{s}^{-1}$; $V_z = 0.0360\,\text{m/s}$).

sediments in suspension injected continuously during a time interval [0, t_d], in which longitudinal mobility is greater than the vertical, are shown in Figure 1. If the grain mobility is not constant, more complex models are generated: Non-homogeneous Poissonian and Non-Poissonian models, for instance. Tests carried out in open channel flows with sediment labeled with radiotracers, permitted the determination of the grain mobility functions and to adjust and validate the resultant $1-D$ and $2-D$ random models.

REFERENCES

Todorovic', P. 1992. *An introduction to stochastic processes*. The Applied Probability Trust. Springler Series in Statistics.

Wilson-Jr., G. 1987. *Etude du transport et de la dispersion des sédiments en tant que Processus Aléatoires*. Thèse de Doctorat d'Etat ès Sciences Physiques. Paris VI, France.

Wilson-Jr.,G. & Monteiro, C.S.G. 2015. $2-D$ Poissonian homogeneous model for suspended sediment and pollutant movements in open-channel flow. *Sedimentation and Hydrology Conferences – SEDHYD*, Nevada, USA.

Experimental study on energy dissipation and beach protection effects of a new type of penetrating frames

Y.F. Xia & H. Xu
Nanjing Hydraulic Research Institute, Nanjing, China
The State Key Laboratory of Hydrology-Water Resources & Hydraulic Engineering, Nanjing, China

Z.M. Fu
Changjiang Waterway Planning Design and Research Institute, Wuhan, China

K.H. Chen
Nanjing Hydraulic Research Institute, Nanjing, China

F. Chen
Changjiang Waterway Planning Design and Research Institute, Wuhan, China

ABSTRACT

Following with the implementation of the 12.5 m Deepwater Channel Regulation Project in the Yangtze River in Nanjing, aiming at the overall stability of the traditional permeable framework is not strong enough, the research of channel engineering has put forward a new type of energy dissipation and beach protection penetrating double H permeable framework (Xia et al. 2015), shown in Figure 1. Unilateral length of the prototype framework is 0.8 m, cross-section of the framework is square, and the side length of it is 0.08 m.

An undistorted model of 1:30, 40 m in length and 1.6 m in width is employed to research on the space frame geometric properties of positive duplex permeable dumped and energy dissipation and beach protection effects. The model design primarily meets the geometric similarity, gravity similarity, weight similarity and sediment incipient motion similarity. The new permeable framework dumping zone is set in the middle section of the flume.

Different layers of porosity changes are studied. When casting layers is less, the porosity rate is greater. Along with the increases of number of layers from 1 to 6, the porosity rates decrease from 89.7% to 87.6%, while the number of layers increases to a certain value, the porosity is stabilized at 87.6%.

The vertical distribution of mean velocity and turbulence intensity in different layers are shown in Figure 2 and 3. Bottom velocity decreases significantly in the casting region, while the upper water body outside the

Figure 1. Sketch of the double H permeable framework.

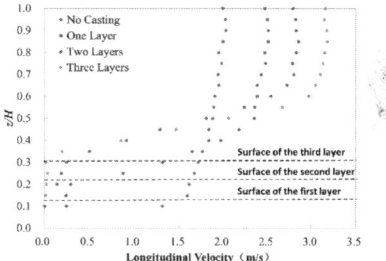

Figure 2. The velocity changes in different layers.

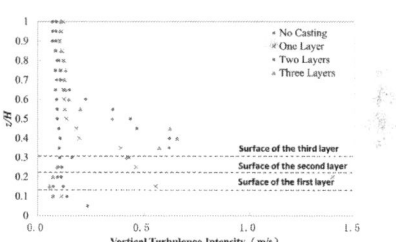

Figure 3. The turbulence intensity changes in different layers.

casting zone experience increasing of flow velocity; and the turbulence intensity is reduced near the bottom.

Beach protection effect: The inside of the entire casting framework is mainly siltation, double H permeable framework played the prevent erosion and promote siltation in guardian area, and served a good overall stability of double H permeable frame group.

REFERENCE

Xia, Y.F., Xu, H. & Wen, Y.C., 2015. Fu Jiang Yangtze River sand, Tongzhou Sand and Baimaosha deepwater channel system of governance key technology research. Nanjing: Nanjing Hydraulic Research Institute.

Suspended sediment dynamics of an allogenic dryland river channel

G.A. Yu
Chinese Academy of Sciences, Beijing, China

M. Disse
Technical University of Munich (TUM), Munich, Germany

Z.W. Li
Changsha University of Science & Technology, Changsha, China

ABSTRACT

Suspended sediment dynamics of the Tarim River, an allogenic river flowing in a very arid environment in China, are analyzed based on flow and suspended sediment data during the last five decades (1960–2011). The runoff of the Tarim River depends on snow and glacier melting water from surrounding mountains, the flow and sediment load are highly concentrated in flood season (June to September, particularly in July and August) when temperature in mountain area is high. The distribution of suspended sediment concentration (SSC) values is non-symmetric and concentrate in lower values (Fig. 1). The runoff and sediment load in the river showed a fluctuating while clear decreasing trend in recent decades, with an obvious increase of occurrence of low flow events while a gentle decrease of moderate-high flow events (Fig. 2). Three forms of hysteresis loop (clockwise, anti-clockwise and figure-eight) are observed, with clockwise loop the major form (Fig. 3). The high possibility of bank collapse near the peak and at the falling limb of the flood hydrograph is the major reason for anti-clockwise and figure-of-eight hysteresis loops.

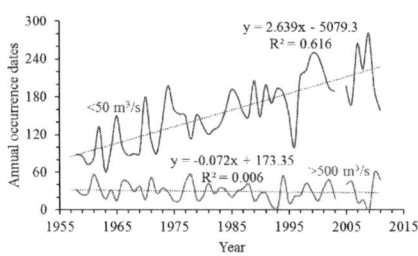

Figure 2. Variations of occurrence dates of low and moderate-high flow at AL station in the past decades.

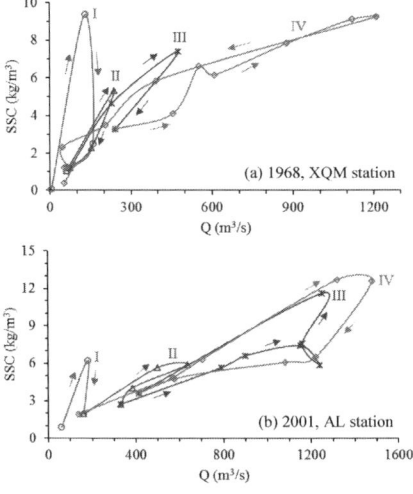

Figure 3. Clockwise loops dominate the suspended sediment concentration variations during flood season for the Tarim River and the loops often move with time from left to right for successive flood events. The rising limb of the first flood (denoted with 'I') has quite steep increase of suspended sediment concentration. The Roman numerals I, II, III, IV show different floods with time sequence.

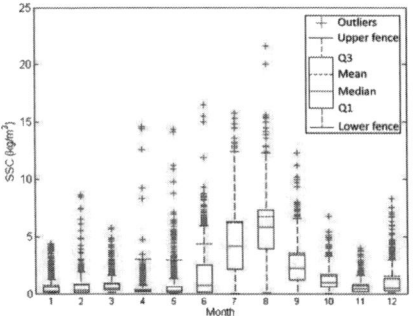

Figure 1. Box plot diagram showing variation of SSC at AL hydrologic station in twelve months over the past five decades.

Sediment transport in the middle reach of the Huaihe River

B. Yu, J. Ni & H. Zhou
Anhui and Huaihe Institute of Hydraulic Research, China

J. Sui
University of Northern British Columbia, Canada

P. Wu
University of Regina, Canada

R. Juepner
Technical University of Kaiserslautern, Germany

ABSTRACT

Sediment transport is an important issue of the Huaihe River. To study the evolution of the riverbed, the relationship between river flow and sediment transport has been done in the paper, based on long-term measured data in the middle reach of the Huaihe River. Both models developed by Han (2012) and Zhang (2000) were used for assessing sediment transport and channel evolution. Field data collected in four gaging stations, namely, Wangjiaba, Lutaizi, Wujiadu, and Xiaoliuxiang were used. In total, 80 sets of data including runoff, water temperature, cross section, sediment and suspended sediment data were chosen. The relationship was developed to indicate the relationship between sediment transport and sediment factor. The relationship between slope and discharge has been developed based on data measured from 1990 to 2007 in Lutaizi and Xiashankou stations. Results indicate that the roughness coefficient decreases with discharge before floodplain got flooded. After floodplain got flooded, roughness coefficient increases with discharge. On the basis of the relationship among gradient and discharge, channel roughness and other hydraulic factors at the Lutaizi Station on the Huaihe River, the models have been validated. The formula for determining sediment-transport capacity which describes the relationship between water flow and sediment transport at the Lutaizi Station has been developed in this paper. By introducing the Han's model into Zhang's formula, the water percentage was included into the sediment transport capacity equation. One suspended sediment transport equation for both coarse and fine sand was developed. By using the data from the Lutaizi Station as the example, the empirical formula for determining transport capacity under equilibrium, erosion and deposition conditions were derived. The annual sediment transport capacity was predicted and validated by using the measured data from 1990 to 2007. Results show that the measured annual sediment transport capacity is 4552×10^4 tons while the calculated annual sediment transport is 4337×10^4 tons, with a difference of 4.7%. From the calculation, it can be concluded that the empirical equation developed for determining sediment transport capacity can reasonably predict the sediment transport in the middle reach of the Huaihe River under different flow conditions.

Sediment transport and silting in Zarzis commercial harbor (Tunisia)

M. Zelleg, I. Said & E. Hamdi
Université de Tunis El Manar, Ecole Nationale d'Ingénieurs de Tunis, LR 14 ES 03 Laboratoire de Recherche d'Ingénierie Géotechnique, Tunisie

Z. Lafhaj
Ecole Centrale de Lille, Villeneuve d'Ascq, France

ABSTRACT

Nowadays, sediment management has an economic and environmental challenge to maintain the activity and movement of trade ships. This study was carried out on the case of a Tunisian harbor suffering from a significant silting which limits its draft and consequently its economic situation. The studied area presents a significant volume of dredged sediment. The objective of this work is to conclude the main source of silting problem in Zarzis harbor. Indeed, the increase of sediment layer prohibits ships movement which arrests the commercial activity of Zarzis harbor.

This paper investigates the influence of hydrodynamic factors, such as waves and currents, on sediment mobility in Zarzis harbor, Tunisia. Furthermore, the major silting observed in the studied area was deduced by bathymetry campaigns carried out over the years. A geographic information system software (ArcGIS 9.3) was used to establish the harbor bathymetric maps for the years 2006, 2008, 2010, 2011 and 2015. Results show an accumulation of 6 meter layer of sediment in the shipping channel. In order to understand the nature of sediment behavior under combined current– wave action at Zarzis harbor, three bed shear stresses were studied: Stress due to waves only (τ_w), Stress due to current only (τ_c) and stress due the combined role of current and waves (τ_{cw}). τ_{cw} values were calculated using the van Rijn (1993) and Soulsby (1995) models. (Porter-Smith, 2004; Zanke, 2003).

Based on Acoustic Doppler Current Profiler (ADCP) results and particle size distribution analysis conducted on marine sediments sampled from

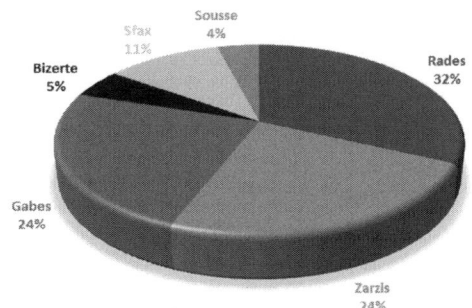

Figure 1. Dredged sediment volumes distribution (in percent) in different Tunisian commercial harbors (2007).

34 sites throughout the studied harbor, the contribution of waves, currents and their combined effect on the generated seabed shear stress was analysed.

The determined seabed shear stress in different zones of the harbor was then compared to the critical threshold value defined by the Shields parameter. Overall, results show the dominant action of waves in sediment mobility, especially in the shipping channel. The stress due to waves only (τ_w) exceeds the threshold value for almost all the studied samples. However, the action of current only does not contribute in the silting problem at the studied harbor.

REFERENCES

Porter-Smith, R., Harris, P.T., Anderson, O.B., Coleman, R., Greenslade, D., Jenkins, C.J., 2004. Classification of the Australian continental shelf based on predicted sediment threshold exceedance from tidal currents and swell waves. Mar. Geol. 211, 1–20.

R.L. Soulsby, R.J.S. Whitehouse, Prediction of Ripple Properties in Shelf Seas, H.R. Wallingford report TR 154, December 2005.

Van Rijn, L.C., 1993. Principles of Sediment Transport in Rivers, Estuaries and Coastal Seas. Aqua Publications, Amsterdam, ISBN 90-800356-2-9.

Zanke, U. C. E. (2003). "On the influence of turbulence on the initiation of sediment motion." Int. J. Sediment Research, 18(1), 1–15.

Table 1. Bathymetry's variation of Zarzis harbor over the years.

Location	Mean depth (m)/year			
	2006	2008	2011	2015
commercial basin	11.25	11.25	10.25	9.5
Swinging circle	10.5	10.25	9.75	9
Shipping channel 1	9.75	9.25	7.75	7

The influences of water-sediment conditions on the sediment delivery rate of the Three Gorges Reservoir

D. Zhandi, L. Qin, H. Haihua & J. Zuwen
China Institute of Water Resources and Hydropower Research, The State Key Laboratory of Simulation and Regulation of Water Cycles in River Basins, Beijing, China

ABSTRACT

The sediment delivery rate is one of the indicators of a reservoir's sediment interception degree. The greater the sediment delivery rate, the smaller the intensity of a reservoir's sedimentation. Conversely, the smaller the sediment delivery rate, the larger the intensity of a reservoir's sedimentation. In this study, based on the measured data of each hydrologic station in the Three Gorges Reservoir area from 2004 to 2012, the sediment delivery rates of the Three Gorges Reservoir's influence factors were analyzed. It was considered that the dam front water level, along with the reservoir's inflow (outflow), were the main factors affecting the sediment delivery rate of the reservoir, followed by the sediment inflow, and particle size. Moreover, it was concluded through further analysis that the sensitivity of the sediment delivery rate to water level changes rapidly increased with the increase of the reservoir's inflow. When the reservoir's inflow was less than 15,000 m³/s, the reservoir's water level changes had little impact on the sediment delivery rate. However, when the reservoir's inflow was greater than 20,000 m³/s, the sediment delivery rate of the reservoir changed greatly, and the reservoir's water level increased from 138 m to 155 m, while the sediment delivery rate was reduced from 62%, to less than 10%. Therefore, in order to improve the sediment delivery rate of the reservoir, an operational mode of "low level for abundant water, high level for scarce water" was recommended for the reservoir's future scheduling. The research results of this study provide references for further study of the sediment delivery rate of the Three Gorges Reservoir, as well as its operation.

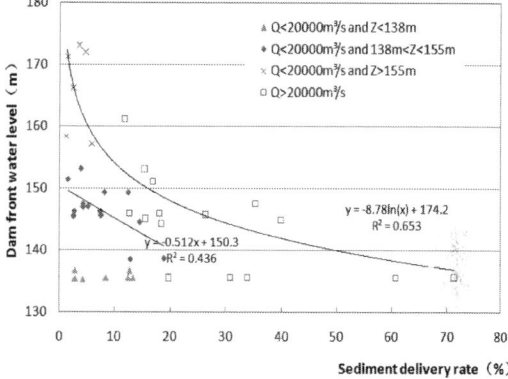

Figure 1. Relationship between the dam front water level and the sediment delivery rate of the reservoir (water level and flow gradation statistics).

REFERENCES

Chen Guiya, Y., Yuan, J. & Xu, Q. 2012. On sediment diversion ratio after the impoundment of the Three Gorges Project. Advances In Water Science, 23(3):355–362.

Lu, J. & Huang, Y. 2009. Study on the problems of long2term use of Three Gorges Project[J]. Journal of hydrauelectric Engineering, 2009,28(6):49–53.

Zhou, J.J. 2004 Schemes of sediment management proposed for the Three Gorges Projet. Proc,9th International Symposium on River Sedimentation.Tsinghua University Press: 250–257.

The response of riverbed erosion and deposition adjustment to the flow and sediment process in the Lower Yellow River

X. Zhang
College of Water Conservancy, North China University of Water Resources and Electric Power, Zhengzhou, China
State Key Laboratory of Water Resources and Hydropower Engineering Science, Wuhan University, Wuhan, China

D.P. Sun & Y. Sun
College of Water Conservancy, North China University of Water Resources and Electric Power, Zhengzhou, China

M.X. Liu
College of Water Resources and Hydroelectric Engineering, Xi'an University of Technology, Xi'an, China

ABSTRACT

In order to investigate the riverbed formation mechanism of alluvial river and the influence of flow and sediment process to alluvial river, it is very necessary to analyze the relationship between the adjustment of channel morphology and the flow and sediment process. Yang (1994) innovatively studied hydraulic characteristics and morphology shaping of alluvial river based on the energy dissipation properties of river system. Sun (1999) explored the riverbed adjustment of Ningxia-Inner Mongolia reach in Yellow River from the perspective of river entropy. In recent years, Chen (2004) adopted the principle of minimum rate of available energy dissipation and the theory of statistical entropy to explore the equilibrium stability mathematical model of riverbed evolution. Xu (2004) thought an alluvial river could adjust itself in accordance with the theory of minimum entropy production or the theory of minimum rate of energy dissipation. Hu (2006) proposed the relation of flow-sediment conditions to cross-sectional geometry in view of the variation of water and sediment in the Lower Yellow River.

The braided reach from Jiahetan to Gaocun of Lower Yellow River is chosen as studying objects in this paper, and studying approaches of observed data analysis and numerical modeling are used. The characteristics of sediment scouring and silting under various runoff-sediment conditions are explored. And the relevance between fluvial facies coefficient and characteristic parameter of water and sediment transport is founded. This reflects the regulation mechanism of river system.

It is shown that different flow and sediment processes have significantly different effects on adjustment of channel morphology, and the scouring intensity of flood closely related with cross-sectional geometry has stage characteristic. The channel shaped by the larger peak discharge is hard to keep its original sediment-transporting intensity during non-peak flood periods. The cumulative channel-scouring effect has a good response relationship to the flow and sediment process. And only the high-efficient sediment-transporting flood process could shape the most appropriate cross-sectional geometry so as to achieve both the optimum sediment-transporting efficiency and the maximum channel-scouring efficiency.

For the different magnitudes of overbank floods, only the flood process ($Q_{max} = 6000$ m^3/s) could shape the channel morphology with the best cumulative channel-scouring effect, and the corresponding synthetical factor $\Phi((B/H)(S/Q)^{0.5})$ is always maximum. Therefore, the discharge corresponding to Φ_{max} could be considered as a regulation index for overbank floods in the Lower Yellow River.

REFERENCES

Chen, X. & Hu, C. 2004. Mathematical model for flow and sediment in alluvial river based on minimum rate of available energy dissipation principle. Journal of Hydraulic Engineering, (8): 38–45.

Hu, C., Chen, J., Liu, D. et al. 2006. Studies on the features of cross section's profile in lower Yellow River under the conditions of variable incoming water and sediment. Journal of Hydraulic Engineering, 37(11): 1283–1289.

Sun, D. 1999. Analysis on energy distribution-dissipation relation on river system. Journal of Hydraulic Engineering, (3): 49–53.

Xu, G. & Lian, J. 2004. Changes of the entropy, the entropy production and the rate of energy dissipation in river adjustment. Advances in Water Science, 15(1): 1–5.

Yang, C.T. 1994. Variational theories in hydrodynamics and hydraulics. Journal of Hydraulic Engineering, ASCE. 120(6): 737–756.

Comparison of capacity and non-capacity sediment transport models for dam break flow over movable bed

J. Zhao, I. Özgen & R. Hinkelmann
Technische Universität Berlin, Germany

F. Simons
Bundesanstalt für Wasserbau, Germany

D. Liang
University of Cambridge, Department of Engineering, Cambridge, UK

ABSTRACT

Sediment transport in flowing waters is an important physical process, as it is one of the main factors of erosion and deposition. The mathematical and numerical modeling of these processes is challenging, because the erosion and deposition processes lead to a time-variable bottom elevation, which in return influences the flow. In addition, the sediment concentration is often considered to influence the momentum of the flow. Furthermore, the erosion and deposition processes are usually described by empirical laws that depend on several parameters.

This study presents a comparison of two different model concepts for the simulation of sediment transport, namely the capacity model with the Exner equation and the non-capacity model with suspended sediment transport and erosion and deposition terms. The Exner equation is modified to account for momentum loss due to sediment particle movement. Both equations are discretized using cell-centered finite-volumes and a second order Godunov-type scheme is applied for the solution. The numerical flux is calculated by a Harten, Lax and van Leer approximate Riemann solver with the contact wave restored (HLLC).

One-dimensional dam break over a horizontal movable bed was carried out as the first test case. Model results are compared with the results obtained in (Cao et al., 2004). Sensitivity study for different model parameters was performed. Results are summarized in table 1.

The capabilities and limitations of both model concepts are demonstrated in the second benchmark experimental test case dealing with dam break flow over variable bed topography. Overall, good agreement between model results and experimental data has been obtained.

The bottom change in the capacity model is calculated by using the sediment flux of the bed load, therefore the morphodynamics is instantaneous. In contrast, in the non-capacity model, the sediment transport is

Table 1. L_1-norm summary of bottom evolution in dependency of model parameters.

Parameters		Non-capacity model	Capacity model
Manning's Coefficient $(s/(m^{1/3}))$	0.015	0.729	0.168
	0.045	0.666	0.227
Sediment Diameter (mm)	0.004	1.130	0.125
	0.012	0.424	0.134
Sediment Porosity (−)	0.2	0.61	0.376
	0.6	0.73	0.483

a process depending on the entrainment and deposition that are calculated separately by using empirical formula. The water carries the sediment concentration and the sediment particle movement influences the water momentum. Thus, the non-capacity model concept is more complex than the capacity model concept. Preliminary research by the authors showed that the capacity model overestimates the flow velocities. Therefore, a simple heuristic modification is proposed that introduces additional source terms in the momentum balance.

Base on the sensitivity test in first case. For the non-capacity model, Manning's coefficient and sediment porosity show an almost linear relationship with the bottom change. For the Exner model, the sediment porosity is considered as the most sensitive parameter.

In the second test case, the modified capacity model and the non-capacity model are compared with experimental data. The capacity and non-capacity models both give good results for the morphodynamics, but for the water surface, the numerical models show higher water depth in the wave front, which might result from non-hydrostatic flow conditions in the experiment.

It is concluded that the modified capacity model can be used for sediment transport modeling for relatively steady flow and the non-capacity can be used in the more rapid flow. However, it is acknowledged that in practice it is difficult to separate these cases.

Numerical modelling of scour – the influence of small scale morphological processes

L. Zhou & R.J. Perkins
Laboratoire de Mécanique des Fluides et d'Acoustique, Université de Lyon, Ecole Centrale de Lyon, France

ABSTRACT

Bed scour may ultimately lead to instability and failure of hydraulic structures, so it is important to be able to predict the location and depth of scour holes that might develop in a range of different flow conditions. Physical modelling of this phenomenon is hampered by problems of scaling, so numerical simulations offer, in principle, an interesting alternative. But numerical modelling of the growth of a scour hole poses its own challenges; the flow itself will be turbulent, often highly three-dimension, with a free surface, and the domain itself will change as sediment is eroded. The flow conditions which produce erosion are usually very different from those studied in standard experiments, so the applicability of the usual laws for sediment transport is questionable. To investigate the influence of these different phenomena, a numerical simulation has been developed which combines both hydrodynamic and morphological processes.

The hydrodynamic module is based on the open source CFD software package OpenFOAM® and solves the Reynolds Averaged Navier-Stokes equations with a $k-\varepsilon$ turbulence closure model. The free surface is tracked using the Volume of Fluid scheme. The morphological module which has been developed consists of three components: a sediment transport module, which includes the suspended load and bed load, obtained by calculating the bed shear stress and solving the sediment continuity equation; the Exner equation, solved by adopting the Finite Area Method on the curved water-sediment interface; and an avalanching mechanism which restricts the bed slope to be smaller than the angle of repose (critical bed slope). The morphological change is incorporated into the hydrodynamic field through automatic mesh deformation.

The suspended load transport was validated independently using the net entrainment test case from Van Rijn (1986). The model was then used to simulate the scour from a submerged jet and the simulation results are compared with the experimental results of Chatterjee et al. (1994).

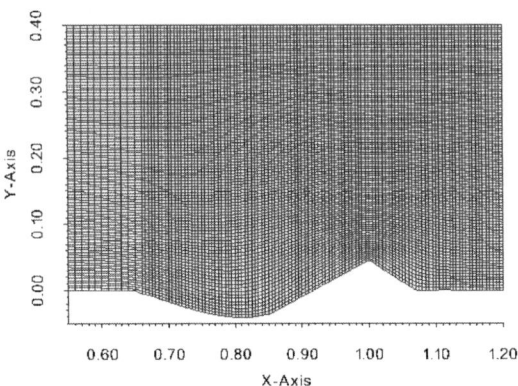

Figure 1. Scour form and mesh deformation from the 2-D simulation of the submerged jet, 5min after the onset of scour.

The flow fields are consistent with the experiments and the evolution of the bed profile agrees well with the experimental data. A detailed analysis of the experimental data shows that when the flow is directed up the bed slope, it can push the sediments and stop them avalanching, creating a bed slope that exceeds the critical slope. This effect should therefore also be included in the model.

REFERENCES

Chatterjee, S., Ghosh, S. & Chatterjee, M. 1994. Local Scour due to Submerged Horizontal Jet. Journal of Hydraulic Engineering, 120(8), 973–992.

Faruque, M., Sarathi, P. & Balachandar, R. 2006. Clear Water Local Scour by Submerged Three-Dimensional Wall Jets: Effect of Tailwater Depth. Journal of Hydraulic Engineering, 132(6), 575–580.

Khosronejad, A., Kang, S. & Sotiropoulos, F. 2012. Experimental and computational investigation of local scour around bridge piers. Advances in Water Resources, 37, 73–85.

Van Rijn, L.C. 1986. Mathematical Modeling of Suspended Sediment in Nonuniform Flows. Journal of Hydraulic Engineering, 112(6), 433–455.

Experimental study on scouring characteristics of cohesive bank soil in the Middle Yangtze River

Q.L. Zong
College of Water Conservancy and Architectural Engineering, Shihezi University, Shihezi, China

J.Q. Xia & Y. Zhang
State Key Laboratory of Water Resources and Hydropower Engineering Science, Wuhan University, Wuhan, China

ABSTRACT

Bank erosion plays an important role in the fluvial processes in the Middle Yangtze River, especially in the Jingjiang Reach. The process of bank erosion is closely related not only to the near-bank fluvial erosion, but also to soil composition and mechanical properties. (Xia et al, 2014). The soil composition of the riverbanks in the Jingjiang Reach is characterized by a kind of typical composite structure, which is composed of the upper cohesive soil and the lower sandy soil. The erosion-resisting capacity of cohesive soil is relatively strong, with its magnitude determining the erosion rate of cohesive riverbank. Therefore, it is necessary to obtain the incipient condition, erodibility coefficient, and the relationship between them in order to investigate the erosion-resisting capacity of cohesive soil.

In this paper, a series of laboratory experiments has been conducted in a closed rectangular flume. According to experimental results, the relationship was obtained between incipient velocity and the ratio of liquid limit to natural water content (Figure 1), as well as the quantitative relationship between critical shear stress and dry density, critical shear stress and liqidity index, which accounted for the effects of various physical properties on the incipient motion of cohesive soil.

In addition, the variation characteristic of erodibility coefficient with the critical shear stress was proposed for the cohesive bank soil in the Jingjiang Reach, and it was found that the obtained erodibility coefficient under the same critical shear stress is much higher, as compared with the value calculated using the existing relations obtained from other river basins, which is attributed to a lower clay content in the study reach and the disturbed soil samples (Hanson and Simon, 2001; Wynn, 2004; Karmaker and Dutta, 2011), shown in Figure 2.

Based on the results of scouring experiment, a quantitative relationship between the erodibility coefficient and critical shear stress of cohesive bank soil was obtained, with a correlation coefficient of 0.90, which can provide a basis for predicting the erosion process of riverbanks in the Jingjiang reach.

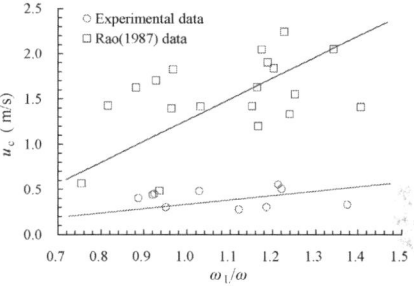

Figure 1. Relationship between incipient velocity u_c and the ratio of liquid limit to natural water content (ω_L/ω).

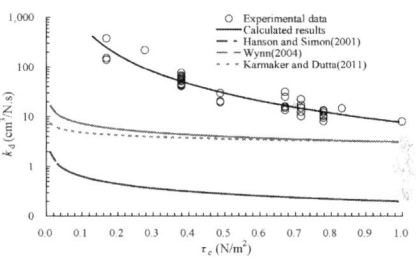

Figure 2. Relationship between erodibility coefficient k_d and critical shear stress τ_c.

REFERENCES

Hanson G.J. and Simon A., 2001. Eodibility of cohesive streambeds in the loess area of the Midwestern USA. Hydrological Processes, 15(1): 23–38.

Karmaker T. and Dutta S., 2011. Erodibility of fine soil from the composite river bank of Brahmaputra in Indi. Hydrological Processes, 25: 104–111.

Wynn T.M., 2004. The effects of vegetation on stream bank erosion. PhD thesis, Blacksburg, Virginia, Department of Biological Systems Engineering, 113–120.

Xia J.Q., Zong Q.L., Deng S.S., et al., 2014. Seasonal variations in composite riverbank stability in the Lower Jingjiang Reach. China, Journal of Hydrology, 519:3664–3673.

C. River morphodynamics

Two-dimensional river bed configuration analysis of the Hii River and diversion channel flood in September 2013

R. Akoh, S. Maeno, S. Hirashita, K. Yoshida & T. Matsumoto
Department of Civil Engineering, Okayama University, Okayama, Japan

ABSTRACT

The Hii River in Shimane Prefecture is well known as a typical sand bed river in Japan and it flows into Lake Shinji. During flood, water level of Lake Shinji is kept rather high because water level difference between the lake and Japan Sea is very small. The backwater from the river mouth considerably increases upstream water level.

To reduce a flood risk, the diversion channel was constructed at 14.4 km from the river mouth in May, 2013. The diversion channel was operated during the flood in September 2013. The measured volume of sand transported to the diversion channel from the Hii River was about 26,000 m³. Therefore, it is very important to get the detailed information of the bed-load transport process around the inlet of diversion channel.

The inlet of diversion channel is shown in Figure 1. The flow dividing weir has 5 openings with a sluice-gate and a relief gate to control the diversion flow rate. Additionally, energy dissipaters are installed behind the weir. The sands through the energy dissipaters are accumulated in a sedimentation basin. In this flood, about 18,000 m³ of sands were deposited in this basin.

In this study, two-dimensional river bed configuration analysis was executed in the section of the Hii River and the diversion channel for flood in September 2013. Unstructured mesh system was used to express the shape of structures at the inlet of diversion channel and bridge piers in the channel (Figure 2).

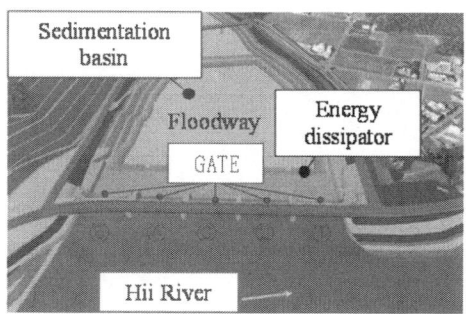

Figure 1. The inlet of diversion channel.

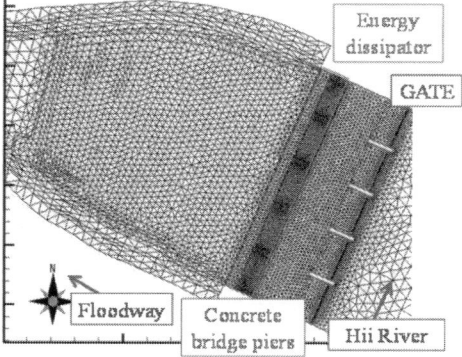

Figure 2. Computational mesh.

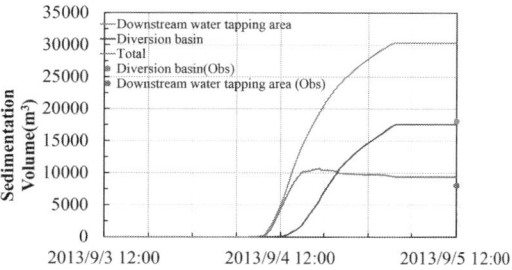

Figure 3. Time series of volume of sands in the sedimentation basin.

A set of shallow water equations and sedimentation transport equation were solved for the triangular mesh system by the finite volume method. Boundary condition was given by the time series of the observed water levels and the flow rate.

Figure 3 shows the time series of simulated volume of sands in the sedimentation basin. It is clarified that the volume of sand in the sedimentation basin is simulated well by the present numerical analysis.

REFERENCE

Maeno, S. & Ogawa, S. 2003. Flow analysis around hydraulic structure based on Finite Volume Method using unstructured grid system. Journal of applied mechanics: JSCE, 6, 857–864.

Morphodynamic modelling of a meandering sand bed river using Delft3D

M.S. Banda & A. Dittrich
Department of Hydraulic Engineering, Leichtweiß-Institute (LWI), TU Braunschweig, Germany

J. Pervez
Bangladesh Water Development Board, Dhaka, Bangladesh

ABSTRACT

A 3D morphodynamic model of a meandering sand-bed river was developed using the Delft3D model package (Lesser et al. 2004). Focus of the study was on the River Dhaleshwari, Bangladesh. The investigations focused mainly on the determination of secondary currents which are characteristic for meanders. For this purpose a boat-mounted acoustic Doppler current profiler (aDcp) at a meander bend section was used to measure the three-dimensional flow field (Fig. 1).

Model parameters were calibrated with the aDcp data for simulating primary and secondary currents (Fig. 2). Roughness and eddy viscosity were used as the main hydrodynamic calibration parameters so as to force agreement between model predictions and field observations. In this context the question occurred if such a calibration might lead to a good estimation of the meandering behavior of the chosen sand-bed river. To answer this question, we aimed to compare the results of one year morphodynamic computation with the field data and adopted the Brier Skill Scores (BSS) (Van Rijn et al. 2003, Sutherland et al. 2004) as quality parameters to assess the ability of model predictions. The model had a skill score of 0.20 and could be qualified as reasonable/fair according to Sutherland et al. (2004). The results of this study suggested

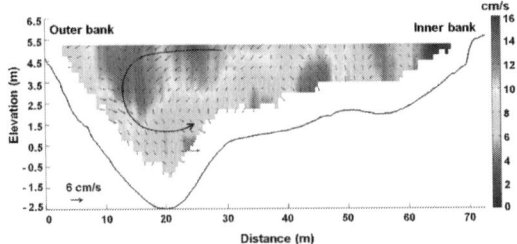

Figure 1. Cross-sectional contour plots of primary flow and vector representation of secondary flow. The cross-section is viewed looking downstream with left bank on the left-hand side.

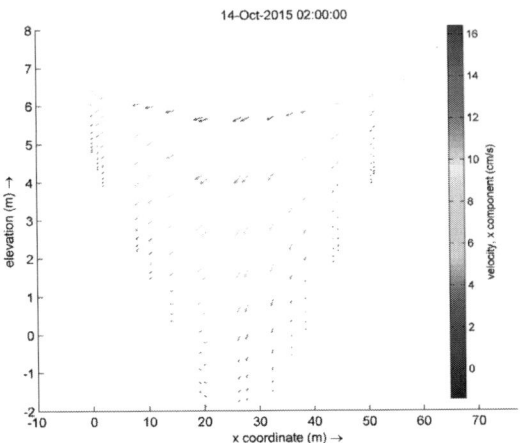

Figure 2. Secondary circulation in the same section as observed (Fig. 1), but obtained from the model. Vector colour scale represents the magnitude of the primary velocity. The cross-section is viewed looking downstream with left bank on the left-hand side.

that once properly calibrated and validated with flow measurements, the modelling approach could provide potentially a sound basis for prediction of bed level changes.

REFERENCES

Lesser, G.R., Roelvink, J.A., Van Kester, J.A.T.M. & Stelling, G.S. 2004. Development and validation of a three-dimensional morphological model. Coastal engineering, 51(8): 883–915.

Sutherland, J., Peet, A.H. & Soulsby, R.L. 2004. Evaluating the performance of morphological models. Coastal Engineering, doi:10.1016/j.coastaleng.2004.07.015.

Van Rijn, L.C., Walstra, D.J.R., Grasmeijer, B., Sutherland, J., Pan, S. & Sierra, J.P. 2003. The predictability of cross-shore bed evolution of sandy beaches at the time scale of storms and seasons using process-based profile models. Coastal Engineering, 47: 295–327.

Laboratory experiments on the influence of the length of a sediment replenishment applied with alternated geometrical configuration

E. Battisacco, M.J. Franca & A.J. Schleiss
École Polytechnique Fédérale de Lausanne, EPFL

ABSTRACT

The natural river conditions are guaranteed by the longitudinal continuity in sediment transport. Regardless their purpose, every dam traps sediment in the upper reservoir, thus the river behavior in the downstream reach is strongly modified in terms of morphology and hydrodynamics. The absence of sediment transport induces many negative changes in rivers. Riverbed incision, bank instability and changes in channel width, together with a loss in the aquatic and riparian habitats, are listed as main consequences. Due to the sediment transport reduction downstream of dams and to the armouring of the riverbed, the possibilities for fish spawning are limited.

The replenishment of sediment downstream of dams is an increasingly common technique to restore the sediment continuum of such disturbed river reaches. At the Laboratory of Hydraulic Constructions (LCH) of the École Polytechnique Fédérale de Lausanne, experimental work is being performed on a flume facility to investigate the river replenishment technique. The model reproduces the typical features of a straight armoured alpine gravel river in terms of slope, cross section and bed grain size. Based on the laboratory tests, an optimised submerged conditon and a geometrical configuration for the replenishment volumes are defined.

The total amount of replenished sediment is placed in four identical volumes following an alternating geometrical configuration. The width occupancy of the replenishment volume is kept constant and equal to 1/3 of the channel width and a completely submergence conditions is applied. The channel bed changes are recorded by photo taken at regular time interval, and the channel bed topography is surveyed by laser scanner at the beginning at the end of each test, after morphological equilibrium is reached in the channel.

The influence on channel bed morphology of three different replenishment lengths is assessed.

The discharge, together with the geometrical placement of volumes, shows a dominant influence on the erosion process. The pattern of the recreated morphological forms, created on the channel and related to the replenishment length, is presented and discussed.

Application of 2D numerical modelling to determination of sediment transport in a Mexican river

G. Cardoso-Landa
Instituto Tecnológico de Chilpancingo, Chilpancingo City, Mexico

ABSTRACT

The main rivers of Mexico and the large dams to operate in the country of Mexico, moves a significant amount of sediment, which have not quantified accurately in our country and causing major problems of flooding, disasters and in general, hydraulic, and hydrological problems, which requiring resolved taking as data, among others, through sediment transport. The amount of sediment transport of the *Papagayo* River was determinate by the CCHE2D model developed by the NCCHE and the results were calibrated with measurements in CFE station 3.

By application of CCHE2D model was obtained the sediment transport in various sections along this river and the results are shown in the next figures 2 (suspended load) and 3 (bed load).

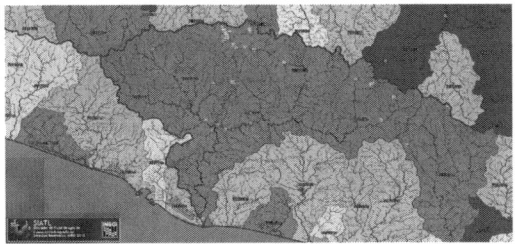

Figure 1. *Papagayo* river basin.

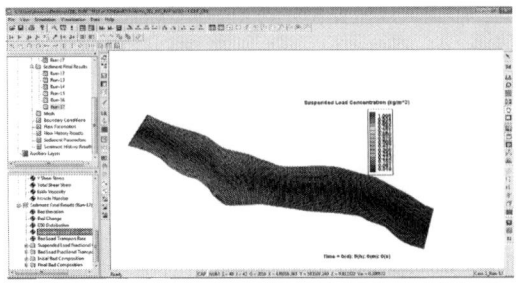

Figure 2. Suspended load of *Papagayo* river.

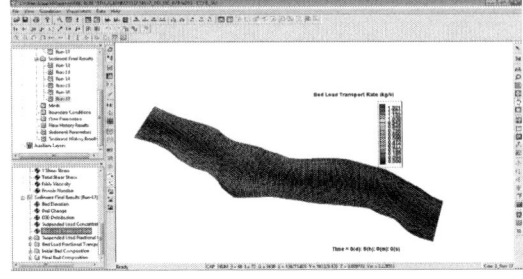

Figure 3. Bed load of *Papagayo* river.

The results reported in this paper suggests that the comparison between the sediment transport along the *Papagayo* river using the CCHE2D and the measuring of sediments realized by Electricity Federal Commission (CFE) at the hydrometric station number 3 *Agua Salada* shown a good accuracy, with a difference of 39% in the mean values for the last year of measurements. It is necessary to work in the another 3 hydrometric station along the *Papagayo* river to obtain more results, which are working at present moment. It is important emphasize that this is the first application of NCCHE models in a Mexican river.

REFERENCES

Duan, J.G., Wang, S.Y. and Jia, Y. 2001. The application of the enhanced CCHE2D model to study the alluvial channel migration processes. *Journal of Hydraulic Research*, 39(5):469–480.

Garcia, M.H. 2008. Sedimentation engineering – processes, measurements, modeling and practice. *American Society of Civil Engineers*, Reston.

Horvat, Z., Isic, M. and Spasojevic M. 2014. Two dimensional river flow and sediment transport model. *Environmental Fluid Mechanics*, 1: 1–31.

Vieira and Wu. 2002. National Center for Computational Hydroscience and Engineering (NCCHE). University of Mississippi, USA. *Technical report No. NCCHE-TR-2002-5*.

Historical and current uses of the Morvan's Rivers (central France): Impacts on bedload transport and fluvial morphology

L. Gilet, F. Gob & E. Gautier
Paris 1 Panthéon-Sorbonne University, Paris, France

C. Virmoux
Laboratory of Physical Geography, CNRS

ABSTRACT

This study focuses on the hydromorphological functioning and the bedload transport characteristics of the Yonne and Cure rivers (Morvan Massif, Central France). These small gravel-bed rivers have undergone two major anthropogenic modifications, one between the mid-16th to the early 20th century for the timber floating industry (Poux *et al.*, 2011), and the second one from the 1920's for hydro-electrical production. For almost four centuries, small dams, artificial cut-offs, straightening and water releases have deeply affected the fluvial dynamics and morphology of the rivers. During the 20th century, the establishment of a small dam (Pierre-Glissotte complex) has locally impacted the upper course of the Yonne River. Almost at the same time, a complex of three large dams has been built on the Cure river (Crescent complex), impacting the sediment flux on several tens of kilometers in the upper part of the catchment.

In order to understand the resulting fluvial forms and the processes at work, two types of influence have to be considered: inherited parameters from former practices (timber floating leading to an incision of the bed) and pressures from remaining activities (dams leading to a sediment deficit). The evolution of the river bed as well as former artificial discharges are first reconstructed from historical archives (old maps, engineer plans, documents for dam management, etc.) and the current morphody-namic processes are then studied through the monitoring of bedload transport using PIT tags (RFID technology). More than 650 particles were marked on 8 study sites on the different reaches of the two hydrological complexes. In this way, the diversity of hydrological conditions regarding dam distribution and functioning is taken into account: reaches located upstream/downstream a dam, reaches receiving water releases, short-circuited reaches, *etc.* The comparison between the results of the different study sites, in the light of their hydrological regime (influenced or not by the dams, type of influence), tends to prove the different impacts of the dams on the bedload transport.

Combined with the historical perspective, current sediment transport characteristics allow a better understanding of the Yonne and Cure response to the construction of the dams. This was completed by fieldwork measurements establishing current cross-section and longitudinal profiles. Identification of morphological readjustments made by the Yonne and Cure rivers will contribute to highlighting the disruptions of sediment transport caused by the dams.

REFERENCE

Poux A-S., Gob F., et Jacob-Rousseau N., 2011. Discharge reconstruction of artificial water releases for timber floating in the Morvan Massif (central France, 16th–19th centuries) from historical archives and geomorphological observations. Géo-morphologie, relief, processus, environnement, 2, 143–156.

Sediment transport and evolution at Pearl River estuary

J. Deng & H. Deng
Key Laboratory of Pearl River Estuarine Dynamics and Associated Process Regulation,
Pearl River Hydraulic Research Institute, Ministry of Water Resources, Guangzhou, China

ABSTRACT

The Pearl River is the third largest river in China. It consists of three rivers, i.e. the West River, the North River and the East River with corresponding lengths 2075 km, 468 km and 520 km respectively. The river Delta consists of more than 200 waterways which form a complex network-river estuary. The Pearl River discharge into the South China Sea through eight outlets named the Humen, Jiaomen, Hongqimen, Hengmen, Modaomen, Jitimen, Hutiaomen and Yamen. Different outlets have different characters of the water flow and sediment transport. The Modaomen has the largest runoff and the Humen has the largest tide. Both the upstream fresh water and the downstream tide are mixed and form back and forth flows in the estuary. Under such flows the sediment movement is extremely complicated.

Under the dual role of both nature and human activities, the evolution of the rivers at the estuary is of complex. Naturally, the network-rivers and the outlets have been in the sediment deposition over a long period of time. This has caused various problems. Due to the sedimentation in the network-rivers, the water level in the Delta center has been getting higher in recent years, which has caused the flood problem. And due to the sedimentation in the outlets, the bars and shoals have been growing and the outlets have been extending to the sea further. The sediment deposition in the harbors or navigational channels has been bothering the navigation development in the estuary, which is always a key problem for the constructions of the new harbors or navigational channels. On the other hand, the extensive human activities have showed an obvious influence on the river water-sediment transport in the estuary. In recent three decades, with the rapid economic development in the estuary area, a lot of infrastructures, such as harbors, navigational channels, bridges, bases for building or repairing ships, and etc. have been built along or in the rivers or outlets. In addition, a large-scale reclamation of the beaches or shoals along coastal areas outside the outlets has been continuously carried out. The uncontrolled sediment mining from the river-bed for the building construction has caused a large amount of river-bed-sediment being excavated. Furthermore, the navigational channels have been dredging deeper and deeper. Many new and large harbors and navigational channels are under construction. These extensive human activities have changed the topographic and hydrodynamic features of the rivers at the estuary and resulted in many problems including the flood-protection, seawater intrusion, fresh-water supply, waterlogged-land drainage, and etc.

Based on analysis of the variations in water-sediment processes both in the network-rivers and the river-outlets, this paper explores the responses of the hydrodynamics, including the flow-diversion ratios at the first- and second-level flow-diversion nodes in the network rivers, and the flow-diversion ratios at outlets. Through the analysis of the variations in water-sediment processes both in the network-rivers and the river-outlets, it is shown that in recent three decades the water-sediment amount diverted towards southeast has get increased in the network-rivers and hence, resulted in an increase of the flow distribution of the Lingding Bay, demonstrating the responses of the human activities to the hydrodynamics of the estuary. Based on the analysis of sediment transport at the estuary, the tidal sediment-laden capacity is set up by taking account for the sediment flocculating settling. The preliminarily analysis of the estuary evolution indicates that the extensive human activities in recent three decades have changed the topographic and hydrodynamic features of the estuary. Especially, the uncontrolled sediment mining from the network rivers has caused a large amount of river-bed-sediment being excavated. And the large-scale reclamation of beaches or shoals along coastal areas outside the outlets has caused a significant influence on the evolution of the estuary.

Conceptual modeling of bank retreat process in the Upper Jingjiang Reach

S.S. Deng, J.Q. Xia, M.R. Zhou & J. Li
State Key Laboratory of Water Resources and Hydropower Engineering Science, Wuhan University, Wuhan, China

ABSTRACT

After the operation of the Three Gorges Project (TGP), the sediment load entering the Upper Jingjiang Reach (UJR) has been reduced greatly, leading to the continuous channel degradation and significant bank retreat processes at local sites of this reach, causing problems to channel stability and flood control management. Bank retreat occurs owing to a combination of hydraulic-induced bank-toe erosion and mass failure. The former is usually controlled by fluvial factors, including the processes of lateral erosion and bed incision; the latter is always affected by non-fluvial factors, covering the variations of in-channel water level, ground water level, and bank soil properties.

In this paper, based on a classical method of bank stability analysis proposed by Thorne & Osman (1988), an improved model for predicting retreat processes is proposed, considering the effects of bank-toe erosion, in-channel water level and ground water level changes. This model consists of three sub-modules: (i) the module for bank-toe erosion calculation in two aspects of lateral basal erosion and near-bank bed evolution, (ii) the module for the one-dimensional simulation of ground water flow instigated by the in-channel water level change, and (iii) the module for evaluating bank stability characterized by the safety factor, covering the terms of pore water pressure and matrix suction for unsaturated soil, which change with the ground water level.

Taking two typical sections of Jing34 and Jing55 in the UJR as examples, the proposed model was tested and applied to simulate their bank retreat processes in 2005. The results indicate that: the calculated retreat widths and bank profiles agreed relatively well with the measurements at these two sections, with six and four bank failures occurred, respectively. Because of the intensive lateral basal erosion and high ground water level, a relatively higher frequency of bank failure was identified during the flood-peak and recessions periods. In addition, the effects of bank material

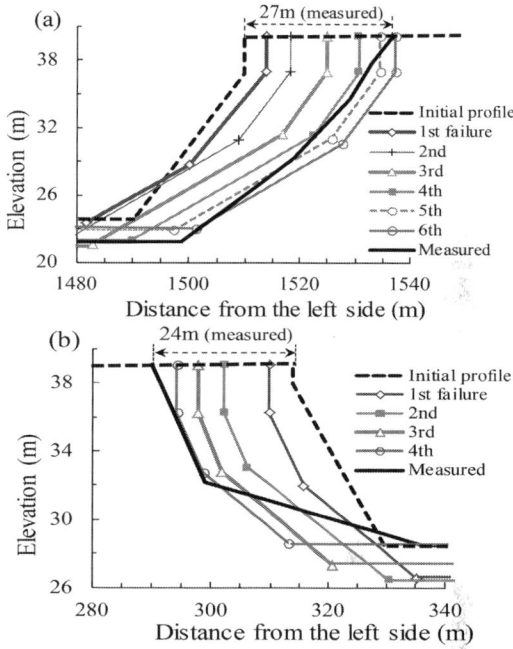

Figure 1. Changes in profiles of (a) the right bank at Jing34 and (b) the left bank at Jing55.

properties on bank retreat were investigated through conducting numerical tests, and the results illustrate that: bank failure frequency and retreat width generally diminished with an increase in bank soil shear strength and permeability, with a unidirectional change tendency for the frequency, while the change in the retreat width with these parameters are more complicated, affected by the bank-toe erosion degree as the corresponding bank failure occurred. Besides, the delayed change in ground water level increased bank failure frequency during the recession stage.

How fast evolve the river-bottom profile and grain-size composition at basin scale

G. Di Silvio & M. Franzoia
Department of Civil, Architectural and Environmental Engineering, University of Padua, Italy

M. Nones
Gerstgraser Ingenieurbüro für Renaturierung, Cottbus, Germany

ABSTRACT

It is known that along any watercourse a clear relation exists between the sediment grain-size in certain reaches and the corresponding bottom slope (Exner 1920, Lane 1955, Egiazaroff 1965, Asselman 2000, Di Silvio & Nones 2014, Franzoia 2014, Horowitz 2003, Nones 2013). At basin scale, if we neglect the localized deviations due to geological constraints, this corresponds to the usual concave profile and progressive sediment fining in the downstream direction (Hirano 1971). Although this configuration is, perforce, not in equilibrium, yet it usually exhibits a quasi-stationary behaviour at very long (historical and even geological) time-scale (Di Silvio & Nones 2014).

The 0-D, two-reaches, two-grain-size hydro-morphological model presented in this paper, based on a number of reasonable and verified simplifications of a 1-D model (Armanini & Di Silvio 1988, Fasolato et al. 2009, Fasolato et al. 2011, Nones 2013), gives reason of the extremely slow evolution of most rivers and provides a quantitative approach for evaluating their "response time". Differently from previous formulations, the response time appears here to be affected, among others, by the granulometry of the sediment input (Franzoia 2014).

Applications of the model to schematic rivers of different sizes and configurations are reported and discussed in the paper.

REFERENCES

Asselman, N.E.M. 2000. Fitting and interpretation of sediment rating curves. *Journal of Hydrology* 234(3): 228–248.
Armanini, A., & Di Silvio, G. 1988. A one-dimensional model for the transport of a sediment mixture in non-equilibrium conditions. *Journal of Hydraulic Research* 26(3): 275–292.
Di Silvio, G. & Nones, M. 2014. Morphodynamic reaction of a schematic river to sediment input changes: Analytical approaches. *Geomorphology* 215: 74–82.
Egiazaroff, I.V. 1965. Calculation of non-uniform sediment concentration. *Journal of Hydraulic Div.* 91: 225–248.
Exner, F.M. 1920. Zur Physik der Dunen. Sitzber Akad. *Wiss Wien*, Part IIa, Bd. 129. in German.
Fasolato, G., Ronco, P., & Di Silvio, G. 2009. How fast and how far do variable boundary conditions affect river morphodynamics?. *Journal of Hydraulic Research* 47(3): 329–339.
Fasolato, G., Ronco, P., Langendoen, E. J., & Di Silvio, G. 2011. Validity of Uniform Flow Hypothesis in One-Dimensional Morphodynamics Models. *Journal of Hydraulic Engineering* 137(2): 183–195.
Franzoia, M. 2014. *Sediment yield in rivers at different timescales*. PhD Thesis. University of Padova, Italy.
Hirano, M. 1971. River bed degradation with armouring. *Trans. of JSCE*.
Horowitz, A.J. 2003. An evaluation of sediment rating curves for estimating suspended sediment concentrations for subsequent flux calculations. *Hydrological Processes* 17(17): 3387–3409.
Lane, E.W. 1955. The importance of fluvial morphology in hydraulic engineering. *Proceedings of the American Society of Civil Engineers* 81: 1–17.
Nones, M. 2013. *Riverine dynamics at watershed scale: hydro-morpho-biodynamics in rivers*. LAP Lambert Academic Publishing, p. 140, ISBN-13: 978-3659367854.

A cellular automata model for riverbed evolvement

M.J. Dong
North China University of Water Resources and Electric Power, Zhengzhou, China

ABSTRACT

Cellular automata are discrete and dynamical systems that are divided up into small cells with each cell taking a certain state. The basic idea of cellular automata is to reduce a complex system of complex rules into something simpler. In the study of cellular automata, some scholars have done the following works: Simulations of the Armaconi Basin, Calabria, Italy showed encouraging results which were in agreement with the findings of other studies (Ambrosio & Gregorio 2001). The cellular model of river avulsion highlights the need to develop models of floodplain evolution at large time and space scales to complement the improving models of river channel evolution (Jerolmack & Paola 2007). The cellular model of braided rivers suggests that the only factors essential for braiding are bed-load sediment transport and laterally unconstrained free-surface flow (Murray & Paola 1994). The simulation of braided river flow using a new cellular routing scheme represents a small and ongoing contribution to the field of numerical simulation of braided river processes (Thomas & Nicholas 2002). The modeling of hillslope runoff and soil erosion at rainfall events suggested that the CA model was an applicable alternate for simulating the hillslope water flow and soil erosion (Ting & Hu 2009).

In this study a novel numerical cellular automata model for riverbed evolvement process was developed. The model was applied to one part catchment of the Wei River the largest tributary of the Yellow River which part is a typical type wandering stream. The dynamics of wandering stream are complex; channels migrate laterally, split, rejoin and develop bars, with the flow shifting unpredictably from one part of the network to another. This model we describe here is a simple, deterministic numerical model of water flow over a cohesionless bed that captures the main spatial and temporal features of real wandering rivers. It involves a number of states including altitude, water depth, total head, infiltration, erosion, sediment transport and deposition. The impact of factors like size of the spatial cell, hydraulic parameters, and the setting of time step and iteration times on the model accuracy was discussed. The comparison of the simulated and measured data suggested that the cellular automata model was an applicable alternate for simulating riverbed evolvement process.

REFERENCES

Ambrosio, D.D. & Gregorio, S.D. 2001. A Cellular Automata model for soil erosion by water. *Physics and Chemistry of the Earth Part B Hydrology Oceans and Atmosphere*, 26(1): 33–39.

Jerolmack, D.J. & Paola, C. 2007. Complexity in a cellular model of river avulsion. *Geomorphology*, 91 (2007): 259–270.

Murray, A.B. & Paola, C. 1994. A cellular model of braided rivers. *Nature*, 371: 54–57.

Thomas, R. & Nicholas, A.P. 2002. Simulation of braided river flow using a new cellular routing scheme. *Geomorphology*. 43 (2002): 179–195.

Ting, M. & Hu, Z.C. 2009. Modeling of Hillslope Runoff and Soil Erosion at Rainfall Events Using Cellular Automata Approach. *Pedosphere* 19(6): 711–718.

Laboratory experiments on gravel deposit erosion

F. Friedl, V. Weitbrecht & R.M. Boes
Laboratory of Hydraulics, Hydrology and Glaciology (VAW), ETH Zurich, Switzerland

ABSTRACT

Many Swiss rivers are heavily affected by various river engineering measures and exhibit ecological deficits. Sediment input from upstream is limited by hydropower plants, sediment retention basins and river training measures, resulting in a lack of bed load and morphological degradation. A possibility to improve the sediment balance is sediment replenish- ment by artificial gravel deposits (Fig 1). Pilot projects have shown that the behaviour of artificial gravel deposits and their impact on the river system are not clearly predictable (Schälchli et al. 2010).

Physical experiments were conducted in a trapezoidal channel with a length of 35 m, a bed width of 2 m and a bed slope of 0.17% at the Laboratory of Hydraulics, Hydrology and Glaciology (VAW). Froude similarity has been used to model the Reuss River at a scale of 1:25. 42 tests were conducted to investigate the influence of the governing parameters, namely the grain size distribution, bulk density, geometry of the deposit and hydraulic load. The investigated hydrological conditions range between a mean annual discharge and a flood with a return period of 30 years (HQ_{30}).

The goals of the model study were to (i) investigate the erosion process of gravel deposits, (ii) improve the understanding of fluvial bank erosion in case of non-cohesive, granular material, (iii) Provide fundamentals

Figure 1. Gravel deposits in the Aare River. View against flow direction (Schälchli et al. 2010).

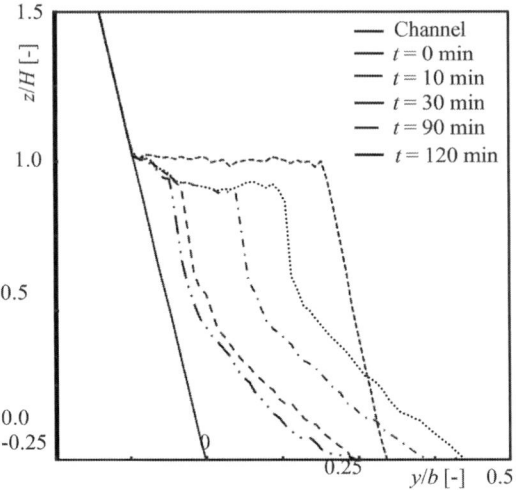

Figure 2. Temporal evolution of the cross section in the middle of a gravel deposit. Bank height H, bed width b, vertical coordinate z and lateral coordinate y.

to implement and improve bank erosion in numerical models.

As we are using a hybrid approach, numerical simulations were performed in parallel. The bank erosion process was implemented and enhanced in the software BASEMENT to simulate the erosion of the gravel deposits (Vonwiller et al. 2016, this conference).

Fig. 2 shows the lateral temporal evolution of a gravel deposit. A significant geometry change in the first time step followed by parallel erosion can be observed. Lateral bank erosion rates and the rate of sediment supply to the downstream river reach were determined.

REFERENCE

Schälchli, U., Breitenstein, M., and Kirchhofer, A., 2010. Kiesschüttungen zur Reaktivierung des Geschiebehaushalts der Aare – die kieslaichenden Fische freuts. *Wasser Energie Luft*, 102(3): 209–213, [in German].

Landslide dam breach during 2015 earthquake in Nepal: Computational modelling of hydraulic and morphological effects

S. Giri & M. Nabi
DELTARES (former Delft Hydraulics), Delft, The Netherlands

J.D. Bricker
Tohoku University International Research Institute of Disaster Science, Tohoku, Japan

B.R. Adhikari
Institute of Engineering, Kathmandu, Nepal

W. Schwanghart
Institute of Earth and Environmental Science, University of Potsdam, Germany

ABSTRACT

Landslides are rather common in Himalayan country Nepal. Almost every monsoon the country suffers significant losses and fatalities due to floods and landslides. A landslide, which occurred on May 24, 2015, had some specific characteristics and aftermath. The landslide occurred during dry period, and in effect, there was no trigger other than the fact that it occurred a month after the 'great' earthquake, which was very destructive. Furthermore, the landslide formed a natural dam in Kali Gandaki River (the length was approximately 700 m and the height was about 30 m), and formed a lake of about 2 km length and more than 30 m deep. Particularly, downstream towns Galeshwor Bazar and Beni Bazar were in threat of inundation in case the breach occurred. The town is located in alluvial deposits, and very vulnerable to extreme flows, propagating from upstream. The landslide dam breached relatively quickly (about 15 hours later), so fortunately there was no fatality due to the propagating flood flow.

In this study, an attempt has been made to reproduce first the flood propagation using 2D Delft-Flexible Mesh (developed at Deltares). Furthermore, a sediment transport and morphological study has been carried out using Delft3D morphological model. The Delft3D model, used herein, is depth-averaged with uniform and non-uniform sediment transport and morphology. The model incorporates all kinds of innovative, recently developed aspects, amongst which domain decomposition, consideration of floodplains including wet and dry processes, sediment transport over non-erodible layers and functionality for sediment management.

The inflow hydrograph was deduced from the BREACH – a model that computes the flow hydrograph, induced by the breach of an earthen dam (in our case a natural landslide dam). The model can be used to compute the breach erosion and sediment transport as well. Since data are sparse in these remote areas, landslide geometry was estimated from

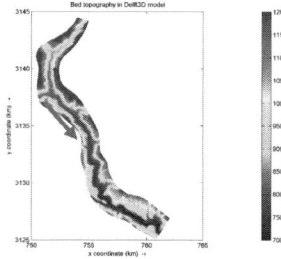

Figure 1. The computational model domain.

available information and Landsat images. Available photos of the landslide allowed gross estimation of geotechnical parameters. Reservoir surface area as a function of stage was determined using SRTM topography data.

The hydrograph, resulting from the BREACH model, was used as upstream condition to simulate the downstream flow propagation. Topography, used in each model, had been processed with hydrological smoothing and incorporated a 1 m deep and 45 m wide channel carved into the thalweg. Hydrologic smoothing and carving of the topography is found to be important, since simulations with a raw topography appear to be causing hindcast flood waves with too little depth, too diffusive, and altered wave propagation speed. High-frequency results were extracted at the locations of the Beni Bridge and Maldhunga Bridge, for comparison between models and with videos recorded by residents. Figure 1 shows the model domain, used for the numerical computations.

2D morphological model will be used to replicate the sediment transport and downstream morphological changes in a relative manner. Also, some synthetic cases will be simulated in order to check some hypotheses and assess their applicability to address such issues and demonstrate the usability of computational models as a part of nonstructural measures in disaster prevention and management.

The work is still under progress.

Analysis of the interaction between the Yangtze River and Poyang Lake, China based on Chaos theory

Jing Hu, Zhi-li Wang & Yong-Jun Lu
Nanjing Hydraulic Research Institute, Nanjing, China

ABSTRACT

Poyang Lake (Fig. 1) is the biggest fresh water lake in China, located in the lower reach of Yangtze River, North of Jiangxi Province. It receives water from five rivers and finally flows into the main stream of the Yangtze River at the outlet, Hukou. The impoundment of Three-Gorges-Dam in 2006 markedly breaks the balance of river-lake system, including water exchange and sediment transport. The process of discharge and sediment transport rate of Hukou represents the interaction between the river and lake.

The behavior of chaotic systems lies between periodical and disorder. Moreover, if a system is sensitive to initial condition, the system is chaotic. River system is a non-equilibrium dynamic system. Every case with different initial condition evolves independently. Thus the river system fits with the definition of chaotic system and can be studied by chaotic theory. Chaotic behavior is a phenomenon which seems irregular and random, appearing in certain non-linear system. The time-delay embedding method and phase-space reconstruction method are used to get the Saturation-Correlation-Dimension. The Saturation-Correlation-Dimension indicates the implicit characteristic of the time series for the dynamic system and predicts future behavior, represents the chaotic degree of the system and moreover, the impact of the operation of the Three-Gorges-Dam. Larger Saturation-Correlation-Dimension means less stable system and more difficulty to describe the system. Smaller means more stable and more regular.

Since the construction of Three-Gorges-Dam is finished by 2006, the year of 1986 and 1987 is chosen as the typical year before the operation, 2011 and 2012 is chosen as the typical year after the operation. The calculation results of Hukou are listed in Table 1.

The Saturation-Correlation-Dimension for the discharge time series of the year 1986–1987 is larger than the year 2010–2011, opposite for the sediment transport rate. The result indicates that the operation of the Three-Gorge-Dam makes the fluvial process more regular, more complicated for the sediment transport, which would be a good reference to investigate the impound effect of the Three-Gorge-Dam.

Table 1. Saturation-Correlation-Dimension of time series.

	discharge		Sediment transport rate	
year	1986–1987	2010–2011	1986–1987	2010–2011
D	1.3749	1.0645	2.2301	3.2136

REFERENCES

Jayawardena, A.W. & Lai, F. 1994. Analysis and prediction of chaos in rainfall and stream flow time series. Journal of Hydrology, 153(1): 23–52.

Sivakumar, B. 2002. A phase-space reconstruction approach to prediction of suspended sediment concentration in rivers. Journal of Hydrology, 258(1), 149–162.

Xu Guo-bin, Yang C.T. 2012. Analysis of river bed changes based on the theories of minimum entropy production dissipative structure and chaos. Journal of Hydraulic Engineering, 43(8): 948–956.

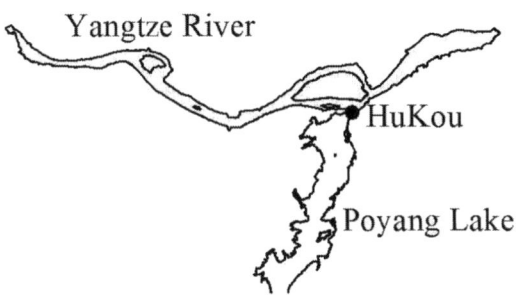

Figure 1. Location of Poyang Lake and Yangtze River.

Dynamics of sediment storage in non-alluvial channels

C.S. James

School of Civil & Environmental Engineering, University of the Witwatersrand, Johannesburg, South Africa

ABSTRACT

The progression of sediment along a river may be disrupted by temporary storage associated with non-alluvial features such as bedrock outcrops, large boulders, woody debris, coarse material substrates and vegetation. Such storages influence the amounts and timing of sediment progression as well as the evolution of sedimentary features in the channel by determining deposition patterns. Conventional sediment routing procedures cannot easily describe the dynamics of such non-alluvial storages, especially under the supply limited conditions that commonly hold in bedrock-dominated channels (James et al. 2001).

An approach is proposed for simulating the dynamics of storage units and hence the progression of sediment through a series of units under varying flow conditions and specified sediment supply rates. The approach is based on the establishment of two relationships: the ultimate equilibrium storage state attained by an indefinitely persisting, steady flow condition, and the time rate of change of storage towards this condition. These relationships can be used to calculate the change in storage under specified flows and sediment supply rates, which can then be included in a simple sediment budget to determine the sediment quantity transferred downstream.

Ultimate equilibrium storage state and storage depletion functions have been determined through laboratory experiments for a variety of storage types, with sediment moving as bed load. Experiments by Harnett (1998) and Odiyo (2005) with pools showed that the ultimate equilibrium storage depends on discharge and sediment size, and that storage depletion under a steady discharge with no sediment supply can be described by an exponential decay function whose exponent depends on discharge and sediment size. Sediment supply gives rise to an "active" storage component (dependent on sediment supply rate and discharge) that is considerably in excess of the storage without sediment supply. This storage persists as long as sediment supply is maintained, but decays according to the no-supply depletion relationship if it is discontinued while discharge is maintained. Similar ultimate equilibrium storage and storage depletion relationships were found for lee bars behind submerged solid objects (Fowler & Wiggett 2015) and emergent porous objects simulating vegetation patches or woody debris accumulations (Brusse 2000).

The storage functions are used in combination with sediment budgeting to simulate sediment storage dynamics and sediment progression, demonstrating, inter alia, Church's (2001) notion of a reservoir cascade for sediment progression, and explaining Thompson's (2008) observations of the time-varying release of bed load from lee deposits.

REFERENCES

Brusse, W. 2000. Investigation of the interaction between sediment and vegetation in rivers, Investigation Project Report, School of Civil & Environmental Engineering, University of the Witwatersrand, South Africa.

Church, M. 2002. Fluvial sediment transfer in cold regions, in Landscapes of Transition, K Hewitt et al. (Eds), Kluwer Academic Publishers, pp 93–117.

Harnett, B.R. 1998. Damming the River Sabie, South Africa: The sedimentation problem. Centre for Water in the Environment, University of the Witwatersrand, South Africa.

James, C.S., Odiyo, J.O. & Lambert, M.F. 2001. Modelling sediment movement and storage in semi-arid, bedrock-controlled rivers, Abstracts, 7th International Conference on Fluvial Sedimentology, Lincoln, Nebraska, 142.

Fowler, N.T.S. & Wiggett, A.R. 2015. Sediment storage dynamics in a lee bar, Investigation Project Report, School of Civil & Environmental Engineering, University of the Witwatersrand, South Africa.

Odiyo, J.O. 2005. Sediment Routing in Bedrock-Controlled Channels, PhD thesis, University of the Witwatersrand, South Africa.

Thompson, D.M. 2008. The influence of lee sediment behind large bed elements on bedload transport rates in supply-limited channels, Geomorphology, 99: 420–432.

Bed variation during floods in the Chikugo River estuary with complex structures of bed layers

Y. Kaneko
Graduate Student of Science and Engineering, Chuo University, Tokyo, Japan

S. Fukuoka
Research and Development Initiative, Chuo University, Tokyo, Japan

ABSTRACT

The Ariake Sea which has the largest tideland in Japan has environmental issues. It is said that one of reasons for issues is the lack of sand supply from the Chikugo River which has the largest catchment area of several rivers flowing into the Ariake Sea. The bed of the Chikugo River estuary is covered with the cohesive sediment such as silty clay called 'gata-soil' which is transported from the Ariake Sea due to large tidal level changes. Therefore it is said that sands may not exist mostly in the Chikugo River estuary. However it is proved that the vertical structure of the river bed is alternate layers consisting of gata-soil and sand, and sands exist sufficiently in the lower layers of the river bed by the core sample survey and supersonic echo sounder carried out in detail around the Chikugo River estuary as shown in Figure 1. In this study, we develop a new method variation analysis of river bed with complex structures of bed layers, and we clarify the mechanism of bed variation and the amount of sand supply from the Chikugo River estuaries to the Ariake Sea during a flood.

Observations of temporal water surface profiles and river bed prifile before and after 2009 flood were carried out in detail from the river mouth to 23 km including the tidal reach of the Ariake Sea. We developed an unsteady quasi-3D flow and 2D bed variation models for the Chikugo River estuary. Structure of bed layers are set in vertical direction at the pitch of 10 cm from river bed and the particle size distribution in each layer is given by the result of the core sampling of river bed materials. In the numerical analysis, when the rate of volume of gata-soil is larger than the porosity of sand, bed variation are calculated by continuity equation for sediment taking into account the sediment discharge of sand (Ashida et al. 1972) and erosion speed of the cohesive material (Nishimori et al. 2009) so as to correspond to the volume of gata-soil and sand because the gata-soil has characteristics of the transportation different from sand and existence of the gata-soil have great influence on the bed variation in such a state that gata-soil contains sand.

The numerical analysis results reproduce the temporal changes of the observed water surface profiles, discharge hydrographs at the observation points and the bed variations before and after the 2009 flood. It is concluded that the present computation method gives better understanding of bed variations in the river estuary which has the alternate bed layers consist of cohesive sediment and sand.

REFERENCES

Ashida, K. & Michiue, M. 1972. Study on hydraulic resistance and bed-load transport rate in alluvial streams, Proceedings, JSCE, Vol.206, pp.59–69.

Nishimori, K. & Sekine, M. 2009. Erosion rate fomura of cohesive sediment, Proceedings, JSCE, Vol.65 No.2, pp.127–140.

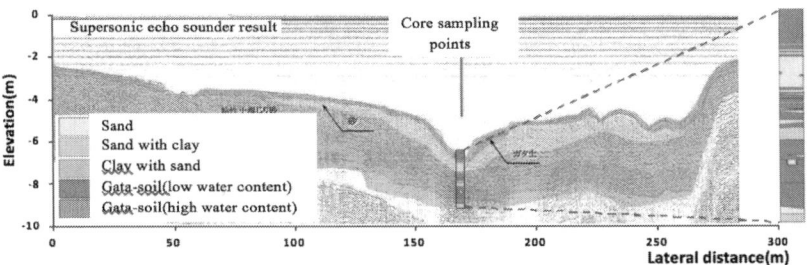

Figure 1. Structure of river bed layer and its materials.

Study for restoring bank protection functions of longitudinal dikes existing in the river with alternate bars

S. Kato
Graduate School of Science and Engineering, Chuo University, Tokyo, Japan

T. Gotoh & S. Fukuoka
Research and Development Initiative, Chuo University, Tokyo, Japan

ABSTRACT

The Kurobe river is a steep-slope river forming alternate bars on the alluvial fan. At the apex of the alluvial fan, Aimoto ground sill was constructed obliquely so as to flow toward the right bank and protect the left bank. The oblique structures control a longitudinal pattern of the alternate bars from downstream of the ground sill (Kinoshita 1957) and flow attacking points to banks. In the reach of alternate bars, longitudinal dikes have been installed at flow attacking points for bank protections since 1991 (see Fig. 1). The longitudinal dikes were designed based on the longitudinal pattern of the alternate bars in 1989. However, 1995's flood which happened in the largest scale in the past 40 years, caused bank erosions and river bed degradations at downstream of the ground sill. Therefore, meandering patterns and flow attacking points of the main channel have changed (see Fig. 1). Most of the longitudinal dikes have not become effective for the bank protection. In this study, we propose a river improvement technique for restoring the patterns of alternate bars and functions of the existing longitudinal dikes.

For restoring longitudinal arrangement of alternate bars, we attempted to control flow attacking points by installing at downstream of the ground sill two sandbars with boulders which had operated well in Jyoganji River as shown in Figure 2 (Koikeda et al. 2012). The space interval between the sandbars with boulders is determined so as to agree with wave lengths of the natural and stable alternate bars. The natural and stable patterns of alternate bars were assumed to be maintained by 1989's channel conditions by investigating the aerial photos and surveying data in the past about 70 years.

In order to investigate the stability of 1989's channels, we carried out two-dimensional flood flow and bed variation analysis method in stony-bed rivers. The numerical analysis method was validated by applying to 1995's flood which was a trigger of the river bed degradation in the main channel. The calculation results showed that installing the two sandbars with boulders were effective for restoring stable patterns of alternate bars. And, the sandbars with boulders

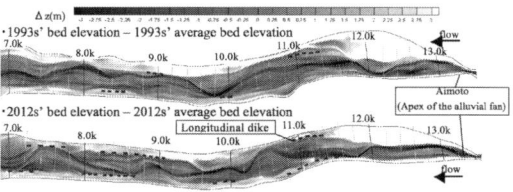

Figure 1. The difference of the bed elevation and average bed elevation.

Figure 2. Sandbars with boulders (Jyoganji River, 2012).

mitigated river bed degradations at downstream of the ground sill. As a result, the numerical calculation showed that the proposed river improvement technique was able to restore the stable pattern of alternate bars which made use of the existing longitudinal dikes against bank erosion.

REFERENCES

Kinoshita, R. 1957. Formation of dunes on river bed, Transactions of Japan Society of Civil Engineers No.42, pp 1–21

Koikeda, S., Ishii, A., Iwai, H., Ishikawa, T. & Fukuoka, S. 2012. Nature friendly bank protection works using sandbars with boulders in the Jyoganji river, Advances in River Engineering, JSCE 18:233–238.

Ikeda, S., Parker, G. & Sawai, K. 1981. Bend theory of river meanders. Part 1. Linear development, Journal of Fluid Mechanics Volume 112, pp 363–377

Long-term numerical investigations of the effects of training structures in a river reach with ongoing river bed deepening

A. Kikillus, L. Seitz, S. Haun & S. Wieprecht
Institute for Modelling Hydraulic and Environmental Systems, University of Stuttgart, Stuttgart, Germany

ABSTRACT

The river Iller can be categorized as significantly modified due to the construction of multiple hydraulic structures. As a consequence of ongoing river deepening processes and a possible drop of the groundwater table ramps and weirs were implemented during the last decades. However, from an evaluation of existing survey data, river sections with still ongoing erosion trends were identified. One of these river reaches is located between km 14.6 and 9.2. Within this section river training structures have been implemented (km 14.6–13.6) over the last couple of years. In addition, further implementations (km 13.6–9.3) will be performed during the next years to stop the progressive erosion tendency.

The aim of this study is to numerically evaluate the effects of the constructed measures to identify long-term trends, as well as to investigate effects of the further planned measures. The numerical sediment transport model BASEMENT developed at the ETH in Zurich is used (Vetsch et al. 2015). Due to the large extent of the investigation area and the long term simulation (50 years), the model requires optimization in order to achieve reasonable computation times. A numerically coupling of the 50 km up-stream reach (1D) and the 5 km long downstream reach, including the training structures (2D), allows reduced simulation times, while maintaining accurate and reliable results in the area of the planned measures. In a second step, the existing mesh, used for hydraulic modelling was coarsened and optimized to additionally reduce the computational time without a loss in accuracy. In the third step, a hydro-graph of the forecast periodic implemented which includes several flood events (e.g. an extreme flood event). The hydrograph was shortened in a way that only bed load relevant discharge rates are included.

As basis for further comparisons and for an analysis of future effects, a long term simulation was performed with measured terrain and bathymetry data before the river training measures were implemented (initial conditions). To evaluate the effectiveness of the already implemented measures, the results of a simulation with the bathymetry data after the construction work are compared with the simulation starting from the initial conditions. Based on these simulations, it is possible to quantify the advantage of the conducted measures. In addition, investigations of the morphological bed changes are per-formed where the newly planned revetments are implemented in the model. Revetments were chosen as mitigation measures to stop ongoing river bed deepening and due to the fact that sediment transport can be assured over the hydraulic structures as well as the fish passability is given. All implemented revetments include a river bed widening in this area.

The results of the investigations show that the already implemented river training structures reduce some of the river bed deepening effects, but equilibrium conditions are not reached so far. However, due to the implementation of new revetments the ongoing erosion processes are likely to be stopped and a possible drop of the groundwater level can be avoided in this reach of the river "Iller"

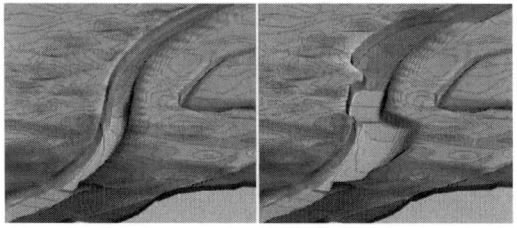

Figure 1. Computational mesh for the status quo (left) and planned measures between km 13.6 and 12.4.

REFERENCE

Vetsch, D., Siviglia, A., Ehrbar, D., Facchini, M., Gerber, M., Kammerer, S., Peter, S., Vonwiller, L., Volz, C., Farshi, D., Mueller, R., Rousselot, P., Veprek, R. & Faeh, R. 2015. Sys-tem Manuals of BASEMENT, Version 2.5, Laboratory of Hydraulics, Glaciology and Hydrology (VAW), ETH Zurich.

Computations on bedform by DEM-URANS coupling with two-way approach

I. Kimura, K. Horiuchi & Y. Shimizu
Graduate School of Engineering, Hokkaido University, Japan

ABSTRACT

In this study, we simulated migrating bedform in open channel flows using advanced numerical techniques. The sediments are considered as group of spherical particles and their motions are simulated by Distinct Element Method (DEM). The flow is calculated by Unsteady Reynolds Averaged Navier-Stokes model (URANS), for which a second order non-liner k−ε model is employed as turbulence closure. A two-way coupling between particles and flow is applied by incorporating a particle-fluid interaction terms of drag force, with and without shield-ing effect. We conducted uumerical simulations under existing experimental conditions, and evaluated the advantage of the inclusion of two-way coupling on replicating the bedform.

The hydraulic conditions in the present study are listed in Table 1. We tested three different computational models listed in Table 2.

The computational results in Run 1 and Run 3 are shown in Figure 1 and Figure 2, respectively. Main findings of our research are summarized as the follows.

(1) Reproducibility of the wave migration velocity, the wave length and the phase of water surface were improved significantly by including the two-way

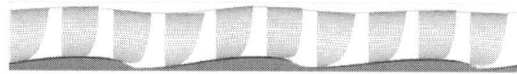

Figure 1. Computational results of Run 1 (t = 120sec).

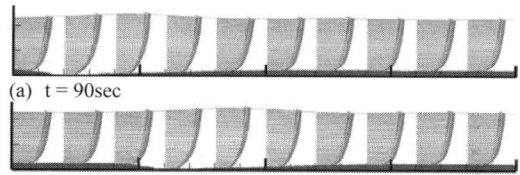

(a) t = 90sec

(b) t = 120sec

Figure 2. Computational results of Run 3.

model with the shield effect in the bed deformation model.

(2) Two-way model without the shield effect could not simulate the formation of dune. The reason seems to be that the particles could not obtain enough momentum from the flow to generate and drive dunes.

(3) Difference of flow velocities in bedload area between each model is small. However, the value of interaction term with the shield effect becomes 1.5 to 2 times larger than the value without the shield effect.

Table 1. Hydraulic conditions.

Slope	Unit flow rate (m^2/s)	Particle diameter (mm)	Froude number
0.0031	0.259	0.89	0.46

Table 2. Tested computational models.

	RUN-1	RUN-2	RUN-3
Shield effect	Without shield effect	Without shield effect	With shield effect
Interaction term	–	include	include
Model	One-way	Two-way	Two-way

REFERENCES

Duff, P.M.D. & Smith, A.J. 1992. Geology of England and Wales. The Geological Society, London.

Guy, H.P., Simons, D.B. & Richardson, E.V. 1966. Summary of alluvial channel data from flume experiments, 1956–61. Proceeding of United States government printing office, Washington.

Kasahara, H., Shimizu, Y., Kimura, I. & Yamaguchi, S. 2012. The bed-deformation calculation by Distinct Element Method. Proceedings of journal of Japan society of civil engineers, Ser. B1, Vol.68, No.4, I_I201-I_I206 (in Japanese).

Kimura, I. & Hosoda, T. 2003. A non-linear k-ε model with reliability for prediction of flows around bluff bodies. International journal for numerical methods in fluids, (DOI:10.1002/fld.540), pp.813–837.

Numerical assessment of the interactions between hydrodynamics, bed morphodynamics and bank erosion

E.J. Langendoen & M.E. Ursic
U.S. Department of Agriculture, Agricultural Research Service, Oxford, MS, USA

A. Mendoza
Department of Basic Sciences and Engineering, Metropolitan Autonomous University, Lerma, Mexico

J.D. Abad
Department of Civil and Environmental Engineering, University of Pittsburgh, Pittsburgh, PA, USA

R. Ata, K. El Kadi Abderrezzak & P. Tassi
EDF R&D, National Laboratory for Hydraulics and Environment & St. Venant Laboratory for Hydraulics, Chatou, France

ABSTRACT

For the meandering Beatton River, Canada, Nanson and Hickin (1983) demonstrated that short-term meander migration rates are not representative of the long-term averages, as two almost identical bends had very different short-term migration rates but similar long-term migration rates. They postulated that this was caused by the asynchronous interactions between erosion of the cut bank along the outside of a bend (or bank pull) and accretion on the point bar along the inside of a bend (or bar push).

The findings of Nanson & Hickin (1983) have important implications for modeling the morphodynamics of meandering streams. For example, most widely-used models of river meandering assume a temporally and spatially constant channel width (Ikeda et al. 1981 & Seminara et al. 2001), and therefore cannot be used to simulate the short-term planform dynamics. In addition, the role of bank pull on scroll-bar formation indicates the importance of incorporating cut-bank erosion processes in numerical models of river meandering to improve the simulation of long term (geologic scale) planform dynamics (Van de Lageweg et al. 2014).

Numerical assessment of river planform morphodynamics requires at least a two-dimensional (2D) depth-averaged model that can adequately simulate the governing processes and their interactions. At present, the simulation of hydrodynamics and river bed morphodynamics using 2D models is relatively straightforward. However, unlike one-dimensional computer models, the implementation of bank erosion processes in multi-dimensional computer models is rather complicated. One-dimensional computer models simulate river morphodynamics using a series of cross sections, and adjust the cross-sectional profile where erosion and deposition occur. These models can handle complex geometry including steep bank sections. Such sections cannot be represented adequately by 2D models, which divide the computational domain into a mesh of elements on which the topography is described. As cut-bank profiles are very steep due to basal erosion, near-bank mesh elements may become too small to perform efficient and numerically stable simulations. Furthermore, the subgrid scale bank topography cannot be represented as modeled bank profiles generally comprise a single, linear segment (or planar surface).

A novel approach to simulate bank erosion in 2D models is presented that combines the TELEMAC2D/SISYPHE computer models of river bed morphodynamics of the TELEMAC-MASCARET Suite of Solvers (EDF-R&D 2015) and the CONCEPTS riverbank erosion algorithms (Langendoen & Simon 2008). The subgrid scale riverbank geometry is represented using a depth-dependent porosity formulation. The new model is used to examine the interactions between hydrodynamics, point bar accretion (bar push) and bank erosion (bank pull) of a bendway on the Goodwin Creek, Mississippi, USA between 1996 and 2007. For this period, a seasonal-resolution time series of detailed bend topography is available.

REFERENCES

EDF-R&D (2015, September). open TELEMAC-MASCARET (http://opentelemac.org).
Ikeda, S., Parker, G. & Sawai, K. 1981. Bend theory of river meanders. Part 1. Linear development. Journal of Fluid Mechanics, 112, 363–377.
Langendoen, E.J. & Simon, A. 2008. Modeling the evolution of incised streams. II: Streambank erosion. J. Hydr. Eng., 134(7): 905–915.
Nanson, G.C. & Hickin, E.J. 1983. Channel migration and incision on the Beatton River. J. Hydr. Eng., 109(3): 327–337.
Seminara, G., Zolezzi, G., Tubino, M. & Zardi, D. 2001. Downstream and upstream influence in river meandering. Part 2. Planimetric development. Journal of Fluid Mechanics, 438, 213–230.
Van de Lageweg, W.I., Van Dijk, W.M., Baar, A.W., Rutten, J. & Kleinhans, M.G. 2014. Bank pull or bar push: What drives scroll-bar formation in meandering rivers? Geology, 42(4): 319–322.

A novel engineering desilting measure – "auto-desilting gallery"

S. Li, Q. Yi, W. Cheng & Q. Liu
School of Water Resources and Hydroelectric Engineering, Xi'an University of Technology, Xi'an, Shaanxi, China

ABSTRACT

Sediment problem is always the important and difficult problem in water conservancy. It has a great practical significance to look for better measures to alleviate the influences on water affairs of river, and to make water conservancy facilities achieving sustainably used. As an engineering method of draining sands, auto-desilting galley is a new-fashioned and efficient desilting facilities with unique structure. This galley has been successfully used in irrigation project channel. The significance of its application is obvious. The objectives of this thesis are three. The first is to confirm the auto-desilting galley on its operating conditions, the second is to search factors influencing desilting and the third is to make the working mechanism of the galley clear. Experiments are taken as a major method for knowing the galley hydraulic characteristics and desilting performance. The main results of this thesis are as follows:

(1) Hydraulic characteristics of the galley. Layout of the galley will increase the flow rate of bottom by about 33%, which is helpful for the transport of sediment at the bottom. The galley inside can produce negative pressure, approximately 1.3 kpa which provides the power to transport sediment out, the mathematical simulation also shows that the flow pattern of the galley is spiral flow.

(2) Desilting performance of the galley. The maximum sediment concentration at the outlet of the galley is 130 kg/m3, which is bigger than other desilting facilities. When the galley is working, the critical gradient decreases with the increase of galley length, but the outlet flow is increased. 1 square water convey 110.1 kg sediment, far higher than the natural river sediment transport. Sediment at the outlet mostly are coarse sand, particle size is 2.3 times than samples.

(3) The length of the galley itself has a certain influence on desilting, and the longer the galley is, the more the flow velocity at the bottom increases, the later outlet concentration peak time appears and the longer the time is. But the length of galley has no effect on sediment particle size at the outlet.

Critical discharge of erosion-deposition process of mid-channel bar head in anabranching channel

Zhi-wei Li
School of Hydraulic Engineering, Changsha University of Science & Technology, China

Guo-an Yu
Institute of Geographic Sciences and Natural Resources Research, Chinese Academy of Sciences, China

Chen-di Zhang
State Key Laboratory of Hydroscience and Engineering, Tsinghua University, China

ABSTRACT

The head zone of mid-channel bar is a dynamic region where water and sediment are divided into two branching channels in anabranching channel. Erosion-deposition process in the bar head relies on upstream water-sediment supply and local topographic condition, i.e. generally experiencing alternate deposition and erosion in non-flood and flood seasons respectively. The geometric shape of mid-channel bar head is simplified by a symmetric triangle with a longitudinal upslope and two lateral downslopes. A conceptual model of bedload transport rate on the bar head is derived based on the concept of upslope facilitating deposition and lateral downslope dividing sediment, meanwhile deriving a critical discharge formula of the erosion-deposition transition point. Along with increasing upstream discharge, the deposition rate in bar head increases up to the maximum and then decreases to reach the critical discharge, and finally occurs to erode. The deposition rate in the bar head is proportionally related to the longitudinal channel gradient and bed sediment diameter, inversely proportional to lateral channel gradient, upslope, and downslope. If other conditions keep constant, the plane shape of bar head is more asymmetric with less sediment load supply, more conducive to the erosion. The critical discharges of erosion-deposition process of bar head zone in the Tianxingzhou of Wuhan reach and the Tiebanzhou of Datong reach in the middle and lower Yangtze River are predicted by using this model since the impoundment of the Three Gorge Reservoir. The predicted critical discharges are roughly in agreement with the observation of the hydrological stations. Since the impoundment of the Three Gorge Reservoir was operated in 2003, the bar head zone of Tianxingzhou and Tiebanzhou has experienced an ongoing erosion process. The erosion volume of the bar head zone (1–7 m contour) of the Tiebanzhou from 2008 to 2011 is 62000 m^3, erosion depth of 1.2 m, erosion length of 910 m. However, the Tianxingzhou bar head appears to somehow deposit when the upstream discharge surpasses the critical discharge according to the hydrologic station. To this particularity, this study provides a new explanation that when surpassing the critical discharge, the regulation project (i.e. revetment and protection) in the Tianxingzhou bar head plays a crucial role on resisting the flow erosion and supporting the sedimentation in higher discharge condition, in particular, the bar head submerged in flood season.

Study on the flow around the Baguazhou Island in the lower reach of the Yangtze River

D. Liang & X. Wang
Collaborative Innovation Center for Advanced Ship and Deep-Sea Exploration, Shanghai Jiao Tong University, Shanghai, China
Department of Engineering, University of Cambridge, Cambridge, UK

P. Yu & H. Tang
State Key Laboratory of Hydrology-Water Resources and Hydraulic Engineering, Hohai University, Nanjing, China

ABSTRACT

The downstream reach of the Yangtze River takes on the characteristics of braided fluvial rivers, and there are many islands inside the river channel. The Baguazhou Island is a large island, which divides the Yangtze River into two branches in the north of the Nanjing city. The ratio of the flow rates through the two branches has been slowly changing in recent years. The flow rate in the left branch is now only around 13% of the total flow rate of the Yangtze River, which causes the gradual degradation and may eventually lead to the disappearance of the left branch. However, the area along the left branch has been experiencing rapid economic development, which has significantly benefited from the presence of the left branch of the river channel through water course navigation and almost unlimited water resources. The decline of the left branch is a major threat to the continuous prosperity in this area. In order to modify the flow conditions and increase the flow rate through the left branch, many hydraulic structures have been constructed in front of the Bagua Island to guide more water into the left branch of the channel, which achieved limited success. In this paper, laboratory experiments and numerical simulations have been carried out to investigate the most effective design of a diversion dike in front of the Bagua Island to regulate the flow. The physical experiments were conducted in the hydraulics laboratory in Hohai University, and numerical simulations were undertaken based on an extensively-verified shallow water solver developed at the University of Cambridge. We first compared the numerical results and the experimental and field measurements under the present scenario, and good agreement was reached. Then, a systematic parametric study was conducted, with the two geometric parameters of the diversion dike being its angle and length. Considering the financial and technical constraints on the building of such a dike, recommendations were provided at the end of the study regarding the shape of the diversion dike and its effectiveness.

A shallow-water model was set up for the flow around the Bagua Island located in the Nanjing reach of the Yangtze River. A high-quality curvilinear mesh was generated to cover the domain, which started at Sihao Wharf and ended at Gangchi. It was first verified by comparing the simulated water depths, velocities and diversion ratios with the field measurements and the physical model experiments for the three hydrological seasons. A diversion dike can modify the flow rates to the left branch of the river. The left branch's diversion ratio varies greatly when the length and the angle of the diversion dike changes. An exception is when the angle of the diversion dike is about 55°, where the diversion ratio barely changes with the length of the dike. Considering the financial and technical constraints on the building of such a dike, a 550 m long and 5° angled diversion dike may be a reasonable option for increasing the flow rate to the left branch. Only the diversion dike option is considered as an engineering measure to modify the flow in the Bagua river reach in this study. To increase the flow through the left branch to ensure the continuing development of area, a number of methods, such as widening and dredging the left branch, can also be considered. Further study is still being carried out to examine the situation more thoroughly.

REFERENCES

Liang, D. Lin, B. & Falconer, R.A. 2007. A boundary-fittednumerical model for flood routing with shock-capturing capability. Journal of hydrology, 332(3): 477–486.

Xiao, Y. & Tang, L. 2012. The hydraulic model test of the river regulation works on the Yangtze River in Nanjing City reach. Technical Report, Hohai University, Nanjing.

Experiment study on sediment control function of river narrow-section

C.H. Lin, C.L. Shieh, C.J. Liu & S.H. Lin
Department of Hydraulic and Ocean Engineering, National Cheng Kung University, Tainan, Taiwan (R.O.C.)

Y.J. Tsai
Disaster Prevention Research Center, National Cheng Kung University, Tainan, Taiwan (R.O.C.)

ABSTRACT

This study conducts a flume experiment to explore the hydraulic and sediment control function in the narrow-section of a river. The laboratory experiment is expected to show how a river's narrow section retains and drainages sand when there is a large scale flood event. A narrow-section scale model was set up for the flume experiments, with parameters including contraction ratio, contraction length, and sediment concentration. The data analysis in sediment deposition, downstream sediment discharge, sediment decrement rate, and sediment trapping efficiency allows to understand the role of a river's narrow section in regulating the amount of sediment. The results show that (1) the contraction ratio is more effective in sediment control than any other parameter; (2) under large supply of sediment cases, the sediment trapping efficiency increases with the contraction ratio; (3) dimensionless sediment concentration decreases with the contraction ratio. Moreover, under different sediment supply cases, high-intensity and short-duration sediment supply improves the sediment trapping efficiency. The results confirm that the natural narrow section of a river has the capacity to trap sediment and detain sediment runoff when a large scale flood event occurs.

Experimental study on velocity pattern and bed morphology around a model patch of vegetation

C. Liu, D. Wang, K. Yang & X.N. Liu
State Key Laboratory of Hydraulics and Mountain River Engineering, Sichuan University, Chengdu, China

ABSTRACT

The aquatic vegetation often grows in natural rivers and plays an important role in the river corridor system. The vegetation is often growing in areas where velocity and turbulent kinetic energy TKE are very low. Within those areas, the vegetation is extending in both lateral and longitudinal directions and traps sediments often combined with organic nutrition. Finally, the vegetation occupies large areas and presents a continuous pattern that locally changes the bed morphology. In wetlands and tidal rivers the vegetation often exhibits a patch with finite width and length, which contributes to a totally different morphology around and within the patch. This is worth to be further discussed because different flow patterns may result in a different distribution of sediment deposition. Besides, in recent years the studies regarding the vegetation patch are rare. Follett and Nepf (2012) have extensively investigated the pattern of sediment deposition inside the patch and also in the patch wake. Follett and Nepf also reported that the depressed sedimentation within a model patch was attributed to the presence of turbulent kinetic energy (TKE). Chen et al. (2012) proposed a method for estimating the wake length behind sparse and dense patches. Ortiz et al. linked the wake sediment deposition to both velocity and TKE, and found that TKE is the dominant factor that contributes to the depressed deposition in the wake. Within the model patch, Liu and Nepf (2016) experimentally determined the threshold of stem turbulence that is expressed using the stem Reynolds number, i.e. $Re_d = 120$, and presented two criteria for determining enhanced deposition inside a vegetation patch in laboratory experiments as well as in a field study. These investigations broaden our understanding for the interaction of vegetation patches and sediment deposition. Meanwhile, we know the flow characteristics upstream and downstream are linked to the patch configuration. In these studies, only one feed condition of upstream sediment was considered, i.e. a recycle sediment feeding. This may not be applicable for some natural rivers where the upstream sediment input is deficient. In a natural scenario, both

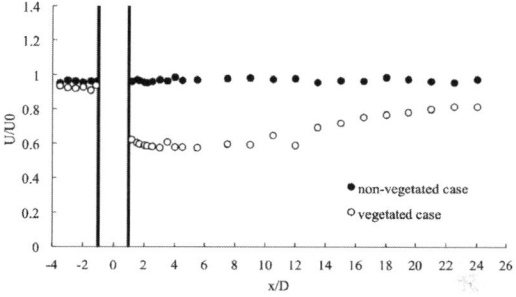

Figure 1. Longitudinal profile of normalized mean velocity at channel centerline in both non-vegetated and vegetated cases.

the enhanced deposition and the depressed deposition behind a patch can be observed. Thus, we are interested in the scenario of reduced or even no sediment input from upstream, which may result in a totally different distribution of sedimentation. Based on the recorded velocity data, the longitudinal velocity profiles in both non-vegetated and vegetated cases are depicted in Figure 1. The model patch ($D = 6$ cm) blocked 20% of channel width (30 cm), resulted in the depressed velocity ($\approx 40\%$ less) directly behind the model patch. In this study, we designed a flume experiment (16 m-long, 30 cm-wide and 40 cm-high) to run detailed investigations on different distributions of sediment depositions in the wake of a vegetation patch. The velocity pattern is also discussed and linked to the change of the channel bed morphology.

REFERENCES

Chen, Z., Ortiz, A., Zong, L. and Nepf, H., 2012. The wake structure behind a porous obstruction and its implications for deposition near a finite patch of emergent vegetation, Water Resour Res, 48(9).

Follett, E. M. and Nepf, H.M., 2012. Sediment patterns near a model patch of reedy emergent vegetation, Geomorphology, 179, 141–151.

Liu, C. and Nepf, H., 2016. Sediment deposition within and around a finite patch of model vegetation over a range of channel velocity, Water Resour Res, 52(1), 600–612.

Effect of the Three Gorges Dam and other upstream factors on the hydrological conditions of Yichang reach, Yangtze River

Huaixiang Liu & Yongjun Lu
State Key Laboratory of Hydrology-Water Resources and Hydraulic Engineering,
Nanjing Hydraulic Research Institute, Nanjing, China

ABSTRACT

As one of the largest hydraulic projects in the world, the Three Gorges project started its operation in 2003. Due to its significant influence, the hydro-logical and bathymetry data near this project was recorded by the government for years.

Yichang-Yangjianao reach locates not far from the Three Gorges dam's downstream. And it is generally considered as the most important representative entrance reach of the Yangtze river's middle stream.

In this paper, the hydrological time series (from the time before the Three Gorges project operation) at the downstream of the Three Gorges project and the bathymetry data of Yichang-Yangjianao reach was collected to study its fluvial morphology process, especially about its tendency after the reservoir's impoundment.

The analysis showed that, the operation of the Three Gorges project has greatly changed the sediment transportation into its downstream river channels. Therefore instability (such as erosion & incision) of Yichang-Yangjianao reach and the deterioration of navigation condition can be expected. The channel's erosion and sedimentation amounts of different river sections in different years were calculated and compared. As a result, the position and movement of main erosion zones were identified and analyzed. Due to this severe erosion, Yichang-Yangjianao river channel was automatically readjusted to suit the new hydrological conditions. The riverbed armoring effect was reflected by the grain sizes of sand and boulder fractions in riverbed composition. Further calculation also proved that the roughness coefficient in some upper river sections had increased more than 30%. This riverbed readjustment had successfully slowed the rate of erosion process and some river sections are approaching a new equilibrium status.

REFERENCES

Harrison, A.S. 1950. Report on Special Investigation of Bed Sediment Segregation in A Degrading Bed. U niversity of California, sept.
Vericat, D.R. Batalla, J. and Garcia, C. 2006. Breakup and reestablishment of the armour layer in a large gravel-bed river below dams: The lower Ebro, Geomorphology, 76(1): 122–136.
Doncker, L. Troch, P. and Verhoeven, R. 2009. Determination of the Manning roughness coefficient influenced by vegetation in the river Aa and Biebrza river, Environmental fluid mechanics, 9(5): 549–567.
Qin, R.L. 1981. Discussion of riverbed scouring and armoring, Engineering Journal of Wuhan University, 3: 43–53(in Chinese).
Gessler, J. 1970. Self Stabilizing Tendencies of Alluvial Channels. J. Waterways and Harbors Division, ASCE, 96(2): 235–249.
Lu,Y.J. and Zhang, H.Q. 1993. Experiment study on armoring mechanism of bed material with wide size distribution by clear water, Journal of sediment research, 1:68–77(in Chinese)

Numerical modelling of the Danube river channel morphological development at the Slovak–Hungarian river section

M. Lukac & K. Holubová
Water Research Institute, Bratislava, Slovak Republic

ABSTRACT

The morphological development of the Danube in Slovak territory was very dynamic in the past. The river transported a huge amount of sediments which accumulated to form thick deposits downstream of Devín (river km 1880) as a result of a drop in the river's gradient. The dynamic development of the river system thus resulted in a massive alluvial fan. The flow conditions and river processes in the Danube were also changed substantially in the 20th century by the construction of flood protection dikes, river regulation works for medium and low discharges, and a cascade of hydropower stations in the upper stretches of the Danube. Further major changes can be attributed to the construction and operation of the Gabčíkovo hydropower station at the Slovak territory, which has accelerated the river processes downstream of Sap (river km 1810). The retention of sediments in the Hrušov reservoir has resulted in a marked sediment transport balance deficit, causing extensive riverbed erosion in the Danube (Holubova et al., 2004).

One-dimensional morphological model MIKE 11 was used to analyse the hydraulic conditions that influence the transport of sediments and to predict how the river channel is likely to change in the near future at the 100 km long Slovak-Hungarian river section downstream from Sap (Holubova et al., 2015). In line with the previous experience of VÚVH in measuring and modelling the transport of sediments, the Engelund & Hansen, Ackers & White and Sato, Kikkawa & Ashida formulae were selected for test calculations. The results of test calculations were compared with the results of direct sediment transport measurements, which were carried out (river km 1795.580) by the Water Research Institute in Bratislava in the years 2000–2002. The comparison has confirmed that the most suitable method for simulating the transport of sediments between Sap and Szob is the Sato–Kikkawa–Ashida formula.

Model simulations with different river channel topographies (1994, 2001, and 2013) provided a picture of the changing dynamics of sediment transport in the area of interest. The changed flow dynamics was documented by the simulated flow velocity and bed shear stress, key hydraulic variables that affect the transport of sediments, as well as with the maximum values of sediment transport simulated with model for different topographies. The most significant morphological changes were simulated in the river section between Sap and the mouth of the Mosoni Danube. The extensive groyne fields and the river's gradient in this section caused concentration of water flow, grain size sorting of the river bed material and intensive sediment transport, even at low discharges.

Using a model schematising the current state, we simulated the probable course of morphological development of the Danube river channel for a time horizon of ten years. The prevailing process is gradual riverbed erosion, the depth of which exceeded 1 m only in isolated areas. Similarly, sediment deposition, rarely exceeding 0.5 m, was predicted only seldom. In general, the prognosis assumes that the intensity of morphological changes in the river channel will moderate in the near future, which corresponds to the previously predicted trends in connection with the river's decreasing gradient.

REFERENCES

Holubova, K., Comaj, M., Lukac, M., Mravcova, K., Capekova, Z. & Antalova, M. 2015. Danube floodplain rehabilitation to improve flood protection and enhance the ecological values of the river in section between Sap and Szob. *Final report of the Slovak partners. Project Reg. Nr. HUSK/1001/2.1.2/0060*. Bratislava: Water Research Institute.

Holubova, K., Capekova, Z. and Szolgay, J., 2004. Impact of hydropower schemes at bedload regime and channel morphology of the Danube River. In: Greco, M., Carravetta, A. and Della Morte, R. (edt.): River Flow 2004. Federico II University of Napoli, Italy. Balkema Publisher. Volume I, pp. 135–142.

Bed-slope-related diffusion of an erodible hump

S. Maldonado
Department of Geophysics, School of Earth, Energy & Environmental Sciences Stanford University, Stanford, CA, USA

M.J. Creed & A.G.L. Borthwick
Institute for Energy Systems, School of Engineering, The University of Edinburgh, Edinburgh, UK

ABSTRACT

Much of the investigation into sediment transport derives originally from study of the fluvial environment. Natural rivers typically present mild slopes whose gradient is of the order of 1:1000 (i.e. $\approx 0.05°$), and so the impact of bed slope on sediment transport is commonly neglected. Yet, the local bed slope has been shown to influence sediment transport phenomena by promoting (inhibiting) the motion of sediment particles in the down(up)-slope direction. Such action translates into a diffusive or smoothing effect, which, in a morphodynamic numerical model, prevents the formation and development of unrealistic oscillations of the bed elevation which would otherwise render the model unstable.

Furthermore, coastal and mountain streams provide examples of environments where neglecting the bed-slope-influence may not be justifiable. Slopes of mountain rivers may be sufficiently large to affect the overall behaviour of the flow, including sediment transport rates (Bayazit 1983).

In the present work, a physics-based model, previously validated against empirically-derived formulae, is used to propose an analytical expression aimed at modifying bed-load formulae (originally derived for nearly horizontal channels) in order to render them applicable to sloping beds. The model is based on the 2-layer Shallow Water Equations (SWE), considering erodible bed and some particular assumptions, which will be discussed in the paper (Maldonado 2015). The slope-related expression is then employed in a Conventional Morphodynamic Model (CMM) (consisting of a coupling between the SWE, the Exner bed-update equation and a sediment-transport formula) to study the evolution of an erodible hump subject to a subcritical steady flow.

Figure 1 illustrates the evolution of a 1D hump predicted by a standard CMM, and a CMM modified through slope-related diffusion i) proposed herein and ii) deduced from theory of Bailard & Inman (1981). Second-order central finite differences and a fourth-order Runge-Kutta method have respectively been

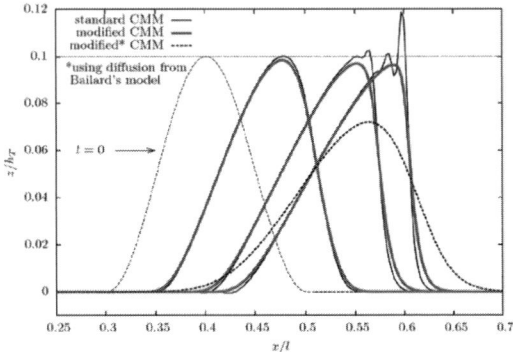

Figure 1. Hump evolution predicted by standard (unmodified) and modified (through two different approaches) CMM.

used for spatial discretisation and time integration of the bed-update equation. A Harten-Lax-Van-Leer Contact (HLLC) Riemann solver is employed to solve the hydrodynamic equations. The standard (unmodified) CMM exhibits the well-documented development of high-frequency oscillations at the crest and base of the hump, which eventually render the model unstable. Such spurious instabilities are not present when Bailard & Inman's (1981) slope-related diffusion is introduced; however, such diffusion appears excessive as it avoids the expected behaviour of the migrating hump (i.e. steepening of its lee) by attenuating its profile. The slope-related diffusion herein proposed prevents unrealistic oscillations from developing while permitting the adequate (from a qualitative and phenomenological viewpoint) migration of the hump, whose lee gradually steepens as it migrates downstream. Therefore, the proposed bed-slope-related diffusive expression can be used to enhance CMMs by improving their numerical stability while retaining close phenomenological representation of the problem. The paper will also present a similar 2D case study.

Restoration of the Eggrank bend at the Thur River in Andelfingen ZH

M. Mende & M. Müller
IUB Engineering Ltd., Bern

P. Sieber & M. Oplatka
Building Department, Canton of Zurich

ABSTRACT

In the framework of the flood mitigation and river restoration project "Hochwasserschutz und Auenlandschaft Thurmündung" the so called Eggrank Bend of the Thur River in Switzerland was restored during winter 2014/15. Initially, this channelized bend was characterized by unnaturally high flow velocities and the lack of flow structuring elements.

A first ecological enhancement consists of a new additional upstream counter bend which slightly elongates the river reach. This new naturalized course leads to an increased dynamic in the riverbed (thalweg formation) and to decreased flow velocities on the erosion bank of the Eggrank Bend. In addition to the new upstream counter bend five micro groins of 30 to 50 m length and a snail shaped groin were implemented. These elements are already overtopped at low flow conditions and result in an additional reduction of flow velocities on the outer bank. Thus, they initiate increased flow variability over the entire cross section, leading to velocity variability along the river depth as well as to substrate sorting.

The third enhancement of the Eggrank Bend consists of an initial widening at the inner bank. Due to the micro groins, the hydraulic forces on the gravel bank on the inner bend are increasing which counteracts the sedimentation tendency of the inner slope in a long-term and sustainable way.

To gather a maximum of knowledge and conclusions about the functionality and efficiency of the restoration measures at the Eggrank Bend, intensive monitoring is carried out. So far, two monitoring campaigns were completed, one before realization and one after finishing the construction works. A third campaign will follow after the next important flood event.

On one hand, monitoring methods comprise a large scale two-dimensional measurement of flow velocities by Surface-Particle-Image-Velocimetry (PIV) from helicopter and drones, respectively (Detert & Weitbrecht 2015). On the other hand, measurements with Acoustic-Doppler-Current-Profilers (ADCP) are carried out to record flow velocities in ten cross sections along the bend. This performance control allows a

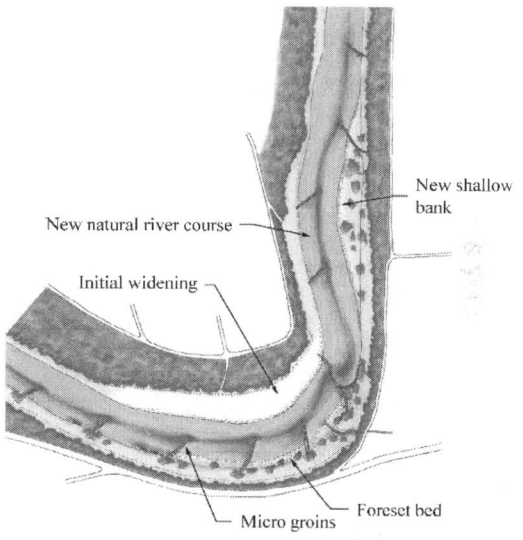

Figure 1. Drawing of restoration measures on the Eggrank Bend of the Thur River / Switzerland (Planergemeinschaft Bachmann, Stegemann + Partner, Staubli, Kurath & Partner AG 2013).

detailed analysis of the micro groin influence on flow characteristics and morphology and aims a better understanding of the hydraulic processes launched by the implemented river restoration measures.

REFERENCES

Detert, M. & Weitbrecht, V. 2015. A low-cost Airborne Velocimetry System: Proof of Concept, J. Hydr. Research, DOI: 10.1080/00221686.2015.1054322

Mende, M. 2015. Naturnaher Uferschutz mit Lenkbuhnen – Grundlagen, Analytik und Bemessung, LWI-Mitteilungen 162/2015, Technische Universität Braunschweig, ISSN 0343–1223

Planergemeinschaft Bachmann, Stegemann + Partner, Staubli, Kurath & Partner AG 2013. HWS und Auenlandschaft Thurmündung,. Technischer Bericht Bauprojekt 2. Etappe, Bericht Nr. 2490

Formation of river dunes by measurement, linear stability analysis and simulation with Bmor3D

P. Mewis
Institute of Hydraulic Engineering and Water Resources Management, TU-Darmstadt, Germany

ABSTRACT

River dunes develop in many sand bed rivers. They are successfully modeled with morphodynamic numerical models that describe a movable river bed (Mewis 2004). The appearance of river dunes may be analyzed using the linear stability analysis of the same mathematical equations (Kennedy 1963, Charru et al. 2013).

Unfortunately, there is no general analytic solution to the linear stability analysis. Instead the perturbation equations have to be derived for a certain model formulation and solved for each parameter set separately. Results are shown of an analysis that differs from that published by Richards 1980 in that a mixing length and a k–ω model are used.

Laboratory measurements at TU-Darmstadt are shown in figure 1. The initial dunes have a relatively short wavelength that is quickly growing by the nonlinear amalgamation process. So while the linear stability analysis may identify the parameter range where dunes develop, it will fail to predict the observed wavelength of the developed dunes in real rivers that are in dynamic equilibrium.

In figure 2 the results of the stability analysis for the phase lead of the shear stress is drawn for different elevations above the bed (y-axis) and different wave numbers of the perturbation (x-axis). It turns out that

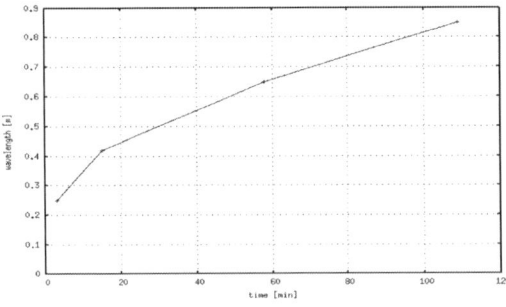

Figure 1. Dune length growth during a laboratory experiment at TU-Darmstadt.

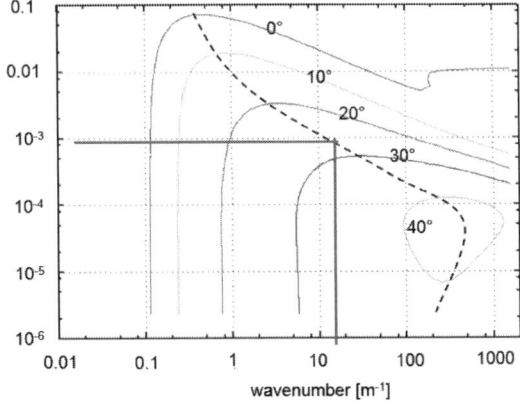

Figure 2. Phase lead of shear stress for different elevations above bed (in m) for the laboratory experiment.

the phase as well as the wavenumber with the maximum phase – that is indicated by the black broken line – is strongly changing with elevation above the bed. The predicted wavelength for a 1 mm particle – indicated by the red lines – is close to the initial wavelength observed in the laboratory.

As long as the same mathematical model formulation is used the same dunes should appear in the early linear stage. In this respect the modelling of dunes may serve as a verification test for numerical models. On the other hand the initial results of the three-dimensional model will be not better than the forecast of the linear stability analysis.

REFERENCES

Kennedy, J.F., The mechanics of dunes and antidunes in erodible-bed channels, *Journal of Fluid Mechanics (JFM)* 16, pp. 521–544, 1963.

Charru F., Andreotti B., Claudin P. 2013 "Sand ripples and dunes", Annu. Rev. Fluid Mech., 45:469–493.

Mewis, P., Are three-dimensional morphodynamic computations without dunes reasonable?" MARID-Workshop 2004, Twente.

Dynamic state of river-mouth bar in the Yuragawa River and its control under flood flow conditions

H. Miwa
National Institute of Technology, Maizuru College, Kyoto, Japan

K. Kanda
National Institute of Technology, Akashi College, Hyogo, Japan

T. Ochi
Kyoto University, Kyoto, Japan

H. Kawaguchi
NTT Infrastructure Network Corporation, Osaka, Japan

ABSTRACT

The Yuragawa River is located in the north of Kyoto Prefecture, the mid-west in Japan. Topographic changes of its river-mouth bar are continuously activated by sediment transport due to river flow and sea wave. On October 2004, a large part of the river-mouth bar was eroded by the huge flood flow due to typhoon. The river-mouth bar has developed along the right bank only since then. This situation may cause some problems such as bank erosion, washout of bank protection works and harmful effects on other coastal structures. Even effects of water discharge during flood periods on responses of the river-mouth bar are not clarified. Therefore, the risk of high water level caused by a river-mouth clogging is high. In order to avoid these problems and risk, it is important to understand the characteristics of the topographic change of the river-mouth bar and its cause, and to propose a control method of the bar geometry.

In this study, the temporal variations in geometrical properties (e.g., bar area and shape) of the river-mouth bar were analyzed on the basis of the hydrological and topographical data in the Yuragawa River estuary. The effects of the river water discharge and the sea wave height on the geometrical properties of the bar were also investigated. In order to investigate responses of area, height and volume of the bar against flood discharges, two-dimensional numerical analysis was conducted next. As for the bar control, the effectiveness of a spur dike and a trench, which can change the flow direction and increase the erosion area respectively, for erosion of the bar was evaluated by means of flume experiments. The two-dimensional numerical model was also applied to further investigate the effects of spur dike and trench on the bar control. The simulation results were verified against the experimental results. The results are summarized as follows:

(1) Although the bar area and volume of the river-mouth bar showed short-term fluctuations due to flooding, they also showed a tendency to increase on a long-term basis. The flush condition of the river-mouth bar sediments due to flooding does not depend on the width of river-mouth channel and the bar area but on flood discharge. The formation of the river-mouth bar may be mainly activated by an increase of longshore sediment transport rate in winter season; the bar area has a strong correlation with wave height.

(2) In the experiment without spur dikes, the scour depth of the river-mouth channel became large because of the flow converging into the channel. On the other hand, in the experiments with the spur dikes, the erosion of the river-mouth bar became active and the scour depth of the river-mouth channel became small because of the spur dikes redirecting the flow toward the river-mouth bar. Sediments deposited at the lower part of the spur dikes because the flow intensity was weak there.

(3) The river-mouth bar can be eroded easily by making a trench. The erosion rate of the bar depends not only on the water discharge and water level at the downstream end but also on the channel width. In particular, the flow over the bar accelerates erosion of the bar.

(4) The numerical simulation results showed that the mechanisms of flow around the spur dikes & trench and river bed variation processes were clarified through the reproduction calculations of the experiments.

(a) Nov. 2011   (b) Sep. 2014

Figure 1. River-mouth of Yuragawa River.

Computational modelling of secondary flow on unstructured grids

M. Nabi, W. Ottevanger & S. Giri
Deltares, Delft, The Netherlands

ABSTRACT

The spiral flow plays a crucial role in the flow structure, sediment transport and morphology. In previous studies, the spiral flow intensity was usually calculated on structured grids. However, the real world applications are not such smooth, and therefore unstructured form of grids are preferable in order to adjust the grid to complex geometries. In the frame of Delft3D Flexible Mesh unstructured hydrodynamic model is implemented. This model is flexible in nature and can be used for a large range of applications for extremely complex geometries. However, calculation of spiral flow intensity on unstructured grids is a challenging task because of irregularity of the grid.

Here, we used a new method based on least-square interpolation to reconstruct the velocity gradients which are necessary for calculation of the streamline curvatures. The calculated streamline curvatures are then used to calculate the spiral flow intensity. The spiral flow intensity is calculated by solving a transport-type equation. The spiral flow intensity is important in estimation of the angle of bed load transport, and hence for calculation of the steady bars. Moreover, it is necessary in calculation of dispersion stresses, which are important in calculation the bedforms under non-linear effects (e.g. sharp bends).

In order to validate the model, we calculated the spiral flow intensity on two different geometries (on smooth grids) and compared the centerline spiral flow intensity with the previously published experimental and theoretical results (Kalkwijk & Booij 1986), and a good agreement was found. Moreover, in order to show the validity of the model on unstructured grids, we repeated the second case on a triangular mesh, and showed that the model accurately reproduced the secondary flow on unstructured grid. Figure 1 shows the grid topography and the velocity contours on an unstructured triangular grid. Figure 2 shows the spiral flow intensity along the bend centerline, together with the previously published theoretical and experimental results.

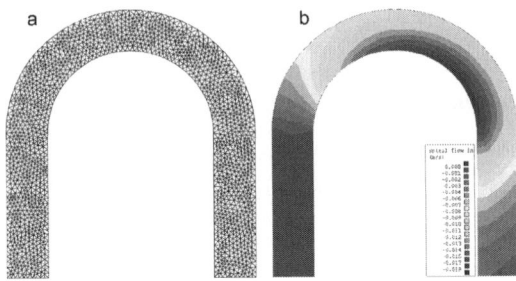

Figure 1. The grid structure for the case LFMII on non-uniform triangular grid (a) and the solution of spiral intensity on this grid (b).

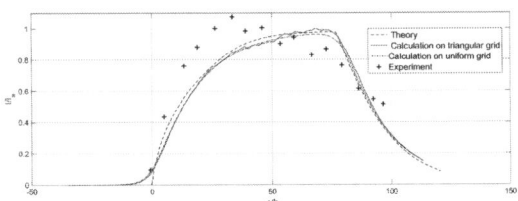

Figure 2. Normalized spiral flow intensity in the centreline for case LFMII on a non-uniform triangular grid, compared with the theoretical and experimental results for Chézy coefficients of $60 \, \text{m}^{1/2}\text{s}^{-1}$.

The model is then applied to simulate the sediment transport and morphological process by applying the transport angle (affected by spiral flow intensity) and bed-slope effect. It is found that the model showed similar morphological development compared to the previous structured version of Delft3D.

REFERENCE

Kalkwijk, J.P.T. & Booij, R. 1986. Adaptation of secondary flow in nearly-horizontal flow. Journal of Hydraulic Research 24(1): 19–37.

Morphological development of tidal tributaries in relation to turbidity and sediment concentration of the main estuary river

E. Nehlsen & P. Fröhle
Hamburg University of Technology, Hamburg, Germany

ABSTRACT

Large estuary systems like the Elbe Estuary usually consist of one main estuary and many tributaries. These tributaries play an important role since they act as transitional waters between marine and fluvial environment. One of the main functions of the tributaries is the drainage of the low-lying marshes in the river catchment. The drainage capability of the tributary depends amongst others on the hydraulically effective cross section, which is influenced by morphological processes. The morphological development mainly depends on the sediment supply from the catchment as well as from the main estuary. In particular, the latter is highly variable due to the non-stationary estuarine turbidity maximum (ETM) in the main estuary (Kappenberg & Fanger 2007). Hence the morphological development of the tributaries – which is for the most parts unknown – can be assumed to be highly variable as well.

In order to analyze the morphological development and its main drivers a field study has been developed and conducted. Within this study the topographies of a tidal tributary of the Elbe has been surveyed in a yearly order for a period of nine years. All measurements were performed with a shipborne multibeam echosounder. A small draft of the research vessel which is only 0.6 m enables the inclusion of the river banks into the survey area. The analysis of the survey data shows that the morphological activity varies along the tributary. The yearly rate of change in bottom elevation varies on average for the whole tributary from −0.15 m up to +0.15 m. Explicit overall trends could not be observed within the considered period since some years show a net deposition and other years a net erosion. However, a correlation could be identified between the morphological development of the tributary and the mean sediment concentration – which may be described as a function of the turbidity – of the main estuary near the junction of the rivers (Fig. 1). A net deposition in the tributary within a period of one year correlates to a high average sediment concentration in the Elbe within the corresponding year. A net erosion in the tributary correlates to a low average sediment concentration in the Elbe, respectively.

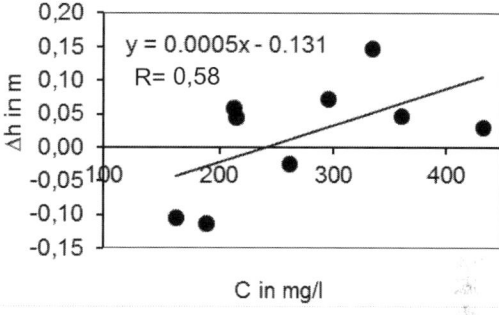

Figure 1. Correlation between yearly morphological development of the tributary Krückau (km 0 – km 7.6) and the corresponding yearly mean sediment concentration (turbidity) in the Elbe near the junction of the two rivers (Nehlsen 2016).

This finding indicates that the turbidity and the position of the turbidity maximum of the Elbe estuary is an important driver of the morphological development of tidal tributaries. However, the mean variation of points indicates that the sediment supply from the main estuary may not be the only relevant driver. Further research should identify other important drivers, including the sediment supply from the catchment, in order to develop elaborated models, that are abled predict the long-term morphological development of the tributaries.

REFERENCES

Kappenberg, J. & Fanger, H.U. 2007. Sedimenttransportgeschehen in der tidebeeinflussten Elbe, der Deutschen Bucht und in der Nordsee. GKSS report 2007/20, ISSN 0344-9629.

Nehlsen, E. 2016: Wasserbauliche Systemanalyse als Grundlage für die Bewertung der Auswirkungen des Klimawandels für tidebeeinflusste Nebengewässer am Beispiel von Este und Krückau. Dissertation. Hamburg: TuTech-Verlag. Hamburger Wasserbauschriften, 19. Submitted for publication.

Numerical modeling of antidune formation and propagation

N.R.B. Olsen
The Norwegian University of Science and Technology

ABSTRACT

The formation of antidunes is an interesting hydraulic phenomenon that has been studied by engineers, geographers and geologists. A number of flume studies with antidunes have been carried out over the years. The current paper present results from numerical modeling of downstream migrating antidunes, replicating two physical model studies (Kennedy, 1960; Núñez-González and Martín-Vide, 2011).

The numerical model solved the 2D Navier-Stokes equations on a non-orthogonal structured width-averaged grid. The convection-diffusion equation for

Figure 2. Longitudinal profile of the computational grid over an antidune in a non-distorted scale. The flow direction is from left to right.

the sediment concentration was solved, including a drift-flux model. Bed elevation changes were computed based on the Exner equation, and the water surface elevations were found from an explicit algorithm based on pressure and continuity. Both the water level and bed level changed over time. The grid was moved vertically accordingly and expanded/contracted as the water depth changed.

The formation and movement of the antidunes are shown in Fig. 1, where longitudinal profiles of the flume at different times. Starting from a flat surface, the dunes start to grow from the upstream end and move in the streamwise direction. The dune lengths become similar to what was observed in the physical models. The water depth above the dunes and the dune celerity were also reasonably well computed. Fig. 2 shows the velocity over an antidune. The velocities are highest close to the water surface and lowest after the dune crest.

Figure 1. Longitudinal profiles of antidune development from a flat bed. The top figure is at the start of the experiment, and there is 10 seconds between each profile. The figures are distorted with a 3x vertical scale. The flow direction is from left to right.

REFERENCES

Kennedy, J. F., 1960. *Stationary waves and antidunes in alluvial channels*, Ph.D. Thesis, California Institute of Technology, USA.

Núñez-González, F. and Martín-Vide, J. P., 2011. Analysis of antidune migration direction, *J. of Geophysical Res. – Earth Surface*, 116, F02004.

On the effect of different upstream schemes on the simulation of the antidunes propagation

E. Rademacher & A. Malcherek
Department of Hydromechanics and Hydraulic Engineering, Institute of Hydro Sciences, University of the German Armed Forces, Munich, Germany

ABSTRACT

The computational simulation of hydro- and morphodynamic processes and the development of adapted numerical methods is currently a wide field of research. Especially morphological models normally assume subcritical flow (Vasquez, Millar & Steffler, 2005). Phenomena like antidunes cause numerical instabilities because the bed celerity is opposite the direction of sediment transport.

(Volp, Prooijen, Pietrzak & Stelling, 2015) presents a new upwind scheme to improve simulation of bed propagation. The presented method provides upwind direction based on the bed celerity c_D (see eq. 1).

$$c_D = \frac{\Delta q_S}{\Delta z_B} = \frac{q_{S,2} - q_{S,1}}{z_{B,2} - z_{B,1}} \quad (1)$$

Δq_S differenz of the transport capacity
Δh_B bottom slope parameter

This improvement ensures numerical stability for both ratios shown in figure 1. In the upper graph, transport rate increases through the shallow area. Therefore transport capacity of the upstream cell has to be taken. Figure 1b shows decreasing transport capacity in the second cell, so upstream direction has to contrary.

The improvement of the outlined scheme has already been demonstrated for sub-critical flow using various test cases (e.g. Gaussian cone). This paper will include the method to the morphological model SediMorph (Malcherek, Piechotta & Knoch, 2005) and expand theory to super-critical flow.

The physical concept is validated by the numerical critical test case of an anti-dune. In result numerical instabilities are prevented and the simulation of propagation will be opposite the direction of flow.

a) bed celerity in the same direction as sediment transport

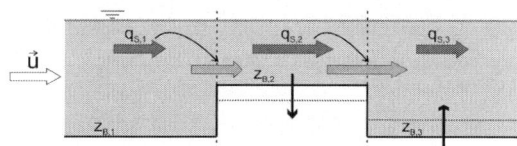

b) bed celerity is opposite sediment transport

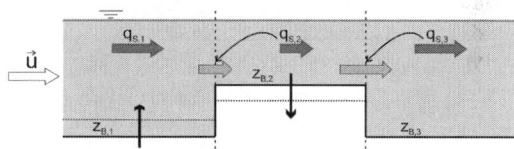

Figure 1. 2D flow through a channel with shallow area.

As a generally valid upwind scheme, this method represents a significant enhancement on convenional schemes and should be standard for morphological models.

REFERENCES

J.A. Vasquez, R.G. Millar, P.M. Steffler, 2005. Two-dimensional morphological simulation in transcritical flow, River, Coastal and Estuarine Morphodynamics.
N.D. Volp, B.C. van Prooijen, J.D. Pietrzak, G.S. Stelling, 2015. A subgrid based approach for morphodynamic modelling, Advances in Water Resources (2015) 1–13.
A. Malcherek, F. Piechotta, D. Knoch, 2005. Technical Report Mathematical Module SediMorph Validation Document.

Quasi-three dimensional computations for flows and bed variations in curved channel with gently sloped outer bank

T. Sasaki
Graduate School of Science and Engineering, Chuo University, Tokyo, Japan

S. Fukuoka
Research and Development Initiative, Chuo University, Tokyo, Japan

ABSTRACT

It is well known that secondary flows cause large velocity and scour around the outer bank and deposition of the inner bank. Various measures for preventing scour and bank erosion around the outer bank have been proposed by researchers (e.g. Roca et al. 2007). Making the slope of the outer bank gentler is one of effective measures for bank protection (Fukuoka et al. 1995). In addition, gently sloped banks have advantages in improving the accessibility of aquatic lives, compared with steep banks. Therefore, it is important to investigate mechanisms of flow structures and bed variation around the gently sloped bank in terms of both flood controls and river environments. From the above reasons, practical calculation model which can estimate three-dimensional flows and bed variations around the banks is required.

Fukuoka and Uchida (2013) have developed a quasi-three dimensional model which can estimate the bottom velocities and three-dimensional velocities (the Bottom Velocity Computation (BVC) method). This study investigates the applicability of the BVC method to flows and bed variations in uniformly curved channels with gently sloped outer bank.

Fukuoka et al. (1995) investigated the effects of the slope of outer banks on the flow structures and bed variation mechanisms for a uniformly laboratory curved channel installing the gently sloped outer bank (see Fig. 1). The experimental conditions are shown in Table 1. The steady discharge and sediment supply were given at the upstream end of the channel. After the equilibrium condition was established, vertical distributions of streamwise velocities and secondary flows and bed topographies were measured. The experiments were conducted in three cases where the outer banks have the slopes 1V:2H, 1V:3H and 1V:4H.

The BVC method was applied to the 1V:2H and 1V:3H experiments. And, we show that the BVC method can almost explain the bed topography at the equilibrium condition and vertical distributions of streamwise velocities and secondary flows in each case (see Fig. 2).

Table 1. Experimental conditions (Fukuoka et al. 1995).

Channel length	24 m
Channel width	1 m

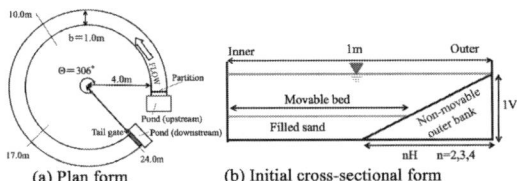

(a) Plan form (b) Initial cross-sectional form

Figure 1. Layout of curved channel with gently sloped outer bank.

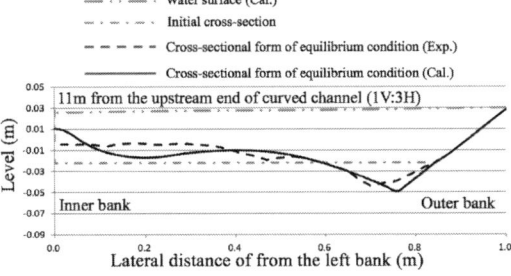

Figure 2. Comparison of cross-sectional form between calculation and experiment.

REFERENCES

Fukuoka, S., Nishimura, T., Sannomiya, T. & Fujiwara, T. 1995. Flow and Bed Profiles in Curved Channels with Gentler Bank Slopes, Proceedings of the Japan Society of Civil Engineers, No. 509 155–167.

Fukuoka, S. & Uchida, T. 2013. Toward Integrated Multi-Scale Simulations of Flow and Sediment Transport in Rivers, Journal of Japan Society of Civil Engineering, Ser. B1, Vol. 69, No. 4, pp. II_1–II.

Roca, M., Martin-vide, J.P. & Blanckaert, K. 2007. Reduction of bend scour by an outer bank footing: Footing design and bed topography. J. Hydraul. Eng., 133(2), 139–147.

Relation between Ishikari River mouth stability and construction of the Ishikari Bay New Port

M. Takezawa, H. Gotoh & R. Hanada
Nihon University

O. Ishikawa & M. Tanaka
Ishikari City

T. Yamamoto
Chuo College of Technology

ABSTRACT

Changes in river mouth morphology are mainly affected by longshore currents and sediment deposition by the river. River mouth closures are important because they can lead to flooding. Although jetties and river mouth excavation are employed as countermeasures for preventing or limiting river mouth closure, such measures are not long-term solutions to the problem. Stabilization of the river mouth of the Ishikari River on the Ishikari Coast is described in this study.

Ishikari Bay is an inlet on the Sea of Japan on the west coast of Hokkaido, Japan. The Ishikari Plain, which has an area $3,800\,m^2$, is drained by the Ishikari River which originates in Mount Ishikari in the Taisetsu Mountain Range and flows through Asahikawa and Sapporo before entering Ishikari Bay. The Ishikari River is 268 km long and drains a catchment area of $14,330\,km^2$. Total discharge from the Ishikari River is $14.8\,km^3$/year, which originate from the major tributaries of the Chubetsu, Uryu, Sorachi and Toyohira rivers. The Ishikari coastline has large areas of quicksand that has been deposited by the Ishikari River, and sandy beaches extend more than 24 km to the north and south of each river bank. Approximately $2,000,000\,m^3$ of sand is discharged by the Ishikari River every year. The large Ishikari River Basin supports a wide variety of life forms, including humans. However, even though flood control measures have been conducted since 1834, several flooding disasters have occurred. The prevention of spits development has been particularly important in terms of preventing river mouth closure of the Ishikari River.

The Ishikari Bay New Port was constructed 7.5 km south of the river mouth of the Ishikari River as part of the Third Hokkaido Comprehensive Development Plan approved by the Japanese Cabinet in July 1970. Under the plan, the Ishikari Bay area just to the north of Sapporo and the surrounding area in western Hokkaido was considered to be well suited for extending the existing function of the area as a distribution and transportation hub for physical goods. To establish a new manufacturing base and distribution hub in the region, construction of the port and development of the area in the vicinity of the port was initiated. After being designated a major port in April 1973, full-scale construction of the port was initiated under the direct control of the national government. In April 1978, the Hokkaido Prefecture Government, Otaru City, and Ishikari City jointly formed the Ishikari Bay New Port Authority to administrator the port. Under the new administration system, the construction of port facilities has been carried out methodically, beginning with the East District. The total area of the Ishikari Bay New Port Area development project is more than 3,000 ha.

The main causes of river mouth closure are a decrease in river discharge due to increased abstraction (i.e. drinking water, irrigation), and the strength of longshore currents. Many countermeasure works against river mouth closure in Ishikari River have been carried out, but no effective solution of river mouth closure could be. Namely, despite implementing numerous countermeasures to prevent river mouth closure of the Ishikari River, none were entirely successful. However, construction of the Ishikari Bay New Port decreased longshore sediment transport, which stabilized the Ishikari River mouth and reduced the incidence of river mouth closure. In this study, the relationship between Ishikari River mouth closure and the construction of Ishikari Bay New Port is described by the authors who have been engaged in the development plan and port construction projects over many years.

A look to valley types developed along the Göksu River (between Mut and Silifke: Southern Turkey)

A. Turan
Faculty Engineering, Geology Department, Selçuk University, Konya, Turkey

ABSTRACT

There are various valley types in the rock groups formed from Ordovician to Miocene at the mouth of the Göksu River. Basement rocks in the area are composed of mainly low resistant Miocene mudstone-shale-marl and high resistant chalk arenites and middle or lesser amount of low resistant Ordovician metamorphic rocks; medium to high resistant Upper Devonian-Upper Cretaceous clastic and carbonaceous rocks. The types of valley with wide and broad floor is dominant in the Langhian-Tortonian low resistant mudstone-shale-marl-clayey limestone cropped out in the Mut town and surrounding area. These rock groups have slightly undulated morphology and low dipping strata. The deposition of the thick alluvial materials at the bottom of the Göksu River has led the developing of the many small planes along the Göksu Valley. A canyon valley in 5 km long was also developed in the Miocene chalk arenites showing plentifull joints and strongly porous nature that were observed in the Evkafçiftliği and Kargıcak villages at the central part of the study area. A very thin alluvial cover composed of pebbles-sands and muds was deposited at the bottom of the canyon. The valleys developed on the Paleozoic-Mesozoic clastic and carbonaceous rocks in the main valley of the Göksu River at the surrounding area of the Deðirmendere and Karkaya villages in the western part of the Silifke are narrow floor, anticline, syncline and cutting through type valleys. Terrace deposits are clearly observed in the main valley slopes while recent alluvial materials deposited in the bottom of the valley are in the shape of narrow and thin strips. The meander valley types have been occurred in the Langhian-Tortonian mudstones-shales-marls and and very thick Holocene alluviums showing a very wide spread at the Göksu delta developed in the center of the Silifke city located in the eastern-southeastern part of the study area.

Annual change of water environment and topographic feature at urban river mouth

K. Uno & S. Kishimoto
The Department of Civil Engineering, Kobe City College of Technology, Japan

ABSTRACT

Generally, it is expected that the micro topography such as bars, tidal flats and pans formed at river mouth have some effect on the water environment and habitat condition for living organism. To support disaster prevention and environment conservation at the same time, it is necessary to understand the topographic dynamics at river mouth. However, it is difficult to approach there due to the restriction of space.

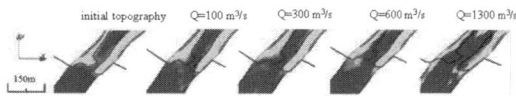

Figure 2. Result of numerical simulations.

In this study, we conducted the field observations and numerical simulations to grasp the water environment and topographic change at river mouth by the flood.

Study site is Akashi River mouth, Hyogo Prefecture in Japan (Fig. 1). The Akashi River is the typical urban river in Japan, its length is 26 km and its basin area is 126.7 square km.

To clarify the characteristics of dynamic of urban river mouth, 2 different observations were carried out. First, to understand the dynamic of river mouth bar, the shape of river mouth bar was recorded by both portable GPS and digital camera. In addition, the grain size distribution on the surface layer at some points was examined. It was conducted at low tide of the spring tide once a month. Second, the continuous monitoring on water level, water temperature and salinity at fixed point was conducted.

From these field observations, it is clarified the river mouth bar and terrace tends to be formed in the winter and be washed away by the flood in rainy season.

Moreover, from the observation by camera monitoring and portable GPS survey, it was clarified the shape of river mouth bar was drastically changed by the flood.

On the other hand, from the result of numerical simulations (Fig. 2), it is clarified that the water level at peak discharge under 300 m^3/s effects on the river bed variation.

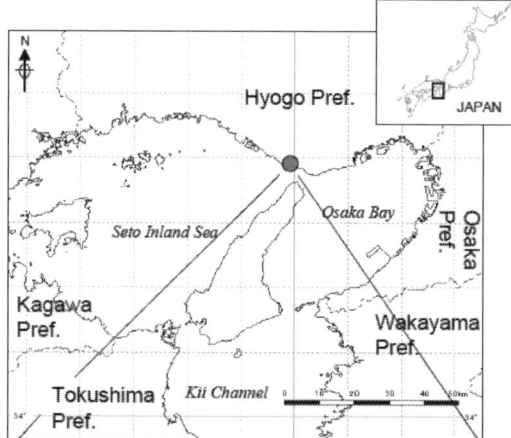

Figure 1. Sample of a figure caption.

REFERENCES

Rubin, D.M. 2004. A simple autocorrelation algorithm for determining grain size from digital images of sediment, Journal of Sedimentary Research, Vol. 74, No. 1: 160–165.

Uno, K. & Arai, J. 2013. A saving, economic and safe measurement on the dynamic state of urban river mouth, Proceedings of ISRS 2013: 1409–1415.

Numerical simulation of gravel deposit erosion

L. Vonwiller, D.F. Vetsch & R.M. Boes
Laboratory of Hydraulics, Hydrology and Glaciology (VAW), ETH Zurich, Switzerland

ABSTRACT

Gravel replenishment by artificial deposits is a potential measure to address sediment deficit in rivers. The main goals of gravel addition are preventing further erosion, reestablishment of natural river morphology, and restoration of spawning grounds for fish. Over the last decade, this technique has gained popularity and is applied in a number of Swiss gravel-bed rivers, such as the Aare, Reuss, Rhine and Limmat Rivers (Fig. 1).

However, the erosion process of artificial gravel deposits and the corresponding impact on river morphology are not yet fully understood. To improve the understanding of the erosion process of gravel deposits, a composite modeling approach was used. Physical experiments of the erosion of gravel deposits were performed for conditions typically found in Swiss lowland gravel-bed rivers (Friedl et al. 2016, this conference). Based on these data, we evaluated the capability of the 2D depth-averaged morphodynamic model *BASEMENT* to reproduce gravel deposit erosion. The included transport model is based on the Hirano-Exner approach for non-uniform sediment,

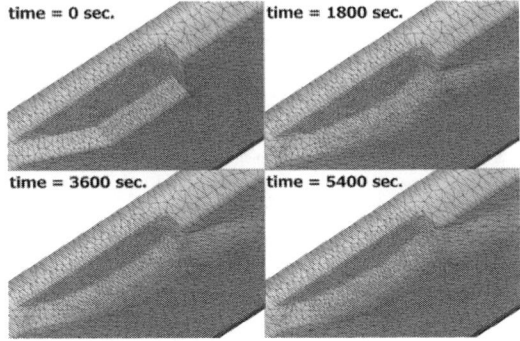

Figure 2. Numerical simulation of gravel deposit erosion.

and considers reduction of the critical Shields parameter due to local slope, correction of bed load transport direction due to lateral bed slope, and bank collapse.

In the present paper, we investigated different configurations with regard to the geometry of the gravel deposit, discharge, and grain size distribution by numerical experiments. The results of the numerical model are in good agreement with the temporal evolution of cross-sectional geometry and erosion rates found in the laboratory experiments. An example of the erosion progress of a gravel de-posit simulated with the numerical model is depicted in Figure 2. The promising results indicate the po-tential use of the numerical model as a planning and decision-making tool for sediment replenishment projects applied on reach-scales.

Figure 1. Gravel deposit in the Limmat River, Sept. 2015.

REFERENCE

Friedl F., Weitbrecht, V., & Boes, R.M. (2016). Laboratory experiments on gravel deposit erosion. Proc. International Symposium on River Sedimentation, Stuttgart, Germany.

Characteristics of flow and sediment at the confluences of mainstream and tributary of the upper reaches of the Yangtze River

P.Y. Wang, L.F. Han, C.Y. Yang & T. Yu
National Engineering Research Center for Inland Waterway Regulation, Chongqing Jiaotong University, Chongqing, China

ABSTRACT

The upper reaches of the Yangtze River is a typical mountain river in the southwest of China, and its tributaries are numerous. This paper deals with the study of the characteristics of flow and sediment accumulation around a confluence of bend-mainstream and tributary of the upper reaches of the Yangtze River. The experiments were carried out in a 180° mobile-bed main channel bend with a by-channel feeding from the top of bend. Uniform sediments having an average diameter of 0.5 mm were used under clear-water scour conditions. Inflow angle and discharge ratios are the main parameters considered in this experiment.

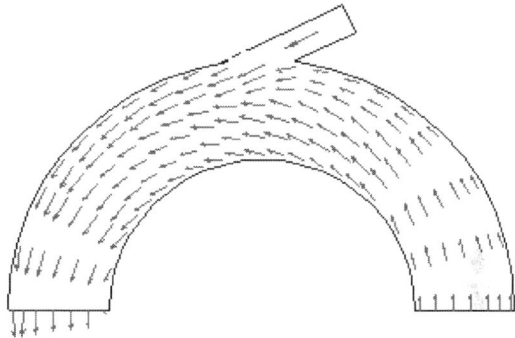

Figure 2. Surface velocity vector horizontal projection in confluences.

Figure 1. Hydraulic experiments.

The results of this paper indicate that:

(1) The flow field around the confluence was divided into five zones: banked-up belt I, banked-up belt II, confluence area, collecting area and separating area;
(2) A new equation for scour parameters around the confluence was developed;
(3) It progressed formula fitting for relationships among the relative length and the relative width of the separating area, discharge ratios, and inflow angles. And it gave formulas for relationships among the relative length and the relative width of the separating area, discharge ratios, and inflow angles in the confluence.

Bedrock channel morphological modeling on the river in Taiwan

K.W. Wu
Water Resources Planning Institute, Water Resources Agency, Taiwan
Department of Civil Engineering, National Chiao-Tung University, Taiwan

K.C. Yeh & C.T. Liao
Department of Civil Engineering, National Chiao-Tung University, Taiwan

Y.G. Lai
Technical Service Center, U.S. Bureau of Reclamation, Denver, CO, USA

ABSTRACT

Rivers in Taiwan typically have steep slopes and subject to rapidly varied transient flood flows. In-stream structures are widely used for flood prevention and water intake in most rivers for limited water resources. Use of these structures, however, leads to both vertical and lateral scour. Furthermore, a 7.3-magnitude earthquake, struck central Taiwan in 1999, caused a differential uplift of river channel, which led to accelerating morphodynamic processes in both stream width and channel slope adjustments. The downstream river reach of Ji-Ji Weir on Cho-Shui River, the longest river in Taiwan with an average channel slope 0.018, is then selected for bedrock channel morphological modeling study river reach (Fig. 1), where a vertical erosion of up to 15 meters was recorded since 1999 (Fig. 2).

A number of engineering schemes have been proposed to stabilize the study river channel. The model simulations have the following three objectives: (1) to assess whether the proposed numerical model is capable for long term simulation on both alluvial and exposed bedrock channel; (2) to evaluate the major agents leading to the severe erosion of the study reach for support of stability project decision; and (3) to identify future research needs for the stability engineering scheme evaluation.

To achieve the above objectives, a two-dimensional mobile bed model, SRH-2D, was selected to simulate both vertical and lateral alluvial erosion, and bedrock incision simultaneously for long-term channel evolution. A validation study was carried out first, covering January 1, 1998 to December 31, 2014. Most model parameters are based on the modeling study on the same river reach (Wu et al. 2015b). The simulated results show that the bedrock hydraulic scour mechanism could reproduce the channel morphological patterns with a satisfied accuracy than the others. Then four selected scenarios, called "Nature", "Earthquake", "Weir" and "Levee", are simulated with hydraulic scour mechanism for exposed bedrock channel and 5 sediment size classes for alluvial channel.

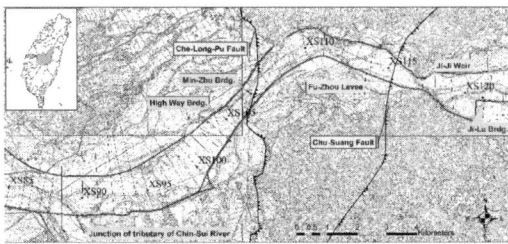

Figure 1. The layout of study area in Cho-Shui River, Taiwan.

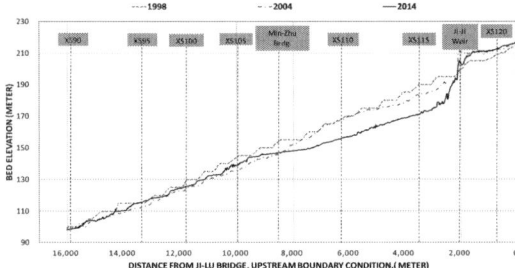

Figure 2. The longitudinal thalweg evolution for the study reach crossing the Ji-Ji weir operation and 921 earthquakes.

The modeling application results shows that the "Weir" scenario dominate the most erosion depth of exposed bedrock reach. And the "Earthquake" scenario, as well as "Levee" scenario, only lead to a minor erosion depth in limited extend. However, the "Levee" scenario could also affect the morphological channel alignment of exposed bedrock reach significantly.

REFERENCE

Wu, K.W., Lai, Y.G., Yeh, K.C. & Liao, C.T. 2015b. Coupled Geo-Fluvial Channel Evolution and Bedrock Erosion Modeling on a River in Taiwan. World Environmental and Water Resources Congress 2015: 1724–1735.

Recent channel adjustments in the Jingjiang Reach controlled by various boundary conditions

J.Q. Xia, M.R. Zhou & S.S. Deng
State Key Laboratory of Water Resources and Hydropower Engineering Science, Wuhan University, China

J.Y. Lu
Changjiang River Scientific Research Institute, Wuhan, China

ABSTRACT

The Jingjiang Reach is located between Zhicheng and Chenglingji in the Middle Yangtze River (MYR), with a length of 347 km, and there are three diversion branches linking the MYR with the Dongting Lake. These branches usually divert water from the main stream during flood seasons but are normally dry during non-flood seasons. At the exit of the Jingjiang Reach, there is a confluence zone at Chenglingji due to a tributary from the Dongting Lake. With the operation of the Three Gorges Project (TGP), the sediment load entering the Jingjiang Reach has been greatly reduced, which leads to remarkable channel adjustments in the reach.

The recent channel adjustments in the Jingjiang Reach are mainly characterised by the variations in the bankfull area and discharge. Measured cross-sectional profiles and hydrological data in the Jingjiang Reach indicate significant variability in the section-scale bankfull variables, and a bankfull variable at a specified section is unrepresentative of the parameter of the whole reach. Using a reach-averaged calculation procedure, the post-flood reach-scale bankfull area

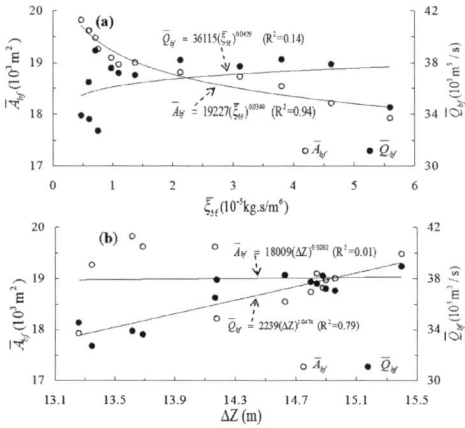

Figure 1. Relationships between the reach-scale bankfull area and discharge and the parameters $\bar{\xi}_{5f}$ and ΔZ.

Table 1. Calibrated parameters for the relations of $\bar{A}_{bf}$ and $\bar{Q}_{bf}$.

	$\bar{A}_{bf}$ (or $\bar{Q}_{bf}$) $= \alpha_1 (\bar{\xi}_{5f})^{\beta_1} + \alpha_2 (\Delta Z)^{\beta_2}$				
Variable	α_1	β_1	α_2	β_2	R^2
$\bar{A}_{bf}$	40.0	−0.347	12914	0.103	0.95
$\bar{Q}_{bf}$	66189	−0.011	45788	0.333	0.91

($\bar{A}_{bf}$) and discharge ($\bar{Q}_{bf}$) in the Jingjiang Reach were calculated annually over the period from 2002 to 2014.

It is found that the recent adjustment in the reach-scale bankfull area or discharge was controlled by three boundary conditions, including the incoming flow and sediment conditions (upstream boundary), the rating curve between stage and discharge at Chenglingji (downstream boundary), and the flow and sediment regime diverted into three branches (lateral boundary). The integrated parameter for the upstream and lateral boundary conditions are represented by the previous five-year average incoming sediment coefficient ($\bar{\xi}_{5f}$) during flood seasons at Zhicheng with the exclusion of three diversion branches, while the downstream boundary condition is given of the difference between mean water levels at Zhicheng and Chenglingji (ΔZ) during a flood season. $\bar{A}_{bf}$ an $\bar{Q}_{bf}$ can respond to the variation in $\bar{\xi}_{5f}$ or ΔZ (Figure 1), with the corresponding empirical relationships between them being developed (Table 1). Therefore, the recent channel adjustments in the Jingjiang Reach were jointly controlled by these boundary conditions.

REFERENCES

Williams, G.P. & Wolman, M.G. 1984. Downstream effects of dams on alluvial rivers. Professional Paper 1286, US Geological Survey, Washington DC.

Xia, J.Q., Li, X.J., Li, T., Zhang, X.L. & Zong, Q.L. 2014. Response of reach-scale bankfull channel geometry in the Lower Yellow River to the altered flow and sediment regime. Geomorphology 213: 255–265.

A comparison of two total sediment transport models for rivers

V.K. Yadav
Department of Civil Engineering, Government Engineering College, Rajkot, India

S.M. Yadav
Sardar Vallabhbhai National Institute of Technology, Icchhanath, Surat, Gujarat, India

S.I. Waikhom
Department of Civil Engineering, Government Engineering College, Surat, India

ABSTRACT

Sediment transport in open channels is a challenge to hydraulics engineer. Despite including many probable significant parameters affecting sediment transport in various relationships proposed by different researchers to predict rate of sediment transport in a stream, the wide and generalized applicability is still a far goal. In pursuit of having a generalized equation for any open channel flow, the dimensionless parameters are selected from some researcher's concepts and are correlated with sediment transport rate parameter of Einstein, in dimensionless form, to develop a regression relation. The two proposed simple models have been developed using four river data sets of Brownlie W. R. (1981), (Red River, Rio-Grande Toffaletti, Rio-Grande Nordin and Trinity River with 365 data points). Einstein's dimensionless sediment transport parameter is tested against a range of dimensionless flow parameters for the best correlation for the selected data sets. Subsequent to analysis, Shield's and Duan's dimensionless flow parameters are selected to finally evolve the models using regression. A performance analysis is carried out to compare the predictability of two models using statistical parameters like MNE, DR, Score, etc. using other data sets. Discrepancy ratio closer to 1 is obtained for Rio-Grande – Toffaletti data for both PM1 and PM2. In general DR is higher for PM2 for all data sets. From MNE, DR, Score, it can be said the models have performed well in every aspect. For other data sets, results are satisfactory, and hence a modification to proposed models is also tested to check increase in their range of application.

REFERENCES

Brownlie, W.R. 1981. "Compilation of fluvial channel data: laboratory and field", Rep. No. KH-R-43B, W.M. Keck Lab. of Hydr. and Water Resources, California. Institute of Technology, Pasadena, Calif.

Duan, J.G. 2013. "A simple total sediment load formula", World Environmental and Water Resources Congress, 2013, ASCE.

Garde, R.J. & Amraei S.R.S. 2009. "Goncharov's total load equation", ISH Journal of Hydraulic Engineering, Vol. 15 (1), No.1, 85–100.

Yang, S-Q. & Lim, S-Y. 2003. Total load transport formula for flow in Alluvial channels, Journal of hydraulic engineering, Vol. 129, pg 68–72, ASCE, ISSN 0733-9429/2003.

Sinnakaudan, S.K., Ab Ghani, A., Ahmad, M.S.S. & Zakaria, N.A. 2006. Multiple Linear Regression Model for Total Bed Material Load Prediction, Journal of Hydraulic Engineering, Vol. 132, No. 5, ASCE, ISSN 0733-9429/2006/5-521–528.

Experiments on the channel plane form with nodes and anti-nodes

S. Yamaguchi
Civil Engineering Research Institute for Cold Region, Sapporo, Japan

Y. Watanabe
National University Corporation Kitami Institute of Technology, Kitami, Japan

K. Sumitomo
Docon Co., Ltd., Sapporo, Japan

ABSTRACT

The widths of natural rivers vary in time and the shape of a channel plane changes mainly with bank erosion. It is essential from the viewpoint of river management to grasp characteristics of channel planes formed by bank erosion with the development of meandering watercourses and sandbars. The process of channel plane formation has been theoretically studied in recent years (e.g. Watanabe & Hasegawa 2014). However, the developmental mechanism of channel planes with changing width, which are observed in such rivers as the Satsunai River shown in Figure 1, has yet to be clarified.

Channel planes with a relatively steep bed slope like in the Satsunai River are characterized mainly by the development of double-row bars and braided streams in the channel. Watanabe & Hasegawa (2014) directed their attention to the structure of a braided channel plane which alternates between a narrow riverbed where watercourses converge (node) and a diverging wide riverbed (anti-node) as is shown in Figure 1, and attempted to theoretically explain its formation mechanism based on the linear stability analysis. Nevertheless, the formation process cannot be fully explained using the theory. Meanwhile, conventional experiments on straight channels with erodible banks (e.g. Ashmore 1982) suggest the appearance of characteristics of the above mentioned nodes and anti-nodes in the process of braiding watercourses along with channel widening. But none of them are experiments for long-term observation of channel planes.

Movable bed experiments using a wide laboratory flume without influence of fixed side walls were carried out in this study. The purpose of this study is to observe the process of river channel plane formation caused by bank erosion along a watercourse in the experiments. The shape of a channel plane changes with bank erosion or formation of braided streams and sandbars in the channel. The stages through which a straight channel, initially built on an erodible bank, develops into a node and anti-node plane were observed as shown in Figure 2.

REFERENCES

Ashmore, P.E. 1982, Laboratory modelling of gravel braided stream morphology, Earth Surface Processes and Landforme, Vol. 7: 201–225.
Kuroki, M. & Kishi, T. 1984. Regime criteria on bars and braids in alluvial straight channels. Proceedings of the Japan Society of Civil Engineers No. 342: 87–96.
Watanabe, Y. & Hasegawa, K. 2014. Stability analysis on periodically changing of channel width. In Proceedings of the International Conference on Fluvial Hydraulics; River Flow 2014: 2273–2281.

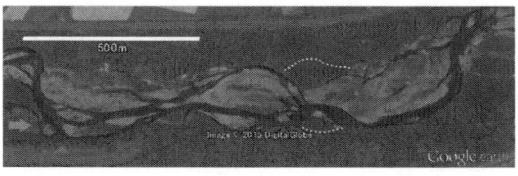

Figure 1. The channel plane form observed in the Satsunai River.

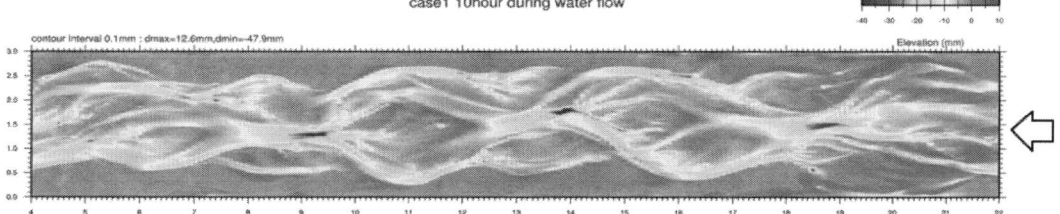

Figure 2. Experimental result. Relative bed elevation. After ten hours constant discharge (Q = 2.76 L/s).

Braided channel evolution in the middle and lower reaches of the Yangtze River after operation of the Three Gorgers Reservoir

S. Yao, G. Qu & H. Wang
Changjiang River Scientific Research Institute, Wuhan City, Hubei Province, China

ABSTRACT

Braided river channel with an island or many islands, which exists extensively in the middle and lower reaches of the Yangtze River, is a typical river pattern. Its fluvial processes are relevant to incoming sediment and flow, riverbed material, boundary condition, and so on. Changes of incoming water and sediment process, erosion-accumulation process, and diversion of flow and sediment on the downstream of the Three Gorges Reservoir (TGR) of typical braided river channel had been analyzed by using the prototype observed data.

Main conclusions made by our research were as follows. (1) After operation of TGR, annual mean time length of high water level and low water level decreased, annual mean time length of middle water level increased at the Jianli and Hankou hydrologic station (Table 1). (2) After operation of TGR, quantity of sediment through the downstream of TGR was greatly reduced. Changes of incoming water and sediment conditions into the downstream of TGR could result in some new changes on evolution of braided river reach such as smaller erosion-and-deposition fluctuation, continuing to erode totally along the time. (3) The operation of TGR took advantage of developing of a branch channel with more energy per unit water body.

Based the theory on river dynamics and riverbed evolution, the relation formula about scouring and silting in braided river channel and change of sediment inflow from upstream was built. The response law of the change of sediment inflow from upstream and diversion of flow and sediment on scouring and silting in braided river channel was analyzed. Main conclusions made by our analysis were as follows. (1) If diversion sediment ratio entering in a branch channel exceeded one of natural condition, it led to deposition totally, or else erosion. (2) If diversion flow ratio entering in a branch channel exceeded one of natural condition, it took advantage of erosion, or else deposition. (3) If diversion flow and sediment ratio entering in a branch channel changed at the same time, it resulted in a change of complicated erosion-and-deposition. The influence mechanism of the braided river channel evolution due to the incoming water and sediment conditions change was explained.

Based on the above analysis of prototype observation data and theory, the trend of typical braided channel evolution in the middle and lower reaches of Yangtze River was forecasted according to incoming flow and sediment condition after operation of TGR. Main conclusions made by our study were as follows. (1) Evolution of the braided channel like goose head shape and bending braided channel could have the trend that main branch channel more easily erode than secondary branch channel. (2) With the continuing erosion of downstream braided channel of TGR, its stability will continue to improve. Research results can provide reference for braided river channel regulation.

Table 1. Statistics of average duration of characteristic discharge level.

	discharge level $\times 10^4$ m^3/s	>3.5	2.0~3.5	1.0~2.0	<1.0
Hankou hydrologic station	1971~1979	70	133	73	102
	1986~2002	46	134	108	77
	2004~2011	48	102	157	52
	discharge level $\times 10^4$ m^3/s	>3.0	1.5~3.0	<1.5	
Jianli hydrologic station	1975~2002	20	106	239	
	2003~2010	9	89	267	

Coastline change of the Yellow River Delta since 1855

S. Yu & S. Tian
Yellow River Institute of Hydraulic Research, YRCC, Zhengzhou City, Henan Province, P.R. China

ABSTRACT

The form of the Yellow River Delta has been analyzed since the great avulsion in 1855 based on coast-line data in 1855, 1937, 1976 and 2013 from site survey and satellite images (Fig. 1). Nearshore wave characteristics were studied to explore the coastline change dynamics.

The Yellow River has the highest suspended sediment concentration in the world and transported mean 0.7bt/a sediment into the estuary recorded at the Lijin Station during the period from 1950 to 2014. The river had been in the state of natural shifting from 1855 to 1976 owing to weak oceanic dynamics with a mean tidal range of 0.4 m–1.8 m and changed its course into the sea 9 times. Thus, sediment load had been nearly uniformly distributed along the whole delta and the coastlines which were developed towards the Bohai Sea with a relatively regular arc shape after two fluctuation periods during 1855–1934 and 1934–1976, respectively.

The coastline before 1855 may be considered the background with no sediment from the river, with an exterior normal direction of 48.1°. The first period of fluctuation completed in 1934, and the coastline developed towards the Sea with 14–25 km. The exterior normal degree was 43.8°. The second period ended in 1976, and the coastline also developed towards the Sea with 14–25 km. The exterior normal direction degree then was 52.9°.

The main wave direction along the offshore is NE, the second strong direction is NNW, and the composite wave energy direction is 46.7°. Hence, exterior normal direction of the coastline is in good accordance with the strong wave direction from north east, which indicates that waves are the main factor controlling the Yellow River Delta coastline development.

The Yellow River has entered to the Bohai Sea through the Qingshuigou Course since 1976. And since then it has been strongly influenced by human interventions for the last 40 years. The only course to the sea produces a continuous extension of depositions in the current estuary, unceasing erosion of adjacent areas and beach degradation of near coastal protection works.

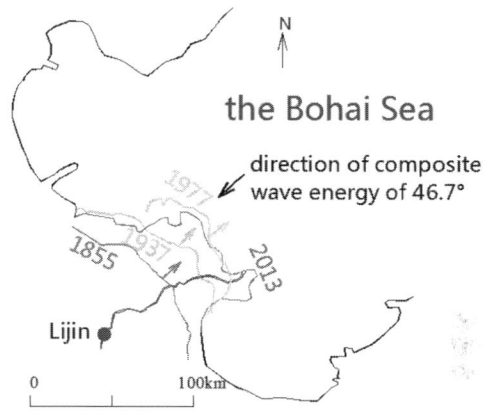

Figure 1. Coastlines of the Yellow River and composite wave direction.

It is suggested that regulation of the Yellow River Estuary should follow the principle of frequent shifting of a single course so that the coming sediment is able to be relatively well distributed along the shore.

REFERENCES

Li, Y., Yu, J., Han, G., Wang, Y. & Zhang, Z. 2012. Coastline change detection of the Yellow River Delta by satellite remote sensing. Marine Sciences, 36(4): 99–106. (in Chinese)

Pang, J.Z. & Jiang, M.X. 2003. On the evolution of the Yellow River Estuary (Part II). Transactions of Oceanology and Limnology, 4: 1–13. (in Chinese)

Peng, J., Ma, S., Chen, H., & Li, Z. 2013. Temporal and spatial evolution of coastline and subaqueous geomorphology in muddy coast of the Yellow River Delta. J. Geogr. Sci., 23(3): 490–502. (in Chinese)

Wang, K., Yu, S. & Ru, Y. 2013. Study on change of flow and sediment entering into the sea and the Yellow River Estuary evolution since 2000. Yellow River, 35(4): 11–13, 126. (in Chinese)

Wang, W. & Zhang, H. 2007. Evolution pattern of the Yellow River Estuary Coast. Yellow River, 29(2): 27–28, 32. (in Chinese)

A physically-based model of individual step-pool stability in mountain streams

C.D. Zhang & Z.L. Wang
State Key Laboratory of Hydroscience and Engineering, Tsinghua University, Beijing, China

Z. Li
School of Hydraulic Engineering, Changsha University of Science & Technology, Changsha, China

ABSTRACT

Step-pool system in high gradient mountain streams stabilize the channel bed by dissipating water energy to a great extent and provide diverse aquatic habitats. However, the failure of step-pools can be triggered by exceptional flood events with a return time of 30–50 years (Lenzi 2001). The destruction of step-pools intensifies sediment transport and channel erosion, and dramatically influences the local fluvial process and aquatic habitats, which leads to great attention to the study of step-pool stability. Combined with effects of water flow, downstream scour of the step, surrounding grains (Zhang et al. 2014) and grain impact, all the forces acting on the keystone were schemed and moment analysis was conducted to advance a new physically-based model of an individual step-pool stability (Fig. 1).

According to the new model, channel gradient, keystone diameter, downstream scour, diameter of the keystone, flow discharge, relative size of the impact grain are the primary factors that affect an individual step-pool's stability. The stability of an individual step-pool decreases with scour angle under a certain discharge. Along with downstream scour angle increasing from 30° to 60°, the critical discharge for keystone instability is diminished over 6 times. Therefore, strong water flow breaches the step as a result of the discharge exceeding the critical discharge reduced by the downstream scour. The grain impact may reduce the critical discharge for step-pool failure and the decreasing effect becomes stronger with the increase of the size and initial velocity of the impact grain and larger scour angle. The large grains with $\eta > 0.4$ ($\eta = D_1/D$, D_1 is the diameter of the impact grain

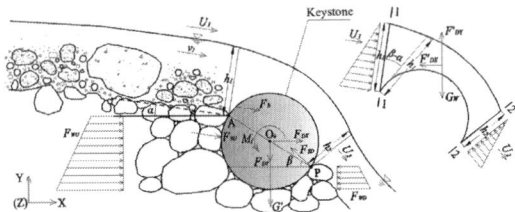

Figure 1. Force analysis of the keystone and the water flowing by.

and D is the diameter of the keystone) supplied by extraordinary floods, debris flows or landslides are able to dislodge the keystone of an individual step-pool directly if they are moving fast or reduce the critical discharge largely to enable the water flow to destabilize the step-pool if with low velocity. The theoretical model was applied to evaluate the stability of both connected and disconnected step-pools to the sediment sources of the mountain river basins and matched well with the field evidence. The model was also successfully to explain the lower stability for connected step-pools with sediment transport in the flood season.

REFERENCES

Lenzi, M.A. 2001. Step-pool evolution in the Rio Cordon, northeastern Italy. Earth Surface Processes and Landforms, 26(9), 991–1008.

Zhang C., Wang Z., Li Z., et al. 2014. A Physically-based Model of Individual Step-pool Failure. Shuili Xuebao/Journal of Hydraulic Engineering, 2014(12): 1399–1409. (In Chinese)

Sensitivity of deposition and erosion to bed composition in the Iffezheim reservoir, Germany

Q. Zhang, T. Speckter & R. Hinkelmann
Chair of Water Resources Management and Modeling of Hydrosystems, Technische Universität Berlin, Germany

G. Hillebrand, T. Hoffmann & H. Moser
Federal Institute of Hydrology, Koblenz, Germany

ABSTRACT

The definition of the initial spatial distribution of the grain size composition is a key challenge when dealing with river morphodynamic modelling. More precisely that means that not only the exact grain size fractions but also the spatial distribution of the material within the model area have to be considered and investigated regarding the effect on the hydraulic roughness and the stability of the river bed. However, the initial grain size distribution is generally not accurately known as a result of the lack of measurements. Furthermore, the grain size distribution is subject to variations during the transport and is sensitive to the history of flow and sediment supply.

The Iffezheim reservoir of the Upper Rhine in Germany has been subject to several studies concerning the mechanisms and prediction of non-uniform sediment transport since its construction in 1977. In this paper several scenarios are performed to determine the sensitivity of the spatially varying grain size characteristics:

i. Simulations with a non-uniform grain size composition distributed uniformly over the whole domain: with a mean grain diameter d_{50} between 0.01 and 19 mm, corresponding to the d_{50} of the fine-grained weir channel fill and the initial river bed before 1977.
ii. Simulations with a variable bed composition: coarser sediments (gravel) from the entrance of the model to the hydropower channel and on the right side of the weir channel; fine sediments (silt and clay) dominate on the left side of the weir channel where the major deposition of fine sediments is observed; nine grain sizes fractions were linearly interpolated at the transition from coarse to fine grains.
iii. Simulation outputs for the initial river bed: grain size distribution of an earlier numerical simulation by Hillebrand et al, 2012.
iv. Long-term (18 years) numerical simulations considering of original river bed material: The simulation was limited to a discharge between 1000 and 2000 m^3/s to avoid erosion during the simulation and to generate a weir channel fill; initial bathymetry of this run was based on the Iffezheim reservoir directly after construction (i.e. without deposition) and a non-uniform grain size composition of the former channel bed (before the construction of the Iffezheim barrage) was used as boundary conditions.

To simulate the morphodynamic processes we use the 3D computational fluid dynamic model SSIIM (Olsen, 2014) with an extension for the erosion process of cohesive sediments. The influence of the initial bed composition was tested by running simulations with identical settings but variable grain size distribution which ranges from 0.4 µm to 90,000 µm in the river bed in each cell.

The bed elevation changes, i.e. the deposition and the erosion patterns within the weir channel of Iffezheim reservoir, are compared to bathymetric surveys after each period.

The quantitative comparison of predicted and measured bed elevation changes indicated the model, with adequate spatially varying grain size distribution of bottom sediment in the Iffezheim Reservoir is generally, in good agreement with the measured data.

REFERENCES

Hillebrand, G., Klassen, I., Olsen, N.R.B. & Vollmer, S. 2012. Modelling fractionated sediment transport and deposition in the Iffezheim reservoir. *10th International Conference on Hydroinformatics,* Hamburg, Germany.
Olsen, N.R.B. 2014. *A three-dimensional numerical model for simulation of sediment movements in water intakes with multiblock option.* Version 1 and 2. User's manual. Department of Hydraulic and Environmental Engineering. The Norwegian University of Science and Technology, Trondheim, Norway.

Features of recent scouring and silting of the river channel of the Jingjiang River downstream of the Three Gorges Project

Y.H. Zhu, X.H. Guo, G. Qu, F. Tang & L.H. Gu
Key Laboratory of River Regulation and Flood Control of MWR, Changjiang River Scientific Research Institute, Wuhan, China

ABSTRACT

Since the operation of the Three Gorges Project (TGP) in the Yangtze River in June 2003, the flow and sediment conditions of the downstream river change significantly, and the channel suffers drastic scouring and silting. The Jingjiang River, part of the middle reach of the Yangtze River and typical of alluvial meandering river, is the most flooding vulnerable reach in the Yangtze River and very closely downstream of the TGP.

By use of prototype observational data, the variation of the flow and sediment conditions of the Jingjiang River and the change of flow and sediment diversion from the river to the Dongting Lake via the three Outlets are analyzed first. Then the fluvial process of the Jingjiang River after the operation of TGP is examined, including the change of shoreline, thalweg, shoals and typical cross-sections, etc. The results of analysis indicate that, though the overall river regime of the Jingjing River is comparatively stable, significant adjustment of the river regime occurs locally after the TGP operation, especially the occurrence of chute cutoff with different degrees at most of the river bends in the lower part of the Jingjiang River (see e.g. Figure 1). As for the extent of river regime variation among different river patterns, the sinuous channel is larger than the anabranch channel, while the straight channel is the smallest.

The results also indicate that the intensity of river channel scouring after the TGP operation ($55.0 \times 10^6 m^3/a$) is much larger comparing with that of $13.7 \times 10^6 m^3/a$ before the project operation.

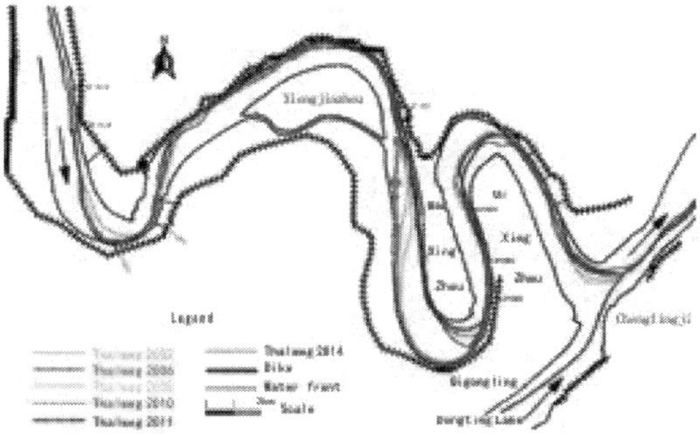

Figure 1. Chute cutoff at the river bed of Oigongling.

REFERENCES

Lu, J.Y. & Zhu, Y.H. 2014. Issues on evolution and regulation of Yangzte River and lakes downstream of TGP. Journal of Yangtze River Scientific Research Institute, 31(2): 98–107. (in Chinese)

Xu, Q.X., Yuan, J., Wu, W.J. & Xiao Y. 2011. Fluvial processes in middle Yangtze River after impoundment of Three Gorges Project. Journal of Sediment Research, 2: 38–46. (in Chinese)

Zhu, Y.H., Guo, X.H., Liao, H.Z., He, G.S. & Qu, G. 2012. Impacts of the Three Gorges Project on the hydrological regime in the Jingjiang reach of the Yangtze River. Proceedings of the 10th International Conference on Hydroscience and Engineering, Orlando, Florida, USA.

D. Hydromorphology meets ecology

Effects of sediment bypass tunnels on sediment grain size distribution and benthic habitats

C. Auel, S. Kobayashi, T. Sumi & Y. Takemon
Water Resources Research Center, Disaster Prevention Research Institute, Kyoto University, Uji, Japan

ABSTRACT

Anthropogenic impacts have altered river systems worldwide and dams interrupt the natural flow and hinder continuous sediment transport resulting in changes in downstream flow regime, bed morphology and ecosystem. A reservoir traps the incoming sediments leading to sediment starving conditions and degradation downstream if no appropriate action is taken. Sediment bypass tunnels (SBT) are one strategy against sedimentation by routing sediment load around reservoirs. Most tunnels are located in mountainous regions where a considerable amount of coarse material is entrained. A SBT has the advantage that only newly entrained sediment from the upstream reach is diverted into the downstream thereby reestablishing the sediment connectivity. The sediment pulse is therefore of natural character as the pre-dam conditions are reestablished during floods. Already accumulated sediments in the reservoir are normally not mobilized. The downstream released sediment lead to morphological changes in the river bed with formerly degraded sections being filled up. Besides morphological changes, also the benthic habitat is largely affected by flushing or bypass operations.

Four reservoirs in Japan and Switzerland with SBTs were monitored in 2014 to analyze their effects in terms of up-to downstream morphological and biotic changes. These were Pfaffensprung and Solis in Switzerland, and Asahi and Koshibu in Japan. The SBTs Asahi and Pfaffensprung are already operated for 17 and 93 years, respectively, whereas Koshibu and Solis are new with no and two years of operation, respectively. Sediment grain size distribution (GSD), local bed characteristics, microhabitat abundance and invertebrate richness were analyzed.

It was found that GSD at new SBTs are fine in the up- and coarse in their downstream due to lack of conveyed sediments in the past. Analysis of biotic data reveal that directly below dams microhabitat richness is low and lentic species abundance is high compared to their upstream, while these differences decrease

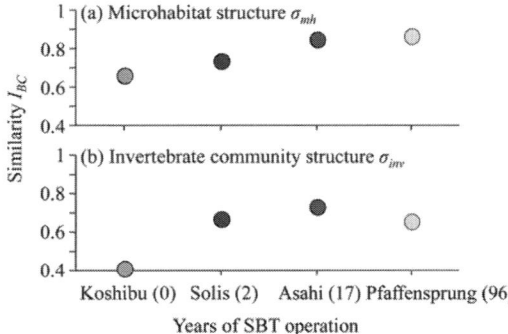

Figure 1. Bray-Curtis similarity index I_{BC} of a) microhabitat, and b) invertebrate community as functions of years of SBT operation.

further downstream. Figure 1 shows the Bray-Curtis similarity index I_{BC} in (a) microhabitat σ_{mh} and (b) invertebrate structure σ_{inv} between up- and downstream. $I_{BC}(\sigma_{mh})$ shows a clear increasing trend with increasing bypass operation (Fig. 1a). This is partly due to similar GSD and less organic microhabitats in the downstream for dams with longer bypass operation. It appears that habitat composition recovers toward the upstream condition with increasing years of bypass operation. Also $I_{BC}(\sigma_{inv})$ between up and downstream increased for dams with longer bypass operation (Fig. 1b). This is partly related to the increase of gliders over net spinners for the Japanese sites and the increase of riffle over pool specialists for both Japanese and Swiss sites. A large difference in similarity between Koshibu and Solis and small one among Solis, Asahi, and Pfaffensprung may suggest that the recovery of invertebrate community towards the upstream condition may occur within a few years of bypass operation. Both, microhabitat and invertebrate community structure tend to be similar between US and DS-SBT for dams with older SBTs. The environment has recovered substantially to a previous state for Asahi and Pfaffensprung and is apparently improving at Solis.

River restoration in sand-dominated lowland streams – a comparison of morphodynamic impacts and response

V. Berger & A. Niemann
Institute of Hydraulic Engineering and Water Resources Management, University of Duisburg-Essen, Essen, Germany

C.K. Feld
Aquatic Ecology, University of Duisburg-Essen, Essen, Germany

ABSTRACT

According to the European Water Framework Directive (WFD) all European water bodies have to reach a good chemical and ecological status by 2027. The ecological status is primarily defined based on ecological conditions of a water body (e.g. a stretch of a stream), taking into account supporting components as physico-chemistry, hydrology and morphology. To achieve this ambitious goal multiple restoration projects are currently carried out throughout Europe. One important aspect of river restoration is the improvement of river morphology, but analysis of morphodynamic processes is currently not the main focus. Yet, there is only few knowledge about the correlation between morphological processes and ecological quality in particular with regard to long-term developments in small rivers (Haase et al. 2013).

In 2012/2013 two river restoration measures have been carried out along the sand-bottom stream "Rotbach" in Dinslaken/Germany. Measures like the reactivation of floodplains, relocation and reshaping of the stream course as well as the placement of large woody debris changed two sections of the former straightened river into morphologically near natural river reaches. Both reaches as well as a natural reach were monitored morphologically with emphasis on capturing and quantifying initial developments and changes (Knighton 1998). The aim of this study is to investigate the temporal and spatial development of the sand-dominated lowland stream and to find linkages between morphodynamic impacts and hydromorphologial development.

Morphological analysis is carried out approx. every three months since spring 2014. Substrate types as well as diversities in flow depths and velocities were recorded for microscale analysis.

For the analysis of the hydraulic impact, discharge dynamics are analyzed as following: First, a statistical analysis of the discharge distribution was conducted; frequency, magnitude, duration of high flow and low flow events as well as diversity and range of flows have been investigated. Additionally, the effective discharge is investigated testing different transport equations (Soar & Thorne 2001). It is shown that the effective discharge is approx. $0.33\,m^3/s$ which is close to the mean discharge of $0.3\,m^3/s$, underlining that dominant discharges in lowland streams are smaller than in mountainous areas. For analyzing the correlation between discharge impacts and morphological response, the temporal development of the coefficient of variation of water levels was chosen as it reflects morphological development best.

The comparison shows that all parameters which depend on the total discharge volume of a period (between two field investigations) do not explain the morphological development. Also periods of respective discharges or the amount of transported sediments do not explain morphological temporal development. Parameters which express the dispersion of discharges can explain the morphodynamic processes. The development of the coefficient of variation of water depth corresponds well with the development of the coefficient of variation of discharge between two field investigations.

It is shown that morphodynamic processes depend more on the variation of discharge than on the maximum discharge or duration of floods.

REFERENCES

Haase, P., Hering, D., Jähnig, S.C., Lorenz, A.W. & Sundermann, A. 2013. The impact of hydromorphological restoration on river ecological status: a comparison of fish, benthic invertebrates, and macrophytes. Hydrobiologia, 704: 475–488.

Knighton, D. 1998. Fluvial Forms & Processes – A new perspective, New York, John Wiley & Sons Inc.

Soar, P.J. & Thorne, C.R. 2001. Channel Restoration Design for Meandering Rivers, Vicksburg, MS, U.S. Army Engineer Research and Development Center, Flood Damage Reduction Research Program.

Application of the hydromorphological assessment framework Valmorph to evaluate the changes in suspended sediment distribution in the Ems estuary

C. Borgsmüller, I. Quick & Y. Baulig
Federal Institute of Hydrology, Koblenz, Germany

ABSTRACT

In the context of the revision of the hydromorphological assessment framework for the environmental impact analysis (UVU) a method was developed to evaluate hydromorphological changes in natural and heavily modified and artificial waterbodies (BfG, 2011). Based on the assessment framework, the module Valmorph was developed in the context of the Integrated Floodplain Model INFORM. This module represents a quantitative measure to describe and evaluate the hydromorphological conditions of federal waterways (Rosenzweig et al., 2012). This is done by using indicator parameters which represent each of the 6 hydromorphological main features used in the UVU assessment framework. The testing, verification and optimization of the described method was done for exemplary investigation areas of the federal waterways. The Ems estuary was chosen to represent heavily modified transitional waters.

In this study, the assessment methodology was tested exemplarily for the tidal Ems taking into account the Valmorph indicator parameter suspended sediment budget.

The suspended sediment distribution of the tidal Ems is characterized by high and increasing sediment concentrations especially in the lower Ems. In the past, many investigations have been conducted evaluating the changes in SSC concentrations close to the surface. In the Ems estuary changes are not only caused by increasing SSC concentrations but also by fundamental changes of the vertical sediment distribution. In former times SSC concentrations in the upper part of the water column could be considered as representative for the entire waterbody. Recent investigations showed that the waterbody is now stratified with very high sediment concentrations close to the river bed and temporarily occurring fluid mud layers of several meters. Measurements from the upper part of the water column might lead to underestimation of the amount of sediment which can be remobilized if fluid mud layers occur.

Additionally this might lead to false estimation of the ability to serve as a habitat for biota or of the consequences for the oxygen budget, phytoplankton and the primary production.

The above mentioned method to characterize and assess hydromorphological changes was applied to describe the reference conditions as well as the present state of suspended sediment distribution in the Ems estuary. For that purpose, historic and current suspended sediment data from long term measurement stations along the tidal Ems were analyzed. Additional vertical measurements of suspended sediment concentration were taken into account to achieve a comprehensive description of the longitudinal and vertical distribution of suspended sediment. Approaches to consider the specific characteristics in the assessment process when fluid mud is present are further discussed.

The results of the analysis can serve as a basis to assess sediment management strategies for heavily modified transitional waters and for decision making processes of federal authorities.

REFERENCES

BfG 2011. Verfahren zur Bewertung in der Umweltverträglichkeits-untersuchung an Bundeswasserstraßen – Anlage 4 des Leitfadens zur Umweltverträglichkeitsprüfung an Bundeswasserstraßen des BMVBS, BfG Bericht 1559. Bundesanstalt für Gewässerkunde, Koblenz.

Rosenzweig, S., Quick, I., Cron, N., König, F., Schriever, S., Vollmer, S., Svenson, C. and Graetz D., 2012. Hydromorphologische Komponenten im Flussauenmodell INFORM – Entwicklung und Anwendung der morphologischen Systemkomponente MORPHO und des Bewertungsmoduls Valmorph zur quantitativen Erfassung und Bewertung hydromorphologischer Veränderungen in Fluss und Aue. BfG Bericht 1657. Bundesanstalt für Gewässerkunde, Koblenz.

Mechanics of biofilm-coated sediment transport

H.W. Fang, H.M. Zhao, W. Cheng, M. Fazeli, Y.S. Chen, Q.Q. Shang, G.J. He & L. Huang
Department of Hydraulic Engineering, Tsinghua University, Beijing, China

ABSTRACT

After the impoundment of the Three Gorges Reservoir (TGR), sediment deposition and related environment problem in the reservoir are issues given more and more attention. High nutrients and weak hydrodynamic condition provide a favorable circumstances for the reproduction and growth of the microbes, most of which prefer to adhere to the sediment along bed. Biofilm, which are the aggregates of microbes and its extracellular polymeric substances (EPS), change the sediment properties, not only the particles itself but also its transport dynamics. Series investigations have been done in the laboratory experiment by our research group. The biofilm was cultivated in the lab with sufficient nutrient, the bulk density of biofilm-coated sediment and the rheological properties of the mixture have been presented. Then changes of the settling velocity of the bio-particles and the incipient of the biofilm-coated sediment were addressed. Moreover, a series of experiments for pure and biofilm-coated sediment have been conducted in a flume, in which the bio-sediment concentration in vertical direction and bedload transport have been proposed. With bio-sediment transport, bedforms induced by the flow and its resistance have also been given, and the effect of the biofilm covered the bed on the turbulent structure of the flow have been explored to be better understanding of the bio-sediment suspension and transport. These studies will make better management of the water and sediment resources in the reservoir and give insight to evaluate the ecological effect in and downstream the reservoir.

Heavy metal concentrations and enrichment of sediment cores: Correlation between geochemistry and geoaccumulation index

F. Fernandes
Centre for Water Resources and Applied Ecology – CRHEA, University of São Paulo, Brazil

C. Poleto
Hydraulic Research Institute – IPH, Federal University of Rio Grande do Sul – UFRGS, Brazil

ABSTRACT

The municipality of Viamão is located in the metropolitan region of Porto Alegre, state of Rio Grande do Sul. The Mãe d'Água dam was built in 1962, in order to attend the demand from the Federal University of Rio Grande do Sul, more precisely the Institute of Hydraulic Research. However, due to a lack of urban planning over the past forty years, the dam has caused several environmental liabilities, such as organic contaminants and/or inorganic contaminants. This manuscript addresses an evaluation of the concentrations of the metals zinc and nickel and Geoac-cumulation Index (I_{geo}), at different depths sampled in sediment cores produced in the watershed that comprises Mãe d'Água. Four sediment core distributed in the dam were collected in June 2014, by "Piston Core" core sampler. Sediments whose fraction were lower than 63 μm were subjected to chemical analyses regarding the presence and concentration of nickel (Ni) and zinc (Zn). The acid digestion methodology employed was EPA 3050, adopted by the US Environment Protection Agency. The analyses were performed in replicates and two USGS (US Geological Survey) reference materials, namely: SGR-1b and SCO-1 were used for the quality control. Geoaccumulation index (I_{geo}) was developed by Müller (1977) and had widely been used in trace metal studies of sediments and soils (Amin et al., 2009; V.K. Singh et al., 2005). The I_{geo} was calculated according to the equation ($I_{geo} = \log2* Ca/1.5*Cp$) and table 1, aiming to quantify the degree of heavy metal pollution on the Mãe d'Água sediments.

All samples showed concentrations of Zn and Ni higher than the local background values and growth patterns, which evidenced an enrichment of the anthropogenic activity. The I_{geo} for most of the metal concentrations at different depths has been characterized as non-polluted. Historically this area is characterized by residential occupation; therefore anthropogenic activity can be considered the main causes of environmental damage to this water body. Sediment contamination was attributed to anthropogenic and natural processes. Therefore, our results showed that the I_{geo} can be used for effective management of fresh water on Mãe D'Água dam.

Keywords: Sediments. Heavy Metals. Trace elements. Geoaccumulation index.

Table 1. Classification of Müller I_{geo}.

Classificação	Classes do IGEO	Intervalos do IGEO
Veryhighly polluted	6	(5–6)
Highly to very highly polluted	5	(4–5)
Highly polluted	4	(3–4)
Moderately to highly polluted	3	(2–3)
Moderately polluted	2	(1–2)
Unpolluted tomoderately polluted	1	(0–1)
Unpolluted	0	(0–0)

REFERENCES

Amin B., Ismail A., Arshad A., Yap C.K. and Kamarudin M.S. 2009. *Anthropogenic impacts on heavy metal concentrations in the coastal sediments of Dumai, Indonesia.* Environ Monit Assess; 148: 291–305.

Müller, G.; Grimmer, G. and Böhnke H. 1977. *Sedimentary Record of Heavy Metals and Polycyclic Aromatic Hydrocarbons in Lake Constance.* Naturwissenschaften. v. 64, pp. 427–431.

Singh, K. P.; Mohan, D.; Singh, V.K. and Maik, S. 2005. *Studies on distribution and fractionation of heavy metals in gomti river sedimentes – a tributary of the Ganges, India.* Journal of Hydrology, v. 312, pp. 14–27.

Microbial biostabilization and flocculation – what can we learn for sediment transport modelling?

S.U. Gerbersdorf, H. Schmidt, M. Thom & S. Wieprecht
Institute for Modelling Hydraulic and Environmental Systems, University of Stuttgart, Germany

ABSTRACT

Over the last years, microbial biostabilization of fine sediments has received great attention since it impacts significantly all parts of the ETDC (Erosion, Transport, Deposition, and Consolidation) cycle (reviewed in Underwood & Paterson 2003, Gerbersdorf & Wieprecht 2015). While the focus has been on marine habitats, the enhanced stability of sediments due to biofilm growth has been reported from freshwaters recently (e.g. Schmidt et al. 2015). However, little is known about the influence of ubiquitously occurring microbes on the characteristics of resuspended sediments. In numerical modelling, the suspended sediment is still regarded as single mineral grains despite better knowledge on the complex composition of natural flocs (Droppo 2001).

Experiments were conducted in mesocosm to grow biofilm under natural but controlled conditions by circulating river water over test sections. The growing biofilms were exposed to varying conditions of hydrodynamic and their development monitored regularly (biomass, species composition, Adhesion and stability). After six weeks, the biofilm – stabilized sediments were eroded in the Gust Chamber (Fig. 1), transferred to a settling column (height of 32.7 cm) and recorded by a Charged Coupled Device CCD camera (photo taken every 12 seconds, Set-up in Fig. 2) during their settling process. Image analysis by Matlab has been used to evaluate geometrical properties of the eroded flocs such as form factor, aspect ratio, Corey shape factor, roundness, solidity, equivalent diameter and perimeter. Additionally, the settling velocity as well as the density, dry weight and organic content of the flocs have been determined.

This presentation will show the results on eroded sediment floc characteristics and relate them to the previous biofilm growth conditions or the "bed history" such as varying hydrodynamics. The data will be opposed to the behavior of single mineral grains.

In particular, the significance of these data will be discussed in terms of natural sediment transport and deposition. Implications for modeling the dynamics of fine sediments will be highlighted.

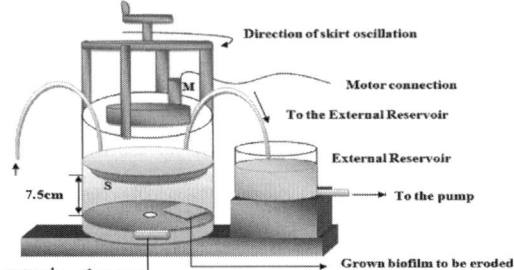

Figure 1. Gust Microcosm.

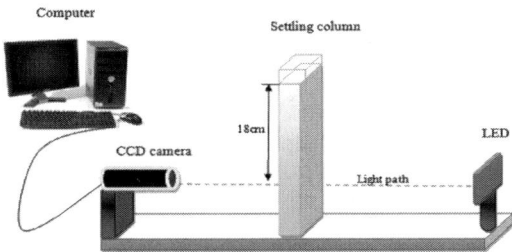

Figure 2. Set-up of settling column and CCD camera.

REFERENCES

Droppo, I.G. 2001. Rethinking what constitutes suspended sediment. Hydrological Processes, 15(9): 1551–1564.

Gerbersdorf, S.U. & Wieprecht, S., 2015. Biostabilization of cohesive sediments: revisiting the role of abiotic conditions, physiology and diversity of microbes, polymeric secretion, and biofilm architecture. Geobiology, 13, 68–97.

Schmidt, H., Thom, M., Matthies, K., Behrens, S., Obst, U., Wieprecht, S. & Gerbersdorf, S.U. 2015. A multidisciplinarily designed mesocosm to address the complex flow-sediment-ecology tripartite relationship on the microscale. Environmental Sciences Europe, 27: 1–11.

Underwood, G.J.C. & Paterson, D.M. 2003. The importance of extracellular carbohydrate production by marine epipelic diatoms. Adv. in Botanical Research, 40, 183–240.

Compensatory measures at a Heavily Modified Waterbody (HMWB) improve the hydromorphological quality, a practical example from the Moselle

D. Gintz & Y. Baulig
Federal Institute of Hydrology, Koblenz, Germany

ABSTRACT

In commission of the Federal Waterways and Shipping Administration (WSA) Trier, the German Federal Institute of Hydrology (BfG) investigates the ecological efficiency of a compensatory measure for the expansion of the outer harbor of Wintrich. Following the landscape conservation support plan, new habitats for wildlife should be created through lengthening of river banks, free landscape succession and the allowance of self-forming processes on unprotected river banks (BfG 2007). For this a new artificial anabranch with a small island was implemented at the left bank of the Moselle on a point bar upstream the village of Minheim (Moselle-km 144.8 to 145.3, area: 6,500 m^2).

The compensatory measure was realized in May/June 2008, the monitoring started in 2010. Currently, two field campaigns have been realized in 2010/11 and in 2015/16. The third and final campaign is planned for 2020. The applied monitoring concept at Minheim (BfG 2010) pays attention to hydromorphology, assuming that the hydromorphological processes influences the hydrodynamic and habitat conditions to support biota (Quick, 2011, Poppe et al. 2015). The concept includes field mapping, GPS survey of cross profiles, turbidity measurements with OBS-probes and sediment samples for grain size analyses to focus on the morphological development of the new river structure and the potential erosion and sedimentation processes.

The Moselle is biocenotical classified as a DE-Typ 9.2: river of the German Central Uplands (c). This type is characterized by a floodplain valley and a sinuate to meandering and braiding channel plan-form (BfN 2009). Due to the fact of recent flow regulation of the Moselle less morphodynamic processes were expected particular for this site in Minheim.

The preliminary results of the monitoring indicate few self-forming processes such as sedimentation and erosion. The grainsize distributions of all sediment samples changed in the observed period of five years to an increased percentage of silt and clay and a respectively decreasing percentage of sand. Also a thin layer of fine sediments and aquatic macrophytes buried recently the in 2011 still visible gravel bars at the shallow banks. A thin layer of fine sediments (silt and clay) and biofilm also cover up the coarse bed sediments of the anabranch channel. The bankside zone was partly covered by a layer of 0.15 m fine muddy sediments, (anthropogenic) floating debris and rising wood followed by a stock of reed bed and alluvial forest species. The preliminary results in Minheim and the results of a decade of monitoring of older compensatory measures (BfG 2007) conforming the research results of Poppe et al. (2015). The measures improved macro- and mesohabitat diversity, promote waves protected areas and minor water flow dynamics and therefore little fluvial processes.

REFERENCES

BFG (2007). Untersuchungen zur ökologischen Wirksamkeit landschaftspflegerischer Kompensationsmaßnahmen an der Mosel. Bericht BfG, 1541. Bundesanstalt für Gewässerkunde. Koblenz.
BFG (2010). Untersuchungen zur ökologischen Wirksamkeit der Ersatzmaßnahme 2. Schleuse Wintrich "Auf der obersten Ray" bei Minheim an der Mosel bei Mosel-km 144,8 bis 145,3 – Untersuchungskonzept für die Jahre 2010 bis 2020.
BFN (2009). Flussauen in Deutschland. – Erfassung und Bewertung des Auenzustandes. BfN-Schriftenreihe "Naturschutz und Biologische Vielfalt". Bundesamt für Naturschutz. Bonn.
Könzen, U. (2005). Fluss- und Stromauen in Deutschland – Typologie und Leitbilder. Angewandte Landschaftsökologie, 65: 1–327.
Poppe, M., J. Kail, J. Aroviita, M. Stelmaszczyk, M. Giełczewski & S. Muhar (2015). Assessing restoration effects on hydromorphology in European mid-sized rivers by key hydromorphological parameters. Hydrobiologia.
Quick, I. (2011). Ermittlung und Bewertung hydrologischer Indikator-Parameter an Bundeswasserstrassen. 13. Gewässermorphologisches Kolloquium – Erfas-sung und Bewertung des hydrologischen Zustandes in Wasserstraßen. 27–40. Bundesanstalt für Gewässerkunde. Koblenz.
Vogel, R.M. (2011). Hydromorphology. Journal of Water Resources Planning and Management, 137: 147–149.

The analysis of sediment diameter with biofilm

G.J. He, H.W. Fang, Q.Q. Shang, F. Mahede & L. Huang
Department of Hydraulic Engineering, State Key Laboratory of Hydroscience and Engineering, Tsinghua University, Beijing, China

ABSTRACT

Sediment particles are often colonized by biofilm in a natural aquatic ecological system, especially in eutrophic water body. A series of laboratory experiments on particle size gradation were conducted after natural sediment was colonized by biofilm for 5, 10, 15 and 20 days. Particle image acquisition and particle tracking techniques were utilized to analyze the changes of these properties. The experimental results indicate that the size gradation of bio-particles underwent significant change due to the growth of biofilm onto the sediment surface. The bio-particle diameter will become 1 to 2 orders of magnitude larger than the initial size. Because of the drag force, in the flowing water, the diameter will decrease with the increasing velocity. These results may be specifically used in the low energy reservoir or lake environment.

Coarse sand as a specific problem for aquatic ecosystems in granite-dominated landscapes

S. Höfler & C. Gumpinger
Consultants in Aquatic Ecology, Austria

C. Hauer
University of Applied Life Sciences Vienna, IWHW, Austria

ABSTRACT

In the past few years problems concerning flood protection and aquatic ecology have increased markedly in most rivers of the Bohemian Massif, mostly due to a special form of "fine sediment" stress. Excessive loads of coarse sand and fine gravel, caused by various anthropogenic impacts, lead to deposits that affect the local flooding situation, form wide areas of mobile substrate and fill up the coarse sediments, with consequences for the habitat suitability of the interstices (Hauer 2015, Scheder et al. 2015).

Complementary to fine sediments, according to the largely accepted definition particles smaller than 2 mm in diameter (e.g. Wood & Armitage 1997, Waters 1995), the fines in the Bohemian Massif range mainly from 1 to 10 mm. This very mobile sediment fraction results from the weathering of the dominant granite and gneiss bedrock that forms the northern part of Austria as well as wide areas in Bavaria and Bohemia.

Depending on hydromorphological conditions, local geology and physico-chemical characteristics, there are variate risks of degradation arising from this sediment class. Furthermore, the mentioned effects overlap with the siltation caused by finer sediments (Höfler & Gumpinger 2014).

Funded by different departments and ministries of the Upper Austrian, Austrian and Bavarian governments a comprehensive study was carried out on this topic.

Investigations included a detailed mapping of more than 450 km in eight different river systems, providing data on the current status and reasons for the degradation and enabling the formulation of improvement approaches. According to this inventory hydraulic experiments were performed; hydrodynamic models on different scales as well as GIS-analyses were used to get a wide understanding of the processes both in detail and on catchment scale.

As the sediment regime is one of the most determining factors for the aquatic fauna an integrated data analysis on the freshwater pearl mussel (*Margaritifera margaritifera*) complement the study. Due to its complex lifecycle and its adaption to crystalline river systems, the freshwater pearl mussel is highly useable as an indicator species for the quality of the hyporheic interstitial and of the composition of the sediments in these ecosystems.

By means of intersecting the population development of relict *M. margaritifera* populations with sediment sieve curves and other collected data, a good understanding of the ecological effects of the high coarse sand loads – also in comparison to the effects of finer sediments – was gained.

Particularly the sediment mobility and the associated lack of structures lead to habitat deficits. This applies to macroinvertebrates, salmonids or different life stages of *M. margaritifera* (L.). While the sand and fine gravel fractions cover necessary structures and form highly mobile habitat patches, the finer particles cause a clogging of the interstitials. Following the results concerning *M. margaritifera*, the fine particles disturb the development of the juveniles, while the mobile sand dunes are responsible for the endangering of both juvenile and adult individuals.

REFERENCES

Hauer, C. 2015. Review of hydro-morphological management criteria on a river basin scale for preservation and restoration of freshwater pearl mussel habitats. Limnol. 50, 40–53.

Höfler, S. & Gumpinger, C. 2014. Erhebung der Feinsedimentbelastung in oberösterreichischen Alpenvorland Gewässern.

Scheder, C. et al. 2015. River bed stability versus clogged interstitial: Depth-dependent accumulation of substances in freshwater pearl mussel habitats in Austrian streams as a function of hydromorphological parameters. Limnol. 50, 29–39.

Waters, T.F. 1995. Sediment in Streams: Sources, Biological Effects, and Control. American Fisheries Society.

Wood, P.J. & Armitage, P.D. 1997. Biological effects of fine sediment in the lotic environment. Environ. Management 21: 203–217.

Current status, sources and effects of fine sediments in Upper Austrian streams

S. Höfler, C. Scheder & C. Gumpinger
Consultants in Aquatic Ecology, Austria

B. Piberhofer
Government of Upper Austria, Directorate Environment and Water Management, Austria

C. Hauer
University of Applied Life Sciences Vienna, IWHW, Austria

ABSTRACT

In the current scientific discussion high loads of fine sediments are considered one of the most important causes of river ecosystem degradation worldwide (see e.g. US EPA 2007 or Zhao et al. 2011, Waters 1997).

Especially in intensively used catchment areas changes in the sediment household must be regarded as a reason, which prevents the achievement of the objectives of the European Water Framework Directive (WFD).

Therefore, the Upper Austrian Water Authorities have launched two comprehensive studies on the topic, one being a survey on the current siltation status in Upper Austria, giving a first assessment of the specific causes (Höfler & Gumpinger 2014). The second, still ongoing study deals with two selected catchments in detail, in order to get a clear picture of the crucial hydrogeological processes and to develop possible role models for measures both in the catchments and in the streams.

For getting a first overview, 78 randomly distributed river sites in the intensively used agricultural areas in the central part of Upper Austria were investigated. In order to collect a wide range of information on the issue, a variety of methods for the classification of fine sediment stress was applied in each sampling site.

Furthermore, a GIS-based analysis was carried out for all examined catchments. The model included data gained from a digital elevation model, land use data and digital soil classification maps. These pieces of information were intersected with the results gained from the field studies.

Based on the results of this preliminary study river sections from two adjacent catchments with different degrees of agricultural use were selected for a more detailed study that is currently carried out: Chemical analyses of fine sediments and attached contaminants (e.g. heavy metals, decomposition products of pesticides, remains of medical drugs, sweeteners) are performed, and the benthic invertebrate fauna is investigated on the microhabitat level. Thereby it will be possible to enhance the understanding of the range of ecological impacts caused by silting-up in different hydro-morphological circumstances and with different fine sediment loads.

The results from the first module show that the watercourses in the intensively used central part of Upper Austria show a significant amount of degradation: About one third of the examined sites are considerably degraded by fine sediments, only about a fifth of the stretches exhibit a natural sediment composition.

In the course of the GIS analysis the following parameters turned out to be the determining factors for the fine sediment situation: local hydro-morphology, topographic catchment metrics, various characteristics of the banks, and the most significant indicators of all land use factors: the traffic-area-ratio, agricultural use and other intensive impacts.

According to these results a risk assessment tool for all watercourses is currently being developed and will be released by 2016, as will be the results of the chemical analyses and the survey on the benthic invertebrates. Preliminary results attest the presence of several kinds of contaminants in nearly all the samples. Due to the necessity of highly integrative improvement measures covering whole catchments, fine sediments must be expected to be one of the most challenging future topics in aquatic ecology.

REFERENCES

Höfler, S. & C. Gumpinger. 2014. Erhebung der Feinsedimentbelastung in oberösterreichischen Alpenvorland-Gewässern.

US EPA 2007. National Water Quality Inventory. Report to Congress. EPA 841-R-07-001.

Waters, T.F. 1995. Sediment in Streams: Sources, Biological Effects, and Control. American Fisheries Society.

Zhao et al. 2011. Macroinvertebrates in the bed sediment of the Yellow River. Int. J. of Sediment Research 26, 255–268.

Reconnection of the Danube floodplain channels as a vital step to restore river morphology and fluvial dynamics

K. Holubová, M. Čomaj & K. Mravcová
Water Research Institute, Bratislava, Slovakia

ABSTRACT

Ecosystems of large rives have been highly altered by regulation and fragmentation by dams. Like many other large rivers the Danube downstream of Bratislava (Slovakia) has been dramatically changed due to multiple pressures (e.g. river regulation, flood protection, commercial dredging, hydropower generation). Disrupted fluvial dynamics has adversely affected habitat diversity and the hydrological connectivity. Consequently, isolation of river and floodplain processes induced successive morphological and ecological degradation of the river ecosystem. Re-establishing of hydrological dynamics is recognized as the most vital step because other processes are influenced by the flow regime and resulting connectivity (Tockner et al. 2000).

The main goal of restoration concept developed for the Slovak-Hungarian Danube segment (∼100 km) is focused on reintroducing of fluvial dynamics and associated lateral connectivity between the main river and floodplain channels to improve habitat diversity preferably for rheophilic species and birds. The first part of this concept developed for one of the most altered reach of the Slovak–Hungarian segment of the Danube river (km 1810-km 1798) is presented in this article. Interaction between ecological restoration of floodplain channels and flood-risk reduction is also addressed.

Historical and present channel/floodplain morphology and related physical processes were considered for the development of restoration scenarios. These include combinations of measures for floodplain (reconnection of side arms, removal or modification of floodplain weirs) and the main river channel (adjustment of in-stream structures) to provide ecological improvement and larger *'room for the river'* to reduce flood risk. Using actual topography (floodplain-LiDAR, channel bathymetry-2013) two dimensional hydrodynamic model CCHE2D was set up to optimize flow conditions, side channels geometry and parameters of measures included in scenarios. 2D hydrodynamic model provides deeper insight into spatial variability of hydraulic characteristics for the main river channel and floodplain (including vegetation) showing the importance of floodplain topography in creating strong lateral gradients in depth, stage and velocity (Jacobson et al. 2015). Channel pattern, flow dynamics and frequency of hydrological connectivity are used as the main indicators to optimize the final scenario.

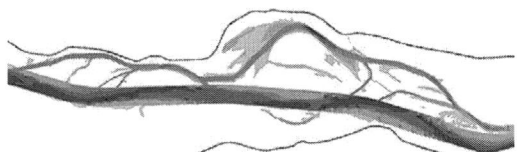

Figure 1. Water depth (Q = 3 000 m^{3}s^{-1}) for final restoration scheme of the Danube river reach (km 1810 – km 1800).

Highly modified large rivers cannot be restored to the pristine state. Nevertheless, reintroducing of fluvial dynamic on degraded reach of the Danube river can substantially contribute to the better hydro morphological and ecological situation that is closer to the natural state. Significant improvement of habitat diversity is already documented by abiotic monitoring from recently reopened smaller side arm in this locality. Reconnection of the floodplain channels in combination with lowering of groyne fields can contribute to the stabilization of incised river bed providing higher capacity for flood discharges. Mean decrease of water level ranges from 22 cm–31 cm for Q_{100} along the Danube reach. Integrated effect of ecological restoration and flood-risk reduction meets requirements of the EU directives for water and nature protection and flood protection.

Modeling of nutrient dynamics and vegetation succession, and comparison between with and without riverbed geomorphological simulation

H. Itoh, S. Yamauchi & M. Sugano
Civil Engineering & Eco-Technology Consultants, Tokyo, Japan

K. Sanjaya & T. Asaeda
Department of Environmental Science and Technology, Saitama University, Saitama, Japan

ABSTRACT

Recently, riparian zone of East Asian rivers are intensively forested, which provides a number of problems in the river management. Under the background, an individual based mechanistic model, Dynamic Riparian Vegetation Model (DRIPVEM) was developed to predict the vegetation succession of the riparian area (Asaeda et al. 2015). The model includes 4 modules, hydrological condition, tree, herb and soil nitrogen to describe the material flows and the influences between each group shown in Fig. 1.

After the validation of the model for the vegetation succession with the observed data of several rivers, the model was combined with the 2-dimensional river bed deformation model to evaluate the effect of the geomorphological change during the flood. For this objective, Nays2DH of the iRIC (International River Interface Cooperative) package (Shimizu et al. 2012, http://i-ric.org) is used.

The model was applied to a real river (Kuzuryu river in Japan) to obtain the effects of the geomorphological change during the flood on the vegetation patterns afterwards. Simulation was conducted either with the simulation of channel bed deformation or using the observed morphological data surveyed before and after the flood. In the former simulations, the riverbed morphology data is inputted at each time step in the DRIPVEM. General curvilinear coordinate system and quadrilateral mesh size of about 10 × 10 meters were used and the total mesh number was 6510. Simulation period was from 2011 to 2014, and 5 times of floods were experienced.

Figures. 2 (a) and (c) show the simulated results compared with the aerial photo (b). *Salix* community on the sand bar was successfully simulated in both cases.

However, with the geomorphological simulation, the distribution of tree area was simulated more accurately than the case of without. This is because the changes in the river bed shape is reflected in detail.

The sandbar was covered with the communities of both *Salix* spp. and *Humulus japonicus*. The simulated

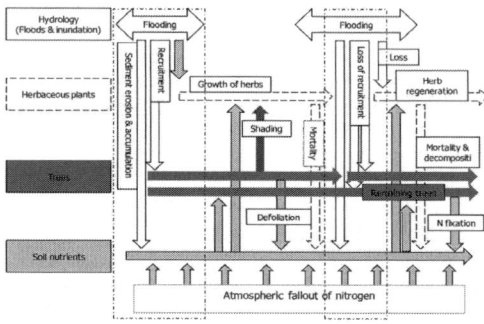

Figure 1. Schematic structure of the model.

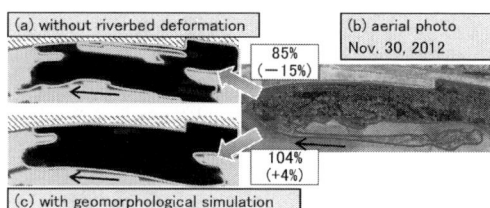

Figure 2. Simulation results compared with aerial photo after the flood.

distribution of *Humulus japonicus* by both models shows *Salix* spp. in the *Humulus* area.

For the practical application, therefore, the geomorphological simulation during the flood period is also required, although it takes longer period.

Moreover, the simulation results of future condition of vegetation indicate the possible application of the present model in the management of rivers.

REFERENCES

Asaeda, T., Rashid, M.H. & Bakar, R.A. 2015. Dynamic modelling of soil nitrogen budget and vegetation colonization in sediment bars of a regulated river, River Research and Applications 31, 470–484.

Shimizu, Y. et al. 2012. http://i-ric.org

Flow patterns, turbidity and sediment size distribution on the Luneplate tidal polder, Lower Weser

E. Kemayou Tchamako & B. Koppe
Institute for Hydraulic and Coastal Engineering, City University of Applied Sciences Bremen, Germany

U. von Bargen
bremenports GmbH & Co. KG, Director Environment & Sustainability at the Ports of Bremen/Bremerhaven, Germany

ABSTRACT

The coastal region of northwestern Germany could be particularly affected by a climate change induced increase in rising sea levels, intensified extreme rain falls and prolonged periods of drought. Adaption strategies to climate change and measures of protection are therefore highly relevant to this region and well supported by local institutions.

An ecological and economical sustainable method for developing lowlands in tidal areas, taking into account climate change effects, is the purpose of tidal polders. Tidal polders have the potential to damp tidal volumes, as well as to reduce suspended sediment concentrations and increase oxygen supply upstream (Donner et al., 2012; Knüppel, 2012). Parallel occurring sea level rise, together with tectonic shifts, can facilitate the accretion of wetlands by sedimentation and subsequently improve flood protection and a long-term increase in agricultural yield. Furthermore, new water management possibilities with flood retention areas and reservoirs against drought can prevent the deterioration of fenlands, thus contributing to NATURA 2000, the European Networking Programme for preserving natural habitat, and the EU Water Framework Directive.

Alongside the benefits and opportunities, which result from the regeneration of wetlands in tidal areas, there are also disadvantages and risks, such as limiting the customary agricultural use, salinization of surface and groundwater and maintenance work involved to ensure the planar accretion of areas by sedimentation. Furthermore, an enrichment of contaminants in the floodplain is possible.

In order to assess whether the advantages of creating tidal polders outweigh the risks of penetrating an existing dyke, it is essential to investigate the local hydraulics and sedimentology. This requires collecting and evaluating data for tide, river discharge, current patterns, salinity, and suspended material. Owing to the complexity of the hydraulic flow processes, calculating flow and morphological conditions in both as-is and as-planned state will be done numerically on the basis of field data.

In 2012, a tidal polder has been opened on the 'Grosse Luneplate' area in the Lower Weser, Germany, to compensate ecological impacts caused by the construction of the 4th Container Terminal of the Port of Bremerhaven. No study on sediment transport processes on this man-made intertidal area has been conducted yet. In cooperation with bremenports, the Institute for Hydraulic and Coastal Engineering of the City University of Applied Sciences Bremen started a monitoring program to understand the correlation between hydrodynamic and sedimentation processes in the Luneplate polder in July 2015.

The paper presents the results of field campaigns set as a first step for the set-up, calibration and validation of the Luneplate sediment transport model. Data of flow velocity and discharge as well as water depth, salinity and turbidity were measured along predefined cross sections both in the Lower Weser River and the tidal polder using an Acoustic Doppler Current Profiler (ADCP) and a Conductivity-Temperature-Depth sonde (CTD). Furthermore, sediment samples were collected to define sediment size distributions.

REFERENCES

Donner, M., Ladage, F., Stoschek, O., 2012: Impact and retention potential of tidal polders in an estuary with high suspended sediment concentrations. Proc. ICCE 2012, Santander.

Knüppel, J., 2012: 'Spadenlander Busch' (Elbe estuary). Measure analysis in the framework of the Interreg IVB project TIDE. Measure 01. Hamburg.

River Sedimentation – Wieprecht et al. (Eds)
© 2017 Taylor & Francis Group, London, ISBN 978-1-138-02945-3

Explicitly salinity and sediment concentration on flocculation processes in estuaries

A. Mhashhash, B. Bockelmann-Evans & S. Pan
School of Engineering, Cardiff University, Cardiff, UK

ABSTRACT

Cohesive sediment is primarily composed of clay minerals and organic matter (Manning et al. 2004), and has an ability to flocculate into large aggregates, namely flocs, which are bigger than individual particles but less dense. According to previous flocculation studies, flocs can be classified according to their size into micorflocs, which have a spherically equivalent diameter of 160 μm and settle at a settling velocity (Ws) of less than 1 $mm{\cdot}s^{-1}$. Furthermore, microflocs combine to form macroflocs with a diameter of more than 160 μm and which settle with settling velocities ranging from 1–15 $mm{\cdot}s^{-1}$ (Eisma 1986, Fennessy et al. 1994, Manning & Dyer 1999, Manning 2001). The flocculation process mainly occurs in the very low salinity region (between 1 and 2.5 ppt) (Wollast 1988), and it is affected by hydrodynamical changes which can alter the suspended sediment particle by modifying its effective particle size, shape, porosity, density and composition (Wollast 1988, Dyer 1994). Laboratory experiments were performed to investigate the effect of key hydrodynamic parameters including salinity (S) and suspended sediment concentration (SSC) on floc size and settling velocity. Experimental research was conducted in a 1 L glass beaker of 11 cm diameter using suspended sediment samples from the Severn Estuary. A non-laser based Particle Image Velocimetry (PIV) system, as shown in Figure 1, and an advanced image processing software were used to measure floc size distribution and settling velocity. This study found that flocculation is enhanced with increasing sediment concentrations from 100 to 250 mg/l at low shear stress of 0.57 $N{\cdot}m^{-2}$. The settling velocity was found to range from 0.2 and 1.3 mm/s^{-1}. Also, it was found that the effect of sediment concentration on settling velocity was controlled by the value of salinity. Where, the faster settling velocity rates occurred within the higher concentration, as well as in the lower S 2.5 ppt. whilst, at high salinity of 20 ppt alongside increasing sediment concentration, the situation was reversed, i.e., the lower sediment concentration the faster the settling velocity was found. A new formula for floc settling velocity was developed including floc size parameters and sediment concentration. This was developed using a Minitab 17 statistical package, and using data gained from the extensive set of experiments.

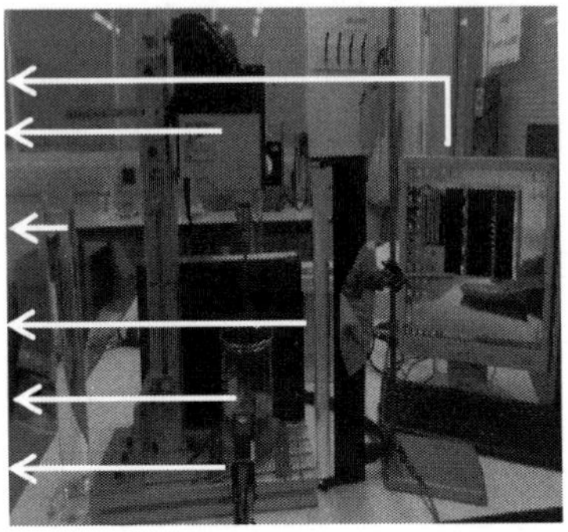

Figure 1. Laboratory setup for settling and flocculation measurements.

REFERENCES

Dyer, K.H. 1994. Estuarine sediment transport. In: PYE, K. (ed.), Sediment transport and depositional processes. Blackwell. Scientific Publications, pp. 193–218.

Eisma, D. 1986. Flocculation and de-flocculation of suspended matter in estuaries. Netherlands Journal of Sea Research 20(2-3), pp. 183–199.

Fennessy, M.J. et al. 1994. inssev: An instrument to measure the size and settling velocity of flocs in situ. Marine Geology 117(1–4), pp. 107–117.

Manning, A.J. 2001. Study of the effect of turbulence on the properties of flocculated mud University of Plymouth.

Manning, A.J. & Dyer, K. R. 1999. A laboratory examination of floc characteristics with regard to turbulent shearing. Marine Geology 160(1–2), pp. 147–170.

Manning, A.J., et al. 2004. Flocculation measured by video based instruments in the gironde estuary during the European commission SWAMIEE project. Journal of Coastal Research (SPEC. ISS. 41), pp. 58–69.

Wollast, R. 1988. The Scheldt Estuary. W. Salomon, B.L. Bayne, E.K. Duursma, U. Förstner (Eds.), Pollution of the North Sea, an assessment, Springer-Verlag, pp. 183–193.

River restoration: The need for a better monitoring agenda

M. Nones
Gerstgraser Ingenieurbüro für Renaturierung, Cottbus, Germany

ABSTRACT

In the last decades, especially after the new impetus given by the European Water Framework Directive (EU 2000, Wharton & Gilvear 2006) and the Directive on the Assessment and Management of Flood Risks (EU 2007, Müller 2013, Nones 2015a), several waterbodies were subjected to restoration measures to improve their ecological status and decreasing the flood risk at the same time (Nienhuis & Leuven 2001, Nones 2015b). Re-engineering channels to reinstate a more natural form and the restoration of water and sediment fluxes can bring multiple benefits, which are achievable because natural rivers are ecosystems that maintain high biodiversity, while the inundation of the floodplains can attenuate flows (Brierley & Fryirs, 2008). Despite numerous restoration projects realized across Europe, only a minor part of them are monitored for sufficient time and at a sufficient large scale in order to give valuable information about the projects outcomes. Moreover, where monitoring programmes of restored reaches are still active, a few quantitative information is available, in favor of more qualitative considerations (Palmer et al. 2007). Comparing different restoration projects, it is evident that monitoring programmes result scarce and not even distributed across the countries (Pedroli et al. 2002, Schiemer 2015), especially with regards to hydromorphological quality elements. The scarcity of monitoring outcomes is generally related to the scarcity of post-projects funds. To overcome this problem, it is necessary to include a long-term monitoring programme in the project agenda (Newson & Large 2006, Slowik 2015), because more outcomes are fundamental to evaluate the quality of the project at large scales, as well as the lessons learnt that can be applied to other river restoration works (Gumiero et al. 2013, Nones & Gerstgraser 2016).

REFERENCES

Brierley, G. & Fryirs, K.A. 2008. *River futures: an integrative scientific approach to river repair*. Society for Ecological Restoration International. Island Press, Washington.
EU, European Union. 2000. Directive of the European Parliament and of the Council 2000/60/EC Establishing a Framework for Community Action in the Field of Water Policy. Official Journal C513, 23.10.2000.
EU, European Union. 2007. Directive 2007/60/EC of the European Parliament and of the Council of 23 October 2007 on the Assessment and the Management of Flood Risks. *Official Journal* L288, 6.11.2007.
Gumiero, B., Mant, J., Hein, T., Elso, J. & Boz, B. 2013. Linking the restoration of rivers and riparian zones/wetlands in Europe: Sharing knowledge through case studies. *Ecological Engineering* 56: 36–50.
Müller, U., 2013. Implementation of the Flood Risk management Directive in Selected European Countries. *International Journal of Disaster Risk Science* 4(3): 115–125.
Newson, M.D. & Large, A.R.G. 2006. 'Natural' rivers, 'hydromorphological quality' and river restoration: a challenging new agenda for applied fluvial geomorphology. *Earth Surface Processes and Landforms* 31: 1606–1624.
Nienhuis, P.H. & Leuven, R.S.E.W. 2001. River restoration and flood protection: controversy or synergism? *Hydrobiologia* 444: 85–99.
Nones, M. 2015a. Implementation of the Floods Directive in selected EU Member States. *Water and Environment Journal* 29(2): 412–418.
Nones, M. 2015b. Sediment management of rivers and Water Framework Directive: the case of the Spree River. *36th IAHR World Congress*, 28 June–3 July, Delft-The Hague, the Netherlands.
Nones, M., and Gerstgraser, C., 2016. Morphological changes of a restored reach: the case of the Spree River, Cottbus, Germany. in *Hydrodynamic and mass transport at freshwater aquatic interfaces*. GeoPlanet Earth and Planetary Sciences Series. Eds. Springer International Publishing.
Palmer, M., Allan, J.D., Meyer, J. & Bernhardt, E.S. 2007. River restoration in the twenty-first century: data and experiential knowledge to inform future efforts. *Restoration Ecology* 15(3): 472–481.
Pedroli, B., de Blust, G., van Looy, K. & van Rooij, S. 2002. Setting targets in strategies for river restoration. *Landscape Ecology* 17(1): 5–18.
Schiemer, F. 2015. Building an eco-hydrological framework for the management of large river systems. *Ecohydrology & Hydrobiology*. in press. doi: 10.1016/j.ecohyd.2015.07.004.
Slowik, M. 2015. Is history of rivers important in restoration project? The example of human impact on a lowland river valley (the Obra River, Poland). *Geomorphology* 251: 50–63.
Wharton, G. & Gilvear, D.J. 2006. River restoration in the UK: Meeting the dual need of the European Union water Framework Directive and flood defence? *International Journal of River Basin Management* 4(4): 1–12.

Reconciling the debate on the impact of vegetation density on river channel braiding

I. Pattison & R. Roucou
School of Civil and Building Engineering, Loughborough University, UK

ABSTRACT

Braided rivers are very dynamic systems which have complex controls over their planform and flow dynamics. Vegetation is an important influence affecting the channel geometry and pattern. However, its effect is uncertain, with multiple previous field based, flume based and numerical modelling studies having been carried out. Previous research in this field has been contradictory; with Gran & Paola (2001) finding that increasing vegetation density decreased the number of active channels. In contrast, Coulthard (2005) observed that as vegetation become denser there was an increase in the number of channels. This was hypothesized to be caused by flow separation around vegetation and the development of bars immediately downstream of the plant. Understanding the effect of vegetation in these highly dynamic systems has multiple consequences for human activity and management. For example, increasing/decreasing channel stability for economic growth/infrastructure on floodplain, flood hazards, and ecological/habitat impacts.

This investigation aimed to assess the impact of vegetation density on channel platform over a wide continuum. A 4 m by 1 m flume was used, with discharge, slope (0.018) and sediment size (D50 = 0.6 mm) kept constant throughout all the experiments. Artificial grass was used to represent vegetation with a density ranging from 100 to 800 plants within the 2 m × 1m focus area of the flume. Digital photographs were taken from above every 30 seconds to observe channel pattern evolution during the 3 hour experiment.

The results of the experiments indicated that a low vegetation density (0–300 plants) the number of channel bars developing doubled from 12 to 24. Above 300 plants there was a decline in the number of bars created to a minimum of 8 for 800 plants.

We develop a simple conceptual model to explain these observations. At low plant densities, each plant acted independently and caused flow separation and convergence around each plant, similar to in the Coulthard (2005) experiment. At medium densities, individual plants start to interact together with

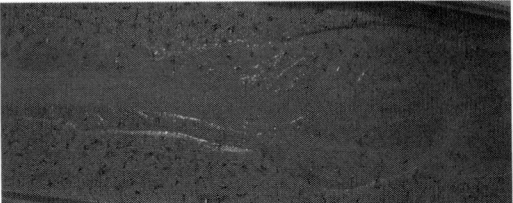

Figure 1. Photograph of example experiment.

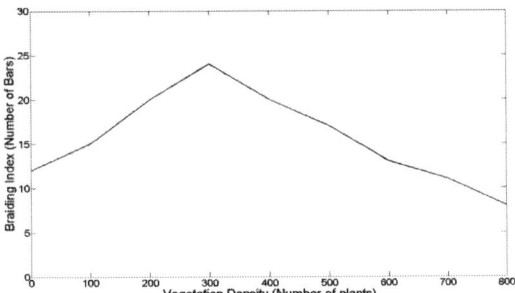

Figure 2. Effect of vegetation density bar development.

narrow channels developing longitudinally between vegetative bars. Finally, at very high densities, there was both lateral and longitudinal interaction between plants meaning that flow was diverted around them forming wandering, meandering channels.

In conclusion these sets of experiments clearly demonstrate that the relationship between vegetation density and channel planform is more complex than previous thought, taking a parabolic shape.

REFERENCES

Gran, K. & Paola, C. 2001. Riparian vegetation controls on braid stream dynamics, *Water Resources Research*, 37, 12, 3275–3283.

Coulthard, T. 2005. Effects of vegetation on braided stream pattern and dynamics. *Water Resources Research*, 41, W04003.

Analysing sediment characteristics of the alpine River Brixentaler Ache (Austria) including *in situ* measurements of dissolved oxygen

L. Seitz, M. Noack & S. Haun
Institute for Modelling Hydraulic and Environmental Systems, University of Stuttgart, Stuttgart, Germany

R. Reindl, G. Senn & M. Schletterer
TIWAG-Tiroler Wasserkraft AG, Innsbruck, Austria

ABSTRACT

The sediment characteristics of a river can indicate the microhabitat availability for different aquatic organisms. In this context, the clogging of riverbeds (colmation) can play a significant role because fine sediments can infiltrate into the riverbed and clog the pore space, which subsequently lead to a reduction in hydraulic conductivity and dissolved oxygen concentrations (Schaelchli 1992). Although colmation is a natural process, the occurrence of colmated riverbeds can be a result of anthropogenic activities such as a non-natural increase of the fine sediment input or the regulation of flow. For the latter, the suppressed peak flows cannot induce bed alterations and thus initiate decolmation processes, resulting in a strengthening of colmation development.

This study aims to investigate the sediment characteristics of a residual flow reach of the Austrian river Brixentaler Ache. Therefore, we investigated the bed sediments at three measuring stations within the residual flow reach and compared them to two measuring stations, which are each located up- and downstream of the residual river reach. In order to evaluate the sediment characteristics, we used three different methods: mapping, sediment sampling and dissolved oxygen measurements from the hyporheiczone.

Mapping allows both a quantitative assessment of the colmation status as well as its spatial distribution. The mapped sites have a length of approx. 200 m and each of them is subdivided into several zones which represent different substrate conditions.

In addition, we took sediment samples using both freeze core and freeze panel technique to ensure a minimal loss of fine sediments during sampling (Carling & Reader 1981). Whereas the sediment cores provide detailed information about the grain size distribution of the subsurface layer and about the vertical stratification, the sediment panels give detailed information about the surface layer. The combination of both methods allows for a proper assessment of the amount of infiltrated fine sediments, which clog the pores and reduce the pore space.

Moreover, we did in situ measurements of dissolved oxygen concentrations (Riss et al. 2008). Therefore, we installed small steel tubes in the riverbed at different depths (5, 10, 15 and 20 cm). The perforated tip of the tubes allows sucking hyporheic water to measure the vertical distribution of dissolved oxygen concentrations. A significant de-crease to a limiting oxygen concentration can serve additionally as an indicator for colmation since the transport of oxygen-rich surface water into the groundwater is disturbed. The tubes remained in the riverbed to observe the dissolved oxygen concentrations in the long-term.

So far, we cannot indicate any influence on the sediment characteristics due to flow regulation. The mapping shows clearly that remarkable depositions of fine sediments occur, predominantly at the riverbanks. However, these depositions occur also in the river reaches up- and downstream of the residual flow reach as result of the catchment characteristics. We found that the deposited sediments contain mainly sand fractions and almost no cohesive material. Thus, almost no colmation can be indicated in this river reach. This is confirmed by the measurements of dissolved oxygen, which show a marginal decrease over depth but not to values that are affecting aquatic organisms

REFERENCES

Carling, P.A. & Reader, N.A. 1981. A Freeze-Sampling Technique suitable for coarser river bed-material. Sediment. Geol. 29: 233–239.

Riss, W.H., Meyer, E.I. & Niepagenkemper, O. 2008. A novel and robust device for repeated small-scale oxygen measurement in riverine sediments - implications for advanced environmental surveys. Limnol Ocean. 6: 200–207.

Schaelchli, U. 1992. The clogging of coarse gravel river beds by fine sediment. Hydrobiologia 235: 189–197.

Correlation between the shelter of juvenile salmonids and bed substrate

M. Szabo-Meszaros, N. Rüther & K. Alfredsen
Norwegian University of Science and Technology, Trondheim, Norway

ABSTRACT

The present study proposes a new approach to evaluate the environment of the juvenile Atlantic salmon (Salmo salar) in the river bed. The goal of the study was to clarify and strengthen the correlation between a physical parameter of the substrate and the natural habitat of the young salmonids in a river section.

The interspaces between particles of the substrate are used by juvenile fish where they hide against predators and seek shelter from velocity. These shelters with different depths could be decreased when fine sediment is accumulated. Finstad et. al. (2007) devised a simple way to measure shelter using quadratic area (50 × 50 cm) with a plastic rubber tube (D = 13 mm). The shelters can be weighted and summarized (Forseth et al. 2014). According to the summarized shelters on an investigated river section the habitat of juvenile salmonids can be classified related to the shelter abundance.

Finstad et. al. (2007) showed that this method is useful to recognize potential locations related to salmons in their first life stage as fry and parr. However, the categorization is static, since a flood event could change the surface of the river bed and alter shelter abundance. To find a link between the potential shelter availability and their gravel environment, site specific samples of the bed surface should be taken, right after each shelter measurement. Hence grain size distribution can be determined in the laboratory and subsequently correlated to the observed shelter availability for that location.

The correlation between D5 and D10 parameters and the number of available shelters showed R2 values of over 0.9. Previous studies showed similar results (Jocham et. al. 2012). Therefore, the existing database was extended to avoid effects of randomness and to strengthen the statistical independence.

Furthermore, the continuation of the recent study is to implement the new relationship into the one-dimensional hydraulic model, in order to predict the development of shelter over time. Here, two main flow scenarios were investigated at two river sites. The first hydrograph consists of an average yearly runoff without the hydropower regulation. The second hydrograph is in correlation with the yearly energy production of the regulation power company.

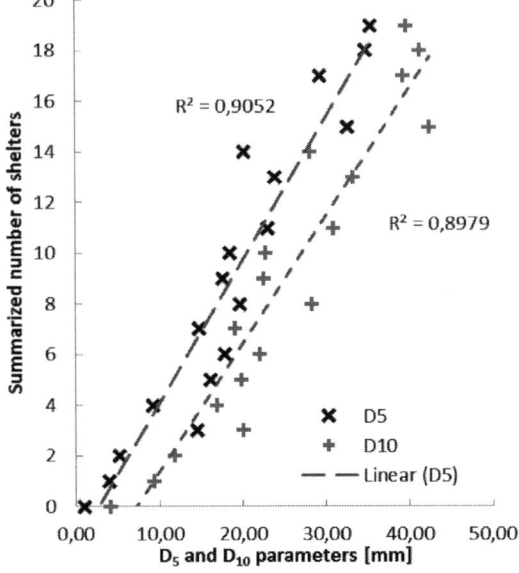

Figure 1. Correlation between D_5 and D_{10} parameters and number of weighted shelters with R^2 coefficients.

As a forerunner before the main simulations, sensitivity tests were run. The comparison at the end of the study contains the main differences between the effects of two types of hydrograph related to chosen and calculated parameters.

REFERENCES

Finstad, A. G., Einum, S., Forseth, T. & Ugedal, O. 2007. Shelter availability affects behavior, size-dependent and mean growth of juvenile Atlantic salmon. Freshwater Biology, 52, 1710–1718.

Forseth, T. & Harby, A. 2014. Handbook for environmental design in regulated salmon rivers. Trondheim, Norway: Norwegian Institute for Nature Research, 48–50.

Jocham, S., Ruther, N., Noack, M. & Sauterleute J.F. 2012. Correlation between shelter abundance and grain size distribution for the assessment of sediment quality for juvenile Atlantic salmon. 9th International Symposium on Ecohydraulics (ISE 2012), sep 17–sep 21.

Analysis of tidal effects on heavy metal transport in coastal aquifers

A. Tao
College of Civil Engineering, Tongji University, Shanghai, China
PowerChina HuaDong Engineering Corporation Limited, HangZhou, Zhejiang Province, China

S.G. Liu, S. Lou, C.M. Dai & B. Tan
College of Civil Engineering, Tongji University, Shanghai, China

R.S. Chalov & S.R. Chalov
Faculty of Geography, Moscow State University, Moscow, Russia

ABSTRACT

Many beaches are formed by thousands of years' sedimentation in the world. With the rapid development of urbanization in coastal areas, many municipal solid waste landfills tend to be built in the suburbs, most of which are near shorelines. Complex hydrodynamic conditions like tidal fluctuations may affect the transport of landfill leachate pollutants in sedimentation beach groundwater system and the seaward pollutants containing heavy metals will exert a profound influence on coastal environment as well as water resources. In order to make an analysis of tidal effects on groundwater movement and heavy metal transport, the author presents a calibrated numerical model of groundwater flow field and heavy metal transport field in a seaside landfill site based on the SEAWAT model.

The conclusions include that both water levels and lead ion concentration in all wells fluctuate with the period about 15 days due to the tidal fluctuations. The water level fluctuations lag time increases from 4.5 h in well 6 to 6 h in well 3 with the offshore distance increasing from 710 m to 990 m correspondingly. Lead ion travels about 34 m farther and 20% more towards the sea when tidal effects are taken into account. The results can provide a theoretical basis for controlling the heavy metal pollution and protecting the coastal resources and environment.

REFERENCES

Asadi-Aghbolaghi, M., Chuang M. H. & Yeh, H. D. 2012. Groundwater response to tidal fluctuation in a sloping leaky aquifer system. Applied Mathematical Modelling. 36(10), 4750–4759.
Ataie-Ashtiani, B., E. Volker, R. & Lockington, D. 2001. Tidal effects on groundwater dynamics in unconfined aquifers.
Chang, S.W. & Clement, T.P. 2013. Laboratory and numerical investigation of transport processes occurring above and within a saltwater wedge. Journal of contaminant hydrology. 147, 14–24.
Galarpe, V.K.R. & Parilla, R. 2014. Analysis of Heavy Metals in Cebu City Sanitary Landfill, Philippines. Journal of Environmental Science and Management. 17(1), 50–59.
Galitskaya, I., Pozdniakova, I. & Toms, L. 2011. Simulation of contaminant transport for contamination risk assessment. IAHS-AISH publication. 341, 172–178.
Hydrological Processes. 15(4), 655–669. Bakis, R. & Tuncan, A. 2011 An investigation of heavy metal and migration through groundwater from the landfill area of Eskisehir in Turkey. Environmental monitoring and assessment. 176(1–4), 87–98.
Li, L., Barry, D.A. & Pattiaratchi, C.B. 1997. Numerical modelling of tide-induced beach water table fluctuations. Coastal Engineering. 30(1), 105–123.
Mastrocicco, M., Colombani, N., Sbarbati, C. & Petitta, M. 2012. Assessing the Effect of Saltwater Intrusion on Petroleum Hydrocarbons Plumes Via Numerical Modelling. Water Air and Soil Pollution. 223(7), 4417–4427.
Motz, L. H. & Sedighi, A. 2013. Saltwater Intrusion and Recirculation of Seawater at a Coastal Boundary. Journal of Hydrologic Engineering. 18(1), 10–18.
Nielsen P. 1990. Tidal Dynamics of the Water Table in Beaches. Water Resources Research. 26(9), 2127–2134.
Nusari, M.S., Soom, M.A.M., Mohammad, T.A., Ghazali, A.H., Wayayok, A. & Zawawi, M.A.M. 2013. Assessment of seawater intrusion in Langat basin, Malaysia. Proceedings of the Institution of Civil Engineers-Water Management. 166(9), 501–15.

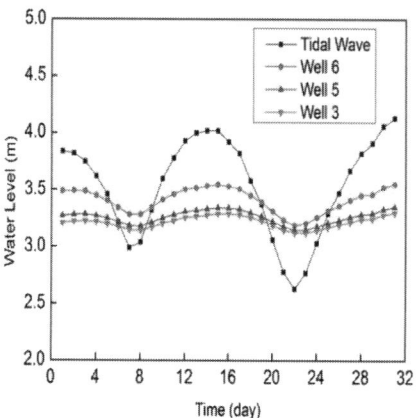

Figure 1. Water levels of observation wells and tide.

The role of surface adhesion in biostabilization processes

M. Thom
Forschungszentrum Küste, Leibniz Universität Hannover/TU Braunschweig, Merkurstrasse Hannover

H. Schmidt, S. Wieprecht & S.U. Gerbersdorf
Institute for Modelling Hydraulic and Environmental Systems, University of Stuttgart, Pfaffenwaldring, Germany

ABSTRACT

The morphodynamics of fluvial fine sediments are especially influenced by biostabilization processes. In this context, biofilms, consisting of algae and/or bacteria and their secreted extracellular polymeric substances, eventuelly glue single sediment grains together to form larger aggregates (or flocs) to impact the resistance towards erosion. Even though it is now well accepted, that adhesion forces (in combination with floc size and density) play a dominant role in this process, only little experimental data can be found in literature. Consequently, current mechanical models could not be validated so far.

In the four year project, "Ecosystem Engineering: Sediment entrainment and flocculation mediated by microbial produced extracellular polymeric substances (EPS)" a method to determine the relative adhesive capacity of surfaces was modified to quantify the surface adhesion forces and applied in four long term experiments on biostabilization under different environmental boundary conditions (light intensity and hydrodynamics). Results of these experiments along with floc size and -density analyses will be discussed in the proceedings and on the conference. These insights may be regarded as valuable information for predicting and understanding the mechanics of biostabilization processes.

Characterizing natural riparian plant stands for modeling of flow and suspended sediment transport

K. Västilä & J. Järvelä
Water Engineering, Aalto University School of Engineering, Espoo, Finland

ABSTRACT

Management of sediment and sediment-bound substances necessitates reliable modeling of in-channel processes, including flow hydraulics within riparian plant stands. For instance, notable amounts of cohesive sediment can be deposited within natural bank and floodplain vegetation. It is obvious that treatment of natural flexible vegetation in numerical models needs elaboration as they mostly parameterize plants as rigid cylinders without taking into account the flexibility-induced reconfiguration.

The purpose of this paper is to improve the parameterization of foliated riparian vegetation and to determine how selected factors affect net deposition of suspended sediment within natural floodplain plant stands. Based on the model of Västilä & Järvelä (2014) and field investigations, we show that the drag of woody plants can be parameterized taking into account the reconfiguration and the complex structure typical of natural riparian vegetation. Flow velocities at the field site were modeled using the two-layer approach of Luhar & Nepf (2013). To explain the net deposition under vegetative conditions ranging from almost bare soil to sparse willows and dense grasses, we constructed a regression model based on the cross-sectional vegetative blockage factor, distance from the suspended sediment replenishment point, and the modeled flow velocity within vegetation (Fig. 1). Net deposition decreased with increasing flow velocity. Further, it decreased with decreasing blockage factor and increasing distance from the sediment replenishment point, indicating that longitudinal advection was the most important mechanism supplying suspended sediment to the floodplain plant stands under real field conditions. The results showed that deposition of fine cohesive sediment can be supply-limited on narrow floodplains having continuous plant stands.

The findings can be used for managing sediment transport. For instance, the results suggested that vegetation dry mass and height above ∼200 g/m² and 0.1 m,

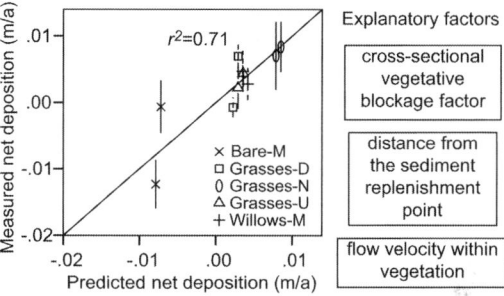

Figure 1. Annual net deposition in differently vegetated sub-reaches: measured vs. predicted by a regression model. The explanatory factors of the regression are shown on right.

respectively, cause deposition within natural grassy stands. To improve water quality, deposition can be enhanced by cutting floodplain vegetation from short, regularly-spaced sub-reaches along the channel while maintaining high vegetation elsewhere. The proposed drag parameterization can be easily implemented into hydraulic and morphological models to improve the description of flexible foliated plants. The vegetative properties needed for modeling can be derived e.g. with terrestrial laser scanning (Jalonen et al. 2015).

REFERENCES

Jalonen, J., Järvelä, J., Virtanen, J.-P., Vaaja, M., Kurkela, M. & Hyyppä, H. 2015. Determining characteristic vegetation areas by terrestrial laser scanning for floodplain flow modeling. *Water* 7(2): 420–437. doi: 10.3390/w7020420

Luhar, M. & Nepf, H. 2013. From the blade scale to the reach scale: A characterization of aquatic vegetative drag. *Adv. Water Resource.* 51: 305–316. doi:10.1016/j.advwatres.2012.02.002.

Västilä, K. & Järvelä, J. 2014 Modeling the flow resistance of woody vegetation using physically-based properties of the foliage and stem. *Water Resourc. Res.* 50(1): 229–245. doi: 10.1002/2013WR013819

The effect to the river environmental preservation of artificial flood in Satsunai River

Y. Watanabe
Kitami Institute of Technology, Kitami, Japan

K. Sumitomo
Docon Co., Ltd., Sapporo, Japan

S. Yamaguchi
Civil Engineering Research Institute for Cold Region, Sapporo, Japan

H. Yokohama
Obihiro Development and Construction Department, Hokkaido Regional Development Bureau, Obihiro, Japan

ABSTRACT

The Japanese government conducts an artificial flood by dam discharge for the purpose of preservation of river environment at the Satsunai river from 2012. The Satsunai river is a braided steep river (bed slope; 1/100-1/250). The gravel bars in this river were covered by riparian forest in recent years and river environment has been changed. At the beginning, artificial flood was generated for the purpose of disturbing a river bed. However, the discharge of the artificial flood is not enough to disserve the whole low water channel. Then, the practical use of the artificial flood to river channel disturbance should be discussed.

Watanabe et al. (2014) pointed out by field surveys on the artificial flood and the 2011 flood (return period; 20 year flood) that although river channel disturbance could not be disturbed by the artificial flood itself, maintenance of anabranch which induces river channel disturbance was possible by artificial flood. In response to this result, Yamaguchi et al. (2015) showed quantitatively and theoretically that maintenance of anabranch is possible by the artificial flood when the inflow part of anabranch is located in the just downstream at the pool of main channel.

This research was conducted using numerical simulation (iRIC Nays2D4.2) for the purpose of understanding quantitatively the effect of the artificial flood to river channel disturbance. The effect to the river channel disturbance of artificial flood was discussed from the following two viewpoints. The 1st viewpoint is maintenance of anabranch which induces river channel disturbance at the time of a large flood. The 2nd viewpoint is a scale of river channel disturbance at the time of a large flood. The difference between the existence or non-existence of the artificial flood has been discussed. Based on past research (Watanabe et al. 2014 & Yamaguchi et al. 2015), the section between the 42 km upstream and the 43km upstream from the

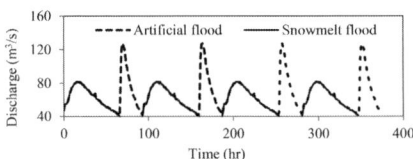

Figure 1. Discharge used for calculation.

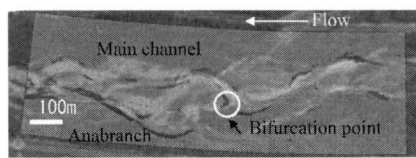

Figure 2. Bed variation after 4 years by snowmelt flood + artificial flood.

Tokachi River junction was selected as the investigation area. The discharge used for the calculation is shown in Figure 1. Example of channel change calculation results is shown in Figure 2. It was evaluated quantitatively that artificial flood is effective in maintenance of distributaries. Moreover, it was found quantitatively that the effect which mainstream alternation at bifurcations during large flood tends to produce is expected by the maintenance of distributaries.

REFERENCES

Watanabe, Y., Kawagishi, H. & Takeda, A. 2014. Restoration of the distributary of Satsunai River by an artificial flood. ISE 2014, Paper No. 90.

Yamaguchi, S., Watanabe, Y., Takeda, A. & Sumitomo, K. 2015. An effective method for restoring the former watercourses of a river where the watercourses are well established. Advances in River Engineering, Vol.21, 217–222. (in Japanese with English abstract).

iRIC Nays2D4.2 http://i-ric.org/en/.

Study on influence of waterway regulation engineering to fish habitat

Yanfen Geng
School of Transportation, Southeast University, Nanjing, China

Zhili Wang
*State Key Laboratory of Hydrology-Water Resources and Hydraulic Engineering,
Nanjing Hydraulic Research Institute, Nanjing, China*

ABSTRACT

As a traditional means of transportation, water transportation has many advantages, such as larger capacity, lesser energy consumption, lower cost and lesser land occupied. In recent years, water transportation in China has been rapidly developing. Among them, waterway regulation engineering plays a very important role in smoothing waterway; ensure waterway security and promoting sustainable economic development. Meanwhile, waterway regulation projects have threatened the ecosystem of rivers. In this paper, the three-dimensional hydrodynamic and sediment transport model (Wang & Geng 2013) combined with fish habitat model (Yi et al. 2014; Yi et al. 2007) are built to simulate the environmental and ecological responses for the waterway regulation engineering. The model was used to perdition the fish habit change by waterway regulation which for improving navigable depth at the Jieshou reach of Xijiang River, China. Figure 2 show the Habitat suitability index of homalopteridae fish, when the river flow is low, normal and flood flow. We compared the habitat area before and after the implementation (Fig. 1). The numerical results show that, waterway regulation increases the diversity of hydrodynamic environment and habitats. The better fish habitat area is increase during drought and flood period, and decrease slightly during middle-flow period.

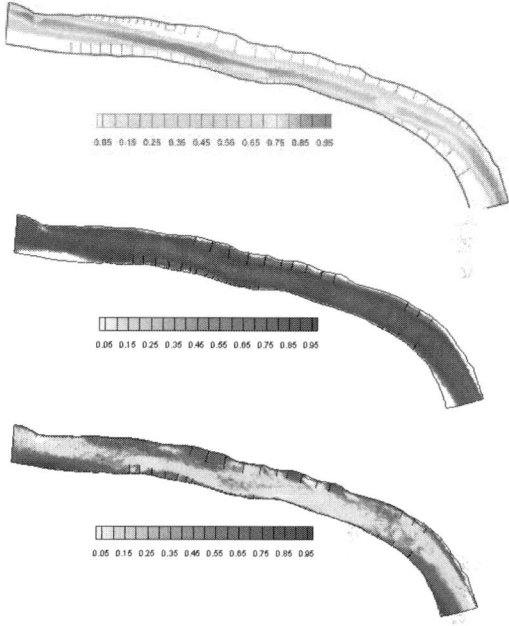

Figure 2. The habitat suitability index of homalopteridae fish Top: Low flow; Mid: normal flow; Bot: flood flow.

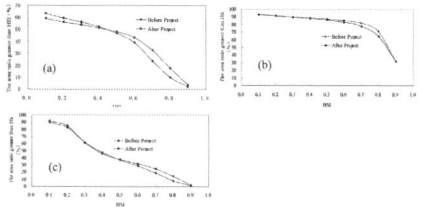

Figure 1. The area ratio that is greater than one HSI (a): Low flow;(b): normal flow; (c): flood flow.

REFERENCES

Yi, Y.J., Tang, C., Yang, Z., et al. 2014. Influence of Manwan Reservoir on fish habitat in the middle reach of the Lancang River Yujun. Ecological Engineering. (69): 106–117.

Yi, Y.J., Wang, Z.Y. & Lu, Y.J. 2007. Habitat suitability index model of Chinese sturgeon in the Yangtze River. Advances in Water Science,18 (4): 538–543. (In Chinese)

Wang, Z.L. & Geng, Y.F. 2013, A three-dimensional semi-implicit unstructured grid finite volume ocean model. Acta Oceanologica Sinica, 32(2): 68–78

Impact of biofilm on the sediment properties and its environmental effects

H.M. Zhao, W.H. Cao, L.Q. Tang, C.H. Wang, Y.H. Wang, D.B. Liu, C.S. Guo & J. Lu
China Institute of Water Resources and Hydropower Research, State Key Laboratory of Simulation and Regulation of River Basin Water Cycle, Beijing, China

Y.F. Zhang
Development Research Center of the Ministry of Water Resources of P.R.China, Beijing, China

ABSTRACT

In recent years, there has been an increasing awareness of the importance of aquatic micro ecosystem on water and sediment environment. Microorganism secrete biofilm on the sediment, forming new bioflocculation sediment and acting an significant impact on sediment properties and its environmental effects, which is important for the study on the interaction mechanism of various factors in water and sediment environment (Decho, 2000; Gerbersdorf et al., 2008; Tolhurst, 2008). In this paper, sediment particles are used as carriers to culture biofilm, and the change of the rheological properties and the adsorption rules on phosphorus element due to biofilm growth were investigated.

The effect of biofilm on rheological property of sediment is first studied by the rheological experiment. The current research demonstrates that biofilm growth significantly affects the properties of the sediment, supporting the field measurements that have shown significant differences in the rheological properties between sediment with biofilm and those without biofilm. It is found that the rheological property of biofilm sediment shows the characteristics of plastic body with thixotropy. Sludge of biofilm sediment has different rheological behaviors, mostly as a consequence of diverse biofilm compositions. In the modeling of biofilm sediment dynamics, bioflocculation sediment is generated by the biofilm acting in the manner of "bridge" on the particle surfaces resulting in the generation of flocculation. The transport mechanisms of different sediment types are somewhat different. The forces acting on sediment particles, including the shear stress, should be calculated to make a judgment of in which type the sediment is transported, which can be calculated based on the rheological equations for the biofilm sediment and the expressions of rheological parameters to be proposed in further study.

Further, phosphorus compound KH_2PO_4 is taken as adsorbent to study the influence of biofilm on phosphorus adsorption rules of sediment particles through the isothermal equilibrium adsorption experiment. The study provides the data evidence for the interaction of sediment particles, biofilm and elements. The Langmuir adsorption isotherm is adopted to analyze the influence of biofilm on phosphorus equilibrium adsorption capacity of sediment particles. The results show that the phosphorus adsorption capacity of sediment particles increase with the growth of biofilm. The changing rules of adsorption isotherm parameters along with the biofilm growth for different time has been obtained. Ongoing studies therefore seeks to extend this data set to quantitative research on interrelation between adsorption rules on phosphorus element of sediment particles and biofilm amount.

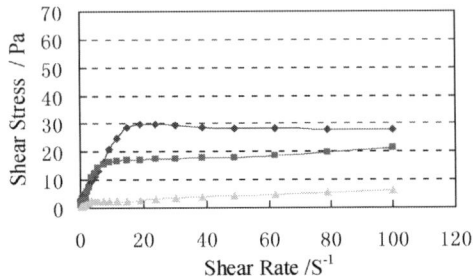

Figure 1. Rheological curves of the sediment with and without biofilm.

Table 1. The Langmuir adsorption isotherm parameters of sediment with and without biofilm

	S_m	K_l	$1/K_l$	$S_m \cdot K_l$
Biofilm sediment	2.791	0.166	6.013	0.464
Original sediment	1.572	2.162	0.463	3.398

REFERENCES

Decho, A. W., 2000. Microbial biofilms in intertidal systems: An overview. *Continental Shelf Research* 20: 1257–1273.

Gerbersdorf, S.U., Jancke, T., Westrich B. and Paterson, D.M., 2008. Microbial stabilization of riverine sediments by extracellular polymeric substances. *Geobiology* 6(13): 57–69.

Tolhurst T.J., Consalvey M. and Paterson D.M., 2008. Changes in cohesive sediment properties associated with the growth of a diatom biofilm. *Hydrobiologia* 596: 225–239.

E. Reservoir sustainability

Impact of a single dam on sediment transport continuity in large lowland rivers

Z. Babiński & M. Habel
Kazimierz Wielki University, Bydgoszcz, Poland

ABSTRACT

According to the research, the waters of the lower Vistula may, on average, transport 2.2 million tons per year of bed load. The lower Vistula currently does not comprise a homogeneous fluvial system. Due to human activity channel lacks its typological continuity – as a result of existing of a dam and channel regulation devices. It was assumed that the reservoir intercepts approximately 42% of suspension and all (100%) of bed load material (Babiński 2005). Disruption of river load movement continuity due to the existence of an artificial reservoir results in increased bottom erosion below the dam, and thus, river tends to replenish the material transported. The process of bottom erosion contributes to the renewal of bed load (partially suspension) accumulated in the upper part of the reservoir, and results in occurrence of an erosion zone (Habel 2013). The zone of intensive bottom erosion below dams occurs within certain reach of the river and relocates in time, at various rate (Williams & Wolman 1984). When the Vistula river reaches its annual mean value of bed load transport (in the case of the lower Vistula the said value amounts to approximately 0.7 mln m^3 – Babiński 1992), which is regarded as a threshold value for the river transport power, the load is "discharged", forming a reach of forced accumulation – functioning of accumulation zone (Fig. 1). For instance, the lower Vistula, within the accumulation reach, takes on the form of a typical braided river with numerous central and lateral bars.

The overriding aim of this paper is to show the process of quantitative and qualitative river load renewal on the reaches below a single dam on alluvial river.

Conducted research on different channel reaches (bathymetric measurements, geodetic measurements of sand bars dynamics, sedimentological analysis and aerial reconnaissance) allowed to get to know different conditions of bed load transport in terms of its size, type of material and mode of transport.

Figure 1. Lower Vistula channel with the accumulation zone, created by alternate and central sand bars, below the erosion stretch, about 60 km downstream from the Wloclawek dam.

REFERENCES

Babiński, Z. 2005. The relationship between suspended and bed load transport in river channels, (in:) Walling D.E., Horowitz A.J. (ed.:) Sediment Budgets 1, IAHS Publication, 291 Proceedings of symposium S1 held during the Seventh IAHS Scientific Assembly at Foz do Iguacu,Brasil): 134–147.

Babiński, Z. 1992. Współczesne procesy korytowe dolnej Wisły, Prace Geograficzne nr 157. Polish Academy of Science, Warsaw.

Habel, M. 2013. Dynamics of the Vistula River channel deformations downstream of Włocławek Reservoir, Kazimierz Wielki University Press, Bydgoszcz, p.144 DOI: 10.13140/2.1.4606.0803

Williams, G.P. & Wolman, M.G. 1984. Downstream effects of dams on alluvial rivers, Geol. Survey Profes. Paper, 1286, Washington.

Experiences of controlled sediment flushing from four alpine reservoirs

M.L. Brignoli & P. Espa
Departmetn of Science and High Technology, University of Insubria, Varese, Italy

S. Quadroni & G. Crosa
Department of Theoretical and Applied Sciences, University of Insubria, Varese, Italy

G. Gentili
Graia s.r.l., Varano Borghi, Varese, Italy

R.J. Batalla
Department of Environment and Soil Sciences, University of Lleida, Catalonia, Spain

ABSTRACT

This paper summarizes the main findings concerning 10 controlled sediment flushing operations carried out between 2006 and 2012 at four Italian Alpine reservoirs located in the Lake Como catchment (Table 1).

The operations displayed some common features as follows. The dose-response model by Newcombe & Jensen (1996) was applied to establish the duration of the works and the suspended sediment concentration (SSC) thresholds of the outflowing waters (typically average values on the entire working period). The respect of these limits was verified through continuous SSC measurements by optical turbidimeters. The reservoirs were completely draw-down and mechanical equipments were employed to dislodge the sediments during daytime. SSC patterns detected downstream were generally characterized by regular daily pulses. The streamflow in the rivers below the reservoirs was increased through regulation of the tributaries intakes to enhance both dilution and transport capacity. The riverbed alteration was estimated through visual inspection and McNeil sampling. The monitored river reaches display as well similar hydromorphologic characteristics. The streamflow is highly regulated due to massive hydropower exploitation and it is generally close to the established minimum flow (5–10% of the mean annual flow), with further contributions due to the residual basins. The investigated channels have slope in the order of few percent, and frequently show morphological alterations (embankments and grade control structures). The riverbed substrate is dominated by pebbles, cobbles and boulders.

On the other hand, the main differences between the analyzed operations are shown in Table 1. They were repeated up to 4 subsequent years and their duration ranged between 3 and 53 days. SSC limits vary in relation to the ecological quality of the impacted rivers and the actual possibility of implementing control measures. Different flushing seasons were scheduled mainly due to the size of the outlet facilities and the needs of other river stakeholders. The flushing season affected both the runoff in the emptied reservoirs and the water availability for the transport and dilution of the evacuated sediments. The grain size of the flushed sediments was almost completely below the sand limit (the sand content varying from negligible to 75%). Therefore, the SSC detected at the downstream monitoring sites were very different, ranging from 0.3 to 4.7 g/l as average values. Sediments deposited along the investigated reaches vary both in terms of grain size and mass per unit area (from few to 300–450 kg/m^2). The cost per unit of evacuated sediment volume ranged between 5 and 45 €/m^3, mainly due to the different technical difficulties occurred during the dislodging works.

Table 1. Capacity (Cap.) of the flushed reservoirs and details of the monitored flushing operations (Y = Years of flushing, D = Duration, SSC$_{lim}$ = SSC limit, L = distance from dam of the SSC monitoring sites where the SSC limits were fixed and measured, SSC$_{ave}$ = average SSC, Q$_{ave}$ = average streamflow, and SSM = SS Mass are reported).

Reservoir	Valgrosina	Cancano	Sernio	Madesimo
Cap. (mm^3)	1.3	124	0.7	0.13
Y (years)	4	3	2	1
D (days)	12–13	40–53	6–16	3
Cost (€/m^3)	35	45	5–15	6–7
L (km)	6.0	22.9	0.9	3.1
SSC$_{lim}$ (g/L)	4.0	3.0	1.5	10
SSC$_{ave}$ (g/L)	3.0–4.7	0.3–0.8	0.7–0.8	2.5
Q$_{ave}$ (m^3/s)	3.2–4.6	3.4–4.8	60–70	10
SSM (10^{3}t)	14–19	15–72	24–75	8

REFERENCE

Newcombe, C.P. & Jensen, J.O.T. 1996. Channel suspended sediment and fisheries: a synthesis for quantitative assessment of risk and impact. North American Journal of Fisheries Management, 16: 693–727.

On the vertical turbulent interaction of non-Newtonian fluid mud

O. Chmiel & A. Malcherek
University of the German Armed Forces Munich, Institute of Hydro Sciences, Hydromechanics and Hydraulic Engineering, Neubiberg, Germany

M. Naulin
Federal Waterways Engineering and Research Institute, Hamburg, Germany

ABSTRACT

Fluid mud can be described as a high concentrated aqueous mixture of fine-grained sediment and organic material (McAnally et al. 2007). In reservoirs where siltation occurs, fluid mud layers can be observed (White 2001). This results in increasing maintenance costs for the reservoir management and has to be considered within different sediment removal operations (van Rijn 2013). During the transport of suspended cohesive sediment, density induced vertical stratification effects occur. Differently concentrated vertical layers can be defined over water depth (Winterwerp & Van Kesteren 2004). The viscous, Newtonian behavior of low concentrated suspensions can be described by turbulence models and the viscoelastic, non-Newtonian behavior of fluid mud by rheological models (Malcherek & Cha 2011, Wehr 2012). However, to characterize the dynamics of muddy reservoirs a continuous modelling approach from damped to free turbulence in suspensions is necessary (Roland et al. 2012). This approach is implemented in an implicit, in stationary and fully coupled 1DV-model. The model solves the one-dimensional momentum equation together with the transport equation for suspended matter and a modified two-equation k–ε turbulence model. To account for the rheological viscosity in the fluid mud layer, the viscosity in the momentum equation is modified as the sum of the turbulent and the rheological viscosity, see Equation 1.

$$v_{eff} = v_t + v_{rh} \quad (1)$$

Accordingly, the shear stress relation including the transition from Newtonian to non-Newtonian flow behavior can be stated, see Equation 2, τ_y being the yield stress of the fluid mud.

$$\frac{\tau_{zx}}{\rho} = v_t \frac{\partial u}{\partial z} + \frac{\tau_y}{\rho} + v_{rh} \frac{\partial u}{\partial z} \quad (2)$$

Results of the 1DV-model, showing the vertical effective viscosity will be presented and discussed at the conference. A continuous transition of a parabolic turbulent viscosity profile in the region of low concentrated suspensions to the point of increasing rheological viscosity in the fluid mud layer is recognizable.

For the validation of this approach, laboratory experiments are conducted. Stratified flow conditions are simulated in a circulating flume at which the density and the turbulent velocity fluctuations will be analyzed by ADV measurements.

REFERENCES

Malcherek, A. & Cha H. 2011. Zur Rheologie von Flüssigschlicken: Experimentelle Untersuchungen und theoretische Ansätze – Projektbericht. Technical report, Universität der Bundeswehr München, Institut für Wasserwesen.
McAnally, W.H., Friedrichs, C., Hamilton, D., Hayter, E., Shrestha, P., Rodriguez, H., Sheremet, A. & Teeter A. 2007. Management of fluid mud in estuaries, bays, and lakes. i: Present state of understanding on character and behavior. J. Hydraul. Eng. 133(1), 9–22.
Roland, A., Ferrarin, C., Bellafiore, D., Zhang, Y.J., Sikric, M.D., Zanke, U. & Umgiesser, G. 2012. Über Strömungsmodelle auf unstrukturierten Gitternetzen zur Simulation der Dynamik von Flüssigschlick.
van Rijn, L.C. 2013. Sedimentation of sand and mud in reservoirs in rivers.
Wehr, D. 2012. An Isopycnal Numerical Model for the Simulation of Fluid Mud Dynamics. Ph. D. thesis, Universität der Bundeswehr München – Institut für Wasserwesen.
White, W.R. 2001. Evacuation of sediments from reservoirs. Thomas Telford. Winterwerp, J.C. & Van Kesteren, W.G. 2004. Introduction to the physics of cohesive sediment dynamics in the marine environment. Elsevier.

Improving the RESCON approach

N. Efthymiou, S. Palt, P. Pintz & P.K. Thapa
Fichnter GmbH & Co. KG, Stuttgart, Germany

G.W. Annandale
Consultant, Denver, USA

P. Karki
World Bank Group, Washington DC, USA

ABSTRACT

The REServoir CONservation (RESCON) approach developed and published by the World Bank (Palmieri et al. 2003) proved a valuable tool for rap-id assessment of expected reservoir sedimentation, and identification of technically feasible and economically optimal reservoir sedimentation management techniques. Since, understanding of reservoir sedimentation management strategies and under-standing of the impact of climate change on the need for storage and reservoir sustainability improved (Annandale 2013 & Xie et al. 2011), prompting the World Bank to update the RESCON software. The updated RESCON model will continue to respect the character of the tool for rapid assessment, based on easily accessible data, in order to identify suitable reservoir sedimentation management strategies at policy and pre-feasibility project level ensuring sustainable development.

The purpose of this paper is to present the new capabilities that have been recently added to the RESCON approach and to provide an illustration of the model performance with regards to a specific case study. The revised model is now capable of per-forming rapid assessment of the technical feasibility and economic optimality of catchment management, sediment removal (drawdown flushing, dry excavation, dredging and hydro-suction) and sediment routing (sluicing, sediment by-pass and density current venting). In addition, the upgraded model is able to allocate the sediment deposits among active and dead storage pools. The issue of climate change and its impact on reservoir sediment management is addressed through a sensitivity analysis, which allows the user to identify possible risks. Finally the economic optimization procedure has been complemented by the concept of a declining discount rate in order to account for the nature of reservoir storage as a renewable resource.

The model is applied for the case study of PB Soedirman reservoir in Java, Indonesia, where sediment management will be applied by means of dredging. The paper presents the very good agreement between the assessment of RESCON and the results of analytic studies for the selection of the most suitable sediment management strategy.

REFERENCES

Annandale, G. 2013. Quenching the Thirst: Sustainable Water Supply and Climate Change, CreateSpace, Charleston, S.C.

Palmieri, A. Farhad, S., Annandale, G.W. & Dinar, A. 2003. Reservoir Conservation – the RESCON Approach, World Bank, Washington D.C.

Xie, J., Annandale, G.W. & Wu, B. 2011. Reservoir Capaci-ty-potential Power Generation-reliability Estimation Based on Gould-Dincer Approach. In: Valentine, EM, CJ Apelt, J Ball, H Chanson, R Cox, R Ettema, G Kuczera, M Lambert, BW Melville, and JE Sargison, Eds. Proceedings of the 34th World Congress of the International Association for Hydro-Environmental Research and Engineering: 33rd Hy-drology and Water Resources Symposium and 10th Confer-ence on Hydraulics in Water Engineering. Barton, A.C.T.: Engineers Australia, 2011: 1811–1817.

Designing reservoir sediment management alternatives with automated concentration constraints in a 1D sediment model

S. Gibson
US Army Corps of Engineers, Hydrologic Engineering Center, Davis, California, USA

P. Boyd
US Army Corps of Engineers, Omaha District, Omaha, Nebraska, USA

ABSTRACT

Sustainable sediment management strategies often have concentration constraints or triggers. Downstream concentration limits often constrain reservoir flushing operations, which draw down reservoir pools to run of river conditions, scouring deposited sediment. Since flushing events release more sediment than the river supplies during the event, they can invoke regulatory standards on releases.

Alternatively reservoir routing (Morris & Fan, 1998) alternatives pass sediment laden flows, (e.g. flood flows) through the reservoir minimizing sediment deposits and extending the reservoir life. Monitoring *inflowing* sediment concentrations and building reservoir alternatives around upstream concentration triggers can optimize these alternatives.

Downstream concentration constraints and upstream concentration triggers can be difficult to implement in sediment transport models. They must be determined iteratively, running the model until it exceeds the trigger threshold, then adjusting the operation and re-running the model. This is time consuming and tedious for short, single event models. It is impossible for long term, period of record simulations with multiple flushing or routing operations.

HEC-RAS version 5.0 includes concentration controls in the operational, making it possible to operate gates and structures based on upstream or downstream concentrations. These rules were applied to a reservoir flushing model of Spencer Dam, on the Niobrara River, a tributary of the Missouri River. First the model was calibrated to a reservoir flush (Fig. 1) and a period of reservoir deposition. Then downstream concentration constraints and upstream concentration triggers were added to the model to simulate alternate flushing (Fig. 2) and routing alternatives respectively. The model used these user defined concentration dependencies and automatically computed dam operations that adhered to defined downstream concentration constraints.

Figure 1. Flushing event at Spencer Dam on the Niobrara River in Nebraska, USA.

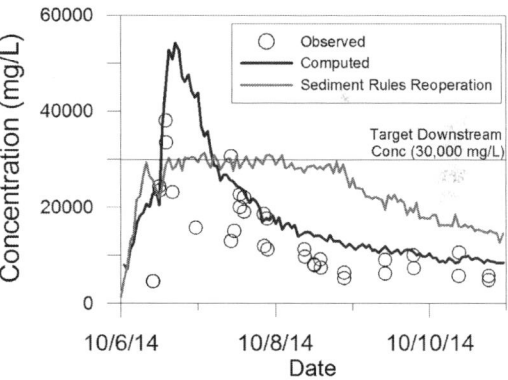

Figure 2. Measured and computed concentration downstream of a flushing event at Spencer Dam.

REFERENCE

Morris, G.L. & Fan J. .1997. Reservoir Sedimentation Handbook: Design and Management of Dams, Reservoir and Watersheds for Sustainable Use, McGraw-Hill, New York.

Reservoir sedimentation issues in India as a part of Dam Rehabilitation and Improvement Project (DRIP): Field reconnaissance and modelling

S. Giri & M. Nabi
DELTARES (former Delft Hydraulics), Delft, The Netherlands

P. Cleyet-Merle
EGIS-Eau, Montpellier, France

B.R.K. Pillai
Central Water Commissions, New Delhi, India

ABSTRACT

Under Dam Rehabilitation and Improvement Project (DRIP) in India, there are concerns from some electricity boards and dam authorities to explore the possibilities for sediment management of some reservoirs, which are not only losing the storage capacity, but also under the threat of malfunctioning of power intakes, sluices and other apparatuses due to rubbles and boulders as well as siltation including consolidation of the deposited materials.

This paper presents our study and assessment on reservoir sedimentation issues as well as desiltation plan and approaches for two reservoirs, namely Pillur in south (Tamil Nadu) and Maneri Bhali I in north (Uttarakhand) part of India. The alteration in nature, magnitude and severity of the problems in these two entirely different reservoirs demonstrates the complexity and uniqueness of the problems, indicating need for distinctive and tailor-made approaches to address them. Proposed desilting approaches for these two reservoirs are distinctive based on: (i) size of the reservoir, (ii) volume of the deposited material, (iii) sediment inflow, (iv) downstream condition, (v) availability of land for disposal, (vi) morphological feature of the reservoir, (vii) minimum generation loss, (viii) environmental and social impacts, and (viii) location and accessibility.

The complexities associated with silt removal approach are not just technical, but also other nuances like the cascade scheme of dams and reservoirs as well as the fact that some of the reservoirs are located in the neighborhood of eco-sensitive area. Besides, apart from electricity generation reservoirs also serve for water supply, irrigation as well as recreational purposes. This implies that the upstream and downstream effects are of the major concerns, and thus the environmental, ecological and social compliances are supposed to be ensured while preparing remedial measures and plan.

In this study, since there are not much data, analyses have been made based on brief field reconnaissance

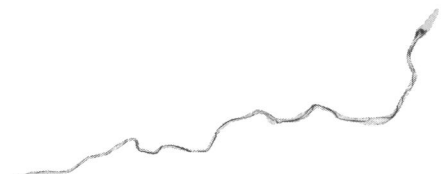

Figure 1. Morphological feature, simulated by the model.

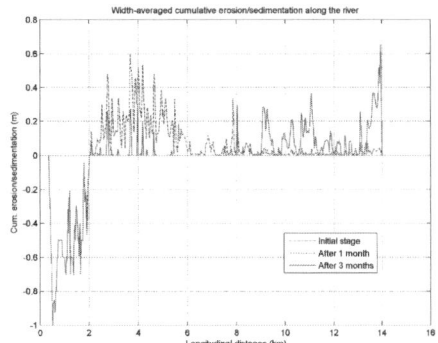

Figure 2. Sedimentation (+ve) and erosion (−ve) of the deposited material and their evolution with time.

and collection of information. Additionally, attempt has been made to demonstrate application of numerical model so as to replicate some synthetic scenarios associated with hydrodynamic and morphological impacts of different interventions and measures, such as morphological evolution along downstream river reach, propagation of replenished sediments, effects of sediment removal from the reservoir and effects of traps. For example, the model has been used to develop qualitatively realistic bathymetry (shown in Fig. 1), which is subsequently used for some scenario simulations and analysis like propagation of deposited silt, which is discharged to downstream (shown in Fig. 2). Based on these studies, recommendation has been proposed.

Sedimentation in rivers and reservoirs following the eruptions of Kelut Volcano, Indonesia

F. Hidayat
Brawijaya University and Jasa Tirta I Public Corporation, Malang, Indonesia

P.T. Juwono, A. Suharyanto, A. Pujiraharjo & D. Sisinggih
Brawijaya University, Malang, Indonesia

D. Legono
Gadjah Mada University, Yogyakarta, Indonesia

ABSTRACT

Volcanic eruptions and subsequent erosional processes deliver enormous quantities of sediment to rivers and reservoirs located near volcanoes. Mobilized eruption material moves rapidly as lahars that can damage structures severely along their way and deposit sediment along a valley floor. Flood-transported sediment can cause similar damage but over a longer period of time, and post eruption sediment transport can cause severe riverbed aggradation and reservoir sedimentation.

Kelut or Kelud Volcano, in eastern Java, is one of the most active and hazardous volcanoes of Indonesia. Historically, Kelut has produced some of Indonesia's most deadly eruptions, which typically include pyroclastic flows and lahars from a crater lake at the summit. During the past century, eruptions occurred in 1919, 1951, 1966, 1990, 2007 and 2014. The eruptions produced devastating lahars, pyroclastic surges and flows as well as ash fall deposits. The volcanic materials produced by an eruption are estimated to be the order of 30 to 160 million cubic meters.

Enormous materials produced by Kelut Volcano's eruptions mixed with flood waters and flowed to the Brantas River subsequently becoming riverbed sediment. Problems associated with riverbed sediment accumulation include reduced channel carrying capacity, resulting in more frequent over-bank flows and greater flood damage to adjacent properties. During the 1951–1970 period, the riverbed of the Brantas River aggraded by about 1.5 m on average (OTCA, 1973). The maximum riverbed aggradation was observed about 1.7 m in 1955 (Nippon Koei, 1961). The channel capacity decreased and sediment had deposited around the intakes and canals of the irrigation systems. Thus difficulties arose in protecting the land from floods in the rainy season and abstracting irrigation water in the dry season.

The eruption of Kelut Volcano in August 1990 caused severe sedimentation in Wlingi and Lodoyo reservoirs, which are located on the skirts of Kelut Volcano. To cope with sedimentation problems in these reservoirs, some strategies have been applied such as reducing sediment inflow from upstream by constructing onstream structures, passing sediment to minimize sediment trapping by constructing sediment bypass channel, and recovering the reservoirs' storage by dredging and empty flushing.

Following the eruption of Kelut Volcano in February 2014, the deposit in the northern skirts of Mt. Kelut became debris flowing to Konto River, triggered by heavy rainfall a few days after the eruption occurred. While the deposit in the southern skirts will potentially become debris flowing to Brantas River through Lekso, Jari, Semut, Putih, Badak Rivers. This paper will discuss the problems on sedimentation in those rivers and reservoirs located near Kelut Volcano i.e. Selorejo, Wlingi, and Lodoyo reservoirs after its eruption. The impact of debris flow to the damages of structures such as bridges, irrigation pond and channels, waterways, hydropower facilities, syphons, check dams, etc. from the results of field observations following eruption of Kelut Volcano in February 2014 will also be discussed.

REFERENCES

Nippon Koei. 1961. Comprehensive Report on the Kali Brantas Overall Project. Ministry of Public Works and Power, Government of the Republic of Indonesia

Overseas Technical Cooperation Agency (OTCA). 1973. Report on the Brantas River Basin Development. OTCA, Tokyo. Japan

Modelling deposition, consolidation and erosion of cohesive sediments in the Upper Rhine

T. Hoffmann & G. Hillebrand
Federal Institute of Hydrology, Koblenz, Germany

M. Noack
Institute for Modelling Hydraulic and Environmental Systems, University of Stuttgart, Germany

ABSTRACT

The Rhine is one of the most important waterways in the world. To secure navigability and to generate energy, 10 dams were built along the upper Rhine between Basel and Iffezheim. Since their construction in the 20th century, the dams disconnect the sediment transport along the Rhine and retain large amounts of sediment in the reservoirs upstream of these dams. Due to historic emissions of particle-bound micropollutants, the fine sediments that are deposited in the reservoirs are highly contaminated increasing the costs of maintenance dredging. Effective sediment management in the Upper Rhine therefore requires a detailed understanding of the sediment dynamics in the Upper Rhine.

Here we present data on the erodibility of cohesive sediments from the impounded Upper Rhine River and a newly developed model of the Upper Rhine sediment cascade, which addresses the sediment transport to and the temporal sediment retention in the reservoirs upstream of the dams (Figure 1). Each reservoir is represented by a 1D sediment budget model, which is coupled to its upstream and downstream neighbour. The budget model considers the (hindered-) settling of sediment in the water column, the consolidation of the deposited sediment and the erosion during discharges that exceed the critical shear stress of the sediments.

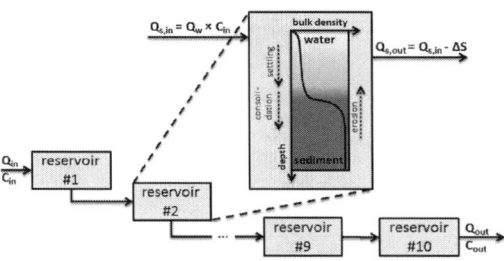

Figure 1. Conceptual framework to model the dynamics of the sediment cascade in the Upper Rhine. Each reservoir is represented by a 1D sediment budget model, which is coupled to the next upstream, through the upstream sediment input, and next downstream model, through sediment output.

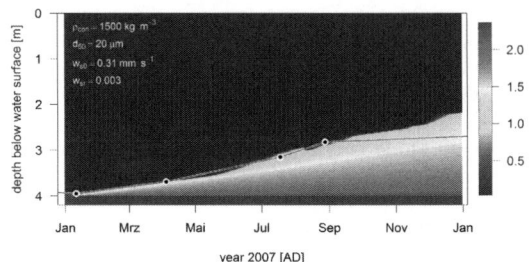

Figure 2. Temporal evolution of the critical shear stress of the Iffezheim reservoir in 2007 modelled using the 1D-sediment budget model. Blue colours represent the water column with low critical shear stress $\tau_c \sim 0\,\mathrm{Nm}^{-2}$. The strong colour gradient from blue to turquois at τ_c between 0.5 and $1\,\mathrm{Nm}^{-2}$ represents the transition from water to sediment. Black dots and lines show the elevation of the echo-soundings and the linear inter-polation between the soundings, respectively.

Critical shear stresses are calculated as a function of the bulk density of the deposited sediment.

The model is calibrated using measurements of sediment erosivity of the Iffezheim reservoir and from a 3D numerical fractionated sediment transport and deposition model of the reservoirs at Iffezheim. Validation of the model is performed using empirical evidences from quarterly echo-soundings of the reservoir. We use Monte-Carlo simulations to evaluate the uncertainty related to our limited knowledge i) of the dynamics of contaminated sediments and ii) of the environmental conditions in the reservoir that control the risk of erosion of contaminated sediments. The Monte Carlo simulations are performed using random variations of the consolidated bulk density, median grain size, hindered settling parameters and erosivity.

Preliminary results (Figure 2) with a special focus on the sensitivity analysis of the sediment cascade model and the resulting implications for the management of contaminated sediments in the Upper Rhine are discussed.

The aging of Japan's dams: Innovative technologies for improving dams water and sediment management

S.A. Kantoush & T. Sumi
Disaster Prevention Research Institute (DPRI), Kyoto University, Kyoto, Gokasho, Ujishi, Japan

ABSTRACT

The aging of Japan's dams and continuous loss of storage capacity due to reservoir sedimentation, coupled with increasing environmental needs, will cause the social, economic, environmental, and political importance of dams to continually increase. Removing such stored sediments is often recommended as a potentially better way to recover reservoirs storage capacities than dam heightening or building new dams. Sediment can be removed by flushing or sluicing it through bottom outlets while lowering the reservoir water levels completely or partially. The rapid reservoir sedimentation not only decreases its storage capacity, but also increases the flood risk in the upstream reaches. Immediately downstream from the dam, the sediment load is greatly reduced in the river.

This paper focuses on one critically important element of these challenges: the need to retrofit or modify the design and operation of existing dams. The retrofitting of dams for modified operation by sluicing refers to cutting dam body and installing (a) New spillways, (b) bottom outlet, (c) bypass tunnel, (d) modify and enlarge existing spillway gates. Moreover, understanding of the necessity of upgrading and retrofitting of aging dams in Japan will be necessary to support the difficult management decisions that ultimately will have to be made. Figure 1 categorizing managing of the aging dams from the standpoint of reservoir recovery, extending, and adding innovative new functions. Possible management issues for aging dams include sediment management techniques to increase the reservoir capacity, extend dam life, dam modification, dam removal, changes in reservoir operational practices, habitat availability and quality, and the environmental consequences of any actions taken. The sustainable development have proliferated, but the following concepts are most relevant from the standpoint of retrofitting dams:

1. Guaranty security and acceptable functioning of the reservoir for its purposes;
2. Today's patterns of infrastructure development should preserve the capacity of the reservoir and maximum possible time of functioning;
3. Produce minimum impact to the environment and maintain biological diversity;
4. Minimize the potential for catastrophic disasters resulting from dam failure;
5. Avoid activities that create a legacy of environmental restoration or dam rehabilitation obligations that disproportional on future generation;
6. Consider all processes: hydrological regime, erosion, climate change, environment and associated ecosystems, water and sediment continuity.

Requiring a reservoir life measured in terms of thousands years instead of decades will demand new methods of analyzing costs and benefits. For all these reasons, developing new techniques to evacuate the fine and coarse sediment to maintain the functionality, and at the same time ecologically rehabilitating the involved landscape would be economically and environmentally beneficial for all types of reservoirs.

REFERENCES

Kantoush, S.A., Sumi, T. & Takemon, Y. 2011. Lighten the load, International Water Power & Dam Construction magazine, May 2011: 38–45.

Sumi, T. & Kantoush, S.A. 2011. Sediment management strategies for sustainable reservoir, 79th Annual meeting of ICOLD, Lucerne, Switzerland.

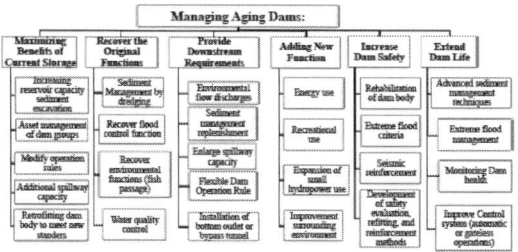

Figure 1. Necessity of upgrading and retrofitting aging dams.

Improvement of a bedload transport rate measuring systems in sediment bypass tunnels

T. Koshiba, C. Auel, D. Tsutsumi, S.A. Kantoush & T. Sumi
Water Resources Research Center, Disaster Prevention Research Institute, Kyoto University, Uji, Japan

ABSTRACT

For long term use of dams, it is urgently required to develop and enhance methods of sediment management in reservoirs. At the same time, they also contribute to appropriate comprehensive sediment management for a basin-scaled sediment routing system. As one effective method, Sediment Bypass Tunnels (SBT) are operated in Japan and Switzerland to prevent reservoir sedimentation. SBT reduces suspended and bed load depositions in reservoirs by routing the incoming sediments around the dam. This system is already applied to several dams and working effectively.

SBT, however, is prone to severe invert abrasion caused by a combination of high velocity and sediment flow. This problem is severe for the long-term maintenance of SBT. Therefore, it is necessary to establish a measurement system of sediment transport rate in the SBT in order to improve maintenance works.

As the measurement systems of bedload transport rate, Hydrophones developed in Japan, and Geophones in Switzerland are examined. A steel pipe of hydrophone detects the sediment rate by the counting number of pulses and sound pressure levels caused by gravel hits. Geophone containing both a steel plate and vibration sensor specially designed for ground motion measures sediment transport rate based on a vibration caused by hitting of gravels. In Europe, Geophone is mainly used. Geophone has, however, several disadvantages such that small particles are hardly detected.

Thus, plate microphones are newly proposed following advantage points based on the geophone and the hydrophone. In this paper, results of calibration experiments and comparison with other devises are discussed in order to verify a validity of the plate microphone. At the Koshibu dam in Japan, a new SBT is supposed to operate from 2016 where plate microphones and other systems are going to be installed on the invert at the end of the tunnel. Plans and preliminary tests for an on-site experiment at the Koshibu dam with the plate microphones are also presented in the paper.

REFERENCES

Hagmann, M., Albayrak, I. and Boes, R.M. (2015). Field research: Invert material resistance and sediment transport measurements. *Proc. Int. Workshop on Sediment Bypass Tunnels*, VAW-Mitteilung 232 (R. Boes, ed.), ETH Zurich, Switzerland, pp. 123–135.

Mizuyama, T., Tomita, Y., Nonaka, M., and Fujita, M. (1996): Observation of sediment discharge rate using a hydrophone. J. Jap. Soc. of Erosion Control Eng, 49(4), 34–37.

Sumi, T., Okano, M. and Takata, Y. (2004). Reservoir sedimentation management with bypass tunnels in Japan. *Proc. 9th International Symposium on River Sedimentation*, Yichang, China, pp. 1036–1043.

Development of a management strategy based on in-situ observation for Agondian Reservoir

C.C. Li & Y.J. Tsai
Disaster Prevention Research Center, Cheng-Kung University, Taiwan

T.H. Wu & H.C. Tai
Southern Region Water Resources Office, Water Resources Agency, Ministry of Economic Affairs, R.O.C, Taiwan

ABSTRACT

Agondian Reservoir with a design storage capacity of 20.5×10^6 m^3 is located in the southwest part of Taiwan near Kaohsiung. It has been in operation since 1953. High sedimentation rate due to strong surface erosion in the watershed by heavy rain resulted in a capacity loss of ca. 71% within 38 years. A renewal project including sediment removing and constructing a spill shaft was then carried out from 1997 to 2005. Empty flushing in the flood period has been taken for desiltation. Since then the sedimentation rate has been decreased (Fig. 1).

Sediment concentrations of inflow and outflow were observed by several typhoons and heavy rain events. Water samples were taken during the whole period of every event to analyze the inflow and out-flow sediment concentration. The desiltation efficiency is about 30% according to the observation data. The expected desiltation efficiency by early stage research, 65%, is still not achieved, therefore further measures has to be taken to reduce sedimentation. The hydrograph of flow and sediment concentration, elevation of storage water level by Typhoon Matmo are plotted in Figure 3. It shows that the observed peak of sediment concentration of inflow will reach the spill shaft in 2 hours. The gate should be opened (indicted by the arrow in Figure 2) to let the high sediment

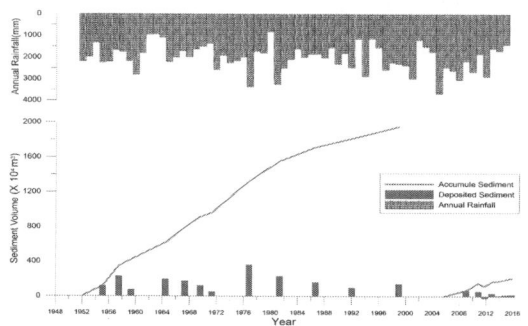

Figure 1. Annual sedimentation of the Agondian Reservoir.

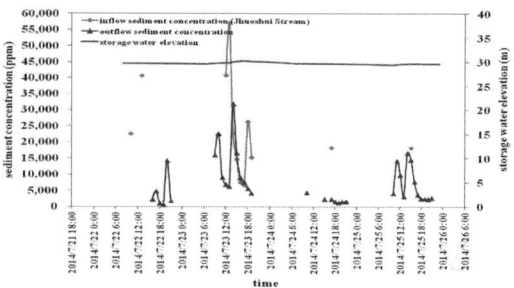

Figure 2. Hydrograph of storage water elevation, inflow and outflow sediment concentration by Typhoon Matmo.

flow through the reservoir. But the sediment concentration of outflow decreases just a few hours after opening of the spill shaft. The storage water level also has influence on desiltation. The desiltation efficiency will become lower if the storage water level is higher than 32 m.

A management strategy is suggested below:

1. to draw down the storage water level soon after the sea alert for a typhoon is issued.
2. to open the spill shaft gate as soon as possible when the sediment concentration in the inflow increases rapidly or exceed 30,000 ppm.
3. to close the spill shaft when the sediment concentration of inflow is lower than 5,000 ppm or the sediment concentration of the outflow is lower than 2,000 ppm.
4. Thus the operation of the Agondian Reservoir could achieve desiltation and other functions such as flood protection, irrigation and sightseeing.

REFERENCES

Atkinson, E. 1996. The Feasibility of Flushing Sediment from Reservoirs, HR Wallingford, HR Wallingford, UK.
Brandt, S. A. 1999. "Reservoir Desiltation By Means Of Hydraulic Flushing, Ph.D. thesis, the Institute of Geography, Faculty of Science, University of Copenhagen.

Density driven underflows with suspended solids in Lake Constance

S. Mirbach & U. Lang
Kobus und Partner GmbH, Stuttgart, Germany

ABSTRACT

The main Inflow of Lake Constance is the River Rhine with a medium discharge of 350 m³/s. The yearly sediment transport is in the range of 2 to 3 m tonnes. The main sediment transport occurs during flood events. Due to the nival runoff regime in the catchment large flood events take place in early summer during snow melt in combination with large rain events.

The high sediment load during flood events of River Rhine leads to a higher density of the river water compared to the lake water. Thus the river water flows along the lake floor. The path length of the underflow is dependent on the impulse of the inflow and sediment concentration. An underflow to the deepest point of the lake has been observed and documented by Eder et al. (2014) during a rain event in August 2005 with a peak discharge of 2500 m³/s in River Rhine. Water temperature increased from 4 to 9 degrees at the deepest point of the lake (250 m). As the underflow came to a halt the larger fractions of suspended solids fell out and the warm riverine water at the bottom of the lake rose together with the finer fractions of the suspended matter. An increased turbidity could be observed in the whole lake at the drinking water intakes.

The hydrodynamic lake model ELCOM-CAEDYM developed by CWR from Hodges & Dallimore (2012) was used to analyze the underflow event. In combination with the ecological model of Hipsey et al. (2012) density driven flow of suspended solids can be simulated. Characteristic grain sizes of suspended solids and relevant resuspension rates in the lake have been identified in a calibration process. The simulation result is shown in Figure 1.

Due to the stratification hypolimnic water is only exchanged in case of total overturns in winter. The ongoing effects of climate change show that the exchange of deep water became more seldom. Only every fifth winter a total water exchange takes place at present. It is expected to become less in future. Using the lake model the influence of flood events to deep water renewal was quantified. Results showed that besides the riverine water the predominant contribution to the renewal is due to epilimnic water, which is entrained into the underflow. Since the occurrence of underflows and their extent greatly vary depending on the flood event, a model study was conducted to quantify the influence of the parameters discharge, suspended sediment concentrations, grain size distribution and water temperature. A clearer picture of the main driving forces was obtained by an analysis of the results in combination with measurement data of the flood events of the last few years. Main factor in the formation of underflows are the concentrations and grain size distribution of the suspended sediments. This is clearly shown by two underflow events in summer 2013, which occurred due to extreme high concentrations of suspended sediments without any increase in discharge.

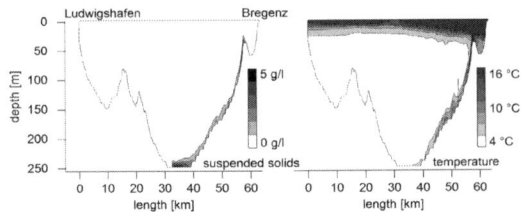

Figure 1. Calculated underflow with suspended solid concentration (left) and water temperature (right) in a vertical cross section of Lake Constance.

REFERENCES

Eder, M., Wessels M. & Dare, J. 2014. Underflows in Lake Constance – Numerical Modeling, Instrumental Observations and Sediment Data. Geophysical Research Abstracts, Vol. 16, EGU2014-10709.

Hipsey, M.R., Antenucci, J.P. & Hamilton, D. 2012. Computational Aquatic Ecosystem Dynamics Model: CAEDYM v3.2 Science Manual. Centre for Water Research, University of Western Australia.

Hodges, B. & Dallimore, C. 2012. Estuary, Lake and Coastal Ocean Model: ELCOM v2.2 Science Manual. Centre for Water Research, University of Western Australia.

Development of oblique flow in barrages due to shoal formation

K. Mishra
Synergy Institute of Technology, Bhubaneswar, Khurda, India

ABSTRACT

A barrage is designed to operate under certain assumed hydraulic conditions. Occasionally, however, unexpected conditions may arise which may be severe enough to threaten the stability of the structure. This paper studies one such condition i.e oblique approaching flow of the water on the upstream pond due to shoal formation which is likely to affect the scour downstream, thus causing concern to the stability of the downstream apron. The oblique approaching flow vectors strike the piers between the gates and are forced to take a turn towards the gates may generate vortices with vertical axes. These vortices generate an additional turbulence which appears to get conveyed downstream with the flow.

The two-dimensional shallow water numerical simulation model, TINFLOW2D was employed to simulate the flow pattern in the pond upstream of Mahanadi Barrage for the 1994 flood, the discharge hydrograph of which is presented in Figure 3. The discharge values, provided by the Mahanadi Barrage Division of the Orissa Irrigation Department, have been computed on the basis of water levels on the upstream and downstream of the barrage structure, openings of the respective gates of the undersluice and weir bays and the standard barrage discharge calculation formulae.

In general, a barrage is designed for operation under certain assumed hydraulic conditions, which are often likely to change during its full service life. Unfortunately, most of these changes are difficult to predict with certainty. For example, the direction of the approaching flow towards the barrage is assumed in the designs to be straight, that is, striking at right angles.

It may be noted that the flood of this year was one of the largest since the completion of the barrage in the 1980s, with the peak reaching 13,200 $m^3 s^{-1}$, about 86 percent of the design flood peak of 15,300 $m^3 s^{-1}$ considered for the barrage. Hence, the determination of the flow pattern during this event seemed relevant.

From a study of the different flow characteristics for the four days of simulation, the following points may be inferred:

The shoals do affect the flow paths but are more prominent for low flows compared to flood flows. Of course, during high flows, there is substantial flow taking place over the shoals as the depth of water over the shoals is relatively more during this time than during low flows.

The gate openings affect the obliqueness of the flows most, especially near the gates, during the low flows with gate controlled discharges. This may be observed from the simulations of 06.07.1994 and 28.07.1994. A comparison of the maximum deviation angles of the flow vectors from the normal to the barrage axis also proves the point. Actually, for the two days of 16.07.1994 and 20.07.1994, all gates were completely withdrawn and hence the flow vectors are seen to straighten out.

The velocities near the gates during the flood flows are generally of the order of 1 ms^{-1} but during the gate controlled low discharge flows it may vary depending upon the gate opening. For example, on comparing the velocities magnitudes near the gates for 06.07.1994 and 28.07.1994, it is observed that the velocity is close to 1 ms^{-1} close to some of the gates for 06.07.1994 but within 0.5 ms^{-1} for 28.07.1994 though the overall discharge of the river is smaller for the former compared to that of the latter.

The four flow conditions, viz., controlled-free, controlled-submerged, uncontrolled-free and uncontrolled-submerged, as investigated by Ghosh (2008) are observed to appear in the simulations. These may be proved with the emerging discharge of the flows through the respective gate bays together with the gate opening and corresponding pond (upstream) and tail (downstream) water levels.

Hydrodynamic instabilities in shallow reservoirs: Implications for sediment management

Y. Peltier, A. de Cuyper, S. Erpicum, P. Archambeau, M. Pirotton & B. Dewals
ArGEnCo Department, Research Group of Hydraulics in Environmental and Civil Engineering, University of Liege, Liege, Belgium

ABSTRACT

Natural or engineered reservoirs are very common structures in hydraulic engineering. They are used as storage reservoirs for flood management or as settling reservoirs for trapping sediments and/or pollutants. For both configurations, controlling sediment transport within these structures is vital to achieve a cost-effective and sustainable management of such structures.

Standard design approaches merely based on the reservoir volume have shown their limitations in terms of sediment management (Dufresne et al. 2012 and Peltier et al. 2013). In addition to the volume, the flow patterns must be taken into account for optimally designing the reservoir (Peltier et al. 2014a, 2014b) and therefore correctly assessing the patterns of sediment deposits (Fig. 1). Indeed, complex flow fields

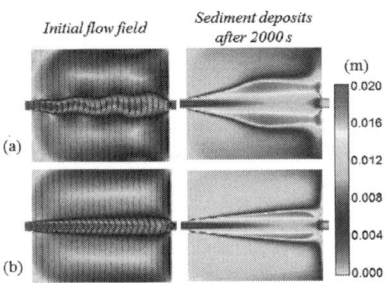

Figure 1. Simulated initial flow field and sediments deposits after 2000 seconds for two types of flow patterns: (a) meandering jet, (b) straight jet (Peltier et al. 2013).

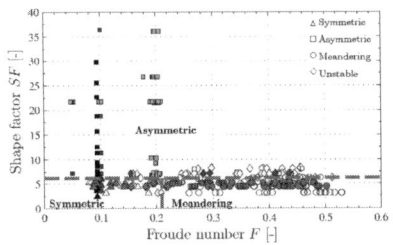

Figure 2. Flow patterns: domains of existence.

develop in shallow reservoirs, such as large scale turbulent gyres and meandering jets, even for in simple geometry (Peltier et al. 2014b). The flow structures are characterised by their streamwise and spanwise lengths, and by their frequencies. Their respective domains of existence are governed by different flow instabilities, and for rectangular reservoir, they are characterised by (Fig. 2):

(i) the shape factor $SF = L/(b^{0.4} \Delta B^{0.6})$, which is a non-dimensional parameter accounting for the length, L, the width of the inlet, b, and the sudden expansion width, ΔB, of the reservoir;
(ii) the flow characteristics (Froude and friction numbers defined at the inlet).

In the present study, we present new experimental data, which clarify the influence on the flow field of ΔB for fixed or for varying SF. New data confirm the contours of the domains of existence (Figure 2). Correlations between the friction number and the newly observed flow patterns highlight that for the considered experimental conditions, the shallowness of the flow influences more the flow structures than the shape factor. In turn, this substantial influence of the friction number on the flow field also affects the resulting sediments deposits.

REFERENCES

Dufresne, M., Dewals, B., Erpicum, S., Archambeau, P. & Pirotton, M. 2012. Flow patterns and sediment deposition in rectangular shallow reservoirs. Water Environmental Journal, 26(4): 504–510.

Peltier, Y., Erpicum, S., Archambeau, P., Pirotton, M., & Dewals, B. 2013. Experimental and numerical investigation of meandering jets in shallow reservoir: potential impacts on deposit patterns. THESIS 2013, Two-phase modelling for sediment dynamics in geophysical flows. SHF – EDF R&D, Chatou, France.

Peltier, Y., Erpicum, S., Archambeau, P., Pirotton, M. & Dewals, B. 2014a. Experimental investigation of meandering jets in shallow reservoir. Environmental Fluid Mechanics, 14: 699–710.

Peltier Y., Erpicum S., Archambeau P., Pirotton M. & Dewals B. 2014b. Meandering jets in shallow rectangular reservoirs: POD analysis and identification of coherent structures. Experiments in Fluids, 55(6): 1–16.

Long term simulation of reservoir sedimentation with turbid underflows

G. Petkovšek
HR Wallingford, Wallingford, UK

ABSTRACT

In some cases sediment laden flow entering a reservoir plunges under the water surface and continues its way towards the dam as submerged turbid density current. If it reaches the dam, it is either vented through a low level outlet or in absence of that, travels up the dam face and then turns back to form an underwater muddy lake.

Numerical modelling of turbid density currents present a particular challenge for engineers who deal with practical reservoir sedimentation problems. In principle, three-dimensional (3D) models are the most suitable to model turbid density currents. These models have been successfully applied to simulate propagation of turbid underflows on the experimental scale as early as in the 1990s and more recently also on a real reservoir scale. However application of 3D models that typically require long run times may not be suitable for practical problems (Cao et al. 2015). Engineers usually resort to 2D or, in particular if the studied reservoir is long and nar-row, they choose 1D models.

This paper presents the development and an example of application of turbid density current module for RESSASS, a one-dimensional quasi-steady reservoir sedimentation model suitable for long term simulations. The module determines the plunging point based on the criterion of densimetric Froude number. Following plunging simulation continues in a supercritical regime as long as the energy of the density flow and slope of bed are sufficient to support the transport in this mode, or until the dam is reached. A muddy lake is then formed. An adapted version of theory of Toniolo et al (2007) is used to compute water and sediment balance in a muddy pond. If the level of muddy pond exceeds the lowest level outlet, sediment is vented from the reservoir.

The model was applied to one of the largest reservoirs in the world, the Nurek reservoir in Tajikistan. Historical data on bed levels and sediment inflow as well as a recent survey from 2015 were used to calibrate and verify the model. The results of verification show promising results in particular in terms of the simulated bed levels.

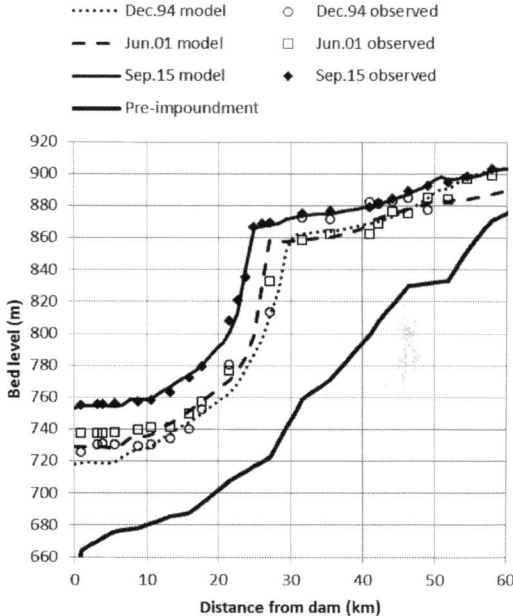

Figure 1. Observed and modelled bed levels.

Finally, re-calibration of the model of Nurek reservoir was attempted without a turbid density current module. However the results were poor, indicating that for simulation of reservoir sedimentation where turbid currents occur, it is necessary to use a model that can simulate this phenomenon properly in order to achieve reasonable predictions of future bed levels and storage volumes.

REFERENCES

Cao, Zh., Pender, G. & Liu, Q. 2015. Whole-Process Modeling of Reservoir Turbidity Currents by a Double Layer-Averaged Model. *Journal of Hydraulic Engineering*, 141(2): 1–19.

Toniolo, H., Parker, G. & Voller, V. 2007. Role of Ponded Turbidity Currents in Reservoir Trap Efficiency. *Journal of Hydraulic Engineering*, 133(6): 579–595.

Controlling sediment flushing to mitigate downstream environmental impacts

S. Quadroni, G. Crosa & S. Zaccara
Department of Theoretical and Applied Sciences, University of Insubria, Varese, Italy

P. Espa & M.L. Brignoli
Department of Science and High Technology, University of Insubria, Varese, Italy

G. Gentili
Graia s.r.l., Varano Borghi, Italy

R.J. Batalla
Department of Environment and Soil Sciences, University of Lleida, Catalonia, Spain

ABSTRACT

This work reports on the main findings concerning the biomonitoring carried out (i) to detect the environmental impacts of ten sediment flushing operations from four Alpine reservoirs, and (ii) to evaluate the suitability of the adopted mitigation measures (mainly SSC thresholds).

Benthic macroinvertebrates and trout were generally surveyed from a year before the flushing operations to a year after the event at different reaches (from 1 to 4) of the downstream watercourses.

In general, the seasonal streamflow increase due to snowmelt (from late spring to summer in the investigated area) seems to superimpose on the flushing disturbance. However, in the first post-flushing sample, collected within a month from the end of the works, a contraction of the benthic community in terms of both density and richness (number of families) was detected (Fig. 1). In the case of density, a similar response to different doses might suggest an asymptotic behavior, i.e. an analogous response of the benthic assemblage once certain SSC thresholds are exceeded. The significant correlation ($R^2 = 0.26$, $p = 0.02$) between sediment dose and richness reduction led us to suppose a taxon-specific response to this kind of perturbation. However, in most cases the benthic community showed a high resilience and recovered to the pre-flushing condition within a year. The comparable recovery can be justified by the low sediment deposition (few kg/m^2). The only exception, in fact, is represented by the benthic community at the monitoring site where the highest sediment deposition (300–450 kg/m^2) was recorded. No cumulative effects, i.e. a progressive decline of the benthic community in the course of the years of flushing, were detected.

Trout data are more difficult to be interpreted, mainly because of the bias due to stocking and fishing in most monitoring sites. However, a substantial

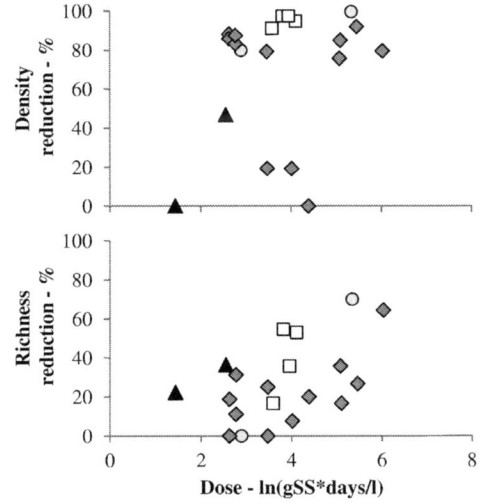

Figure 1. Density and richness reduction of the macroinvertebrates in the first post-flushing sample compared to the pre-flushing one in relation to the sediment dose, computed as natural logarithm of the average SSC (g/l) times the duration of the flushing event (days) according to Newcombe & Jensen (1996). Different symbols indicate the four analyzed reservoirs: square = Valgrosina, diamond = Cancano, triangle = Sernio, circle = Madesimo.

agreement with the Newcombe & Jensen model (1996) was found along with a heavier impact on juveniles.

The reported experiences could be precious to adequately sustain decision making, thus supporting future flushing activities.

REFERENCE

Newcombe, C.P. & Jensen, J.O.T. 1996. Channel suspended sediment and fisheries: a synthesis for quantitative assessment of risk and impact. *North American Journal of Fisheries Management* 16: 693–727.

Economic assessment of the effects of sediment replenishment to rivers and the effectiveness of sediment management

K. Tomita & T. Homma
Civil Engineering And Eco-Technology Consultants Co., Ltd., Tokyo, Japan

T. Sumi
Water Resources Research Center, Disaster Prevention Research Institute, Kyoto University, Kyoto, Japan

ABSTRACT

Japan is characterized by steep mountain topography and weak geology as well as high precipitation and seasonal drastically changing temperature. As a result, sediment produced in mountain areas is frequently transported downstream and supplied to coastal areas through the river mouth which has formed many alluvial fans. In these upstream mountain rivers, many storage dams have been constructed for the purpose of flood control and water utilization. Nowadays, sediment accumulation in their active storage volume is a crucial issue both for reservoir and river basin sustainability.

In Japan, several countermeasures such as excavation, dredging, draw down flushing and bypassing have been introduced to solve this issue. This paper studies the effects of sediment replenishment with a sediment bypass project which restores sediment movement to the original condition and evaluates the project quantitatively by converting the effects to economic value to the extent possible. Additionally, the paper examines optimization of sediment replenishment by conducting sensitivity analysis to know how the effects change through implementation of sediment management.

As a case study, we have selected the Yahagi River where integrated sediment management in a river basin with a proposed sediment bypass tunnel at the Yahagi dam is undergoing. Based on several scenarios, we have conducted economic analysis considering both advantage and disadvantage points of sediment replenishment. We have also studied the effects of sediment management to control shape of reservoir sediment accumulation in the reservoir regarding the quality control in the reservoir. Additionally, based on the amount of necessary sediment excavation volume for maintenance the river channel and downstream reservoir areas after the operation of sediment flushing BP, we studied measures for effective use of sediment resources by sediment nourishment for river-mouth, estuary and coastal areas and recycling for construction materials. In the case, we considered possible site of stockyard where receiving unstable volumes of excavated sediment and delivering these recycling sites continuously.

In this study, historical study is done over dam and rivers and added with cost-benefit analysis of sediment BP over the estuary and coastal zone. Furthermore, in order to enhance the effectiveness and accuracy, estimation of sediment volume inflow of the dam using rainfall data, landslide data, etc. is included into analyzing process.

In the normal basin, sediment is produced in the bare land of landslide trace and supplied to the river channel by the action of gravity, wind, surface flow, etc. When sediment supply into river channel is abundant, talus is formed. Then, erosion of talus supplies sediment volume to the river channel. During heavy rain, bare land gets wider, then, sediment production from a new collapse area occurs rapidly which increases sediment production and supply system. These processes are added into previous model, then, applied on river case study.

Model is calculated with assumption of 50 years stabilized sediment inflow into the dam and sediment supply from eroded area along the way. Then, the result is verified with the real sediment inflow state, which shows good correlation. This includes sediment production results in the rivers, dams region, estuary to the coastal area, and sediment influx from terrain, geology, from rainfall conditions. Moreover, cost-effectiveness calculation of sediment BP can be constructed. It should be noted that aerial photographs, geological data, hydrological data for other dam are also available, thus, study to deter- mine the operation of sediment BP is possible.

From now on, the review and implementation propriety of sediment BP of dam both inside and outside Japan and case study to determine economic effect, etc. are increasing.

The monitoring of empty flushing operation at Agondian Reservoir, Kaohsiung, Taiwan

Y.J. Tsai & C.C. Li
Disaster Prevention Research Center, Cheng-Kung University, Taiwan, R.O.C

ABSTRACT

Agondian Reservoir, located in Kaohsiung, south part of Taiwan (Fig.1), has been operated since 1953 for the flood control. 71% of the total storage volume was filled with deposited sediment in 1991 because of the mud rock in the upstream of the reservoir. To recover flood control ability, a reservoir renewal project was started in 1997. The idea of empty flushing was applied in Agondian Reservoir. The intake was re-construct as the spillway to release the water and sediment during rainfall season. All the re-construct engineer were done in 2004, and the empty flushing operating was started from this year. To understand the effect of the empty flushing, a monitoring system were setup from 2009. There are three station set at the inflow and outflow of the reservoir, which included rainfall gage, water level meter, flow velocity meter, water sampler (Fig.2). There are 9 events were recorded from 2009-2015, included the rainfall, inflow and outflow discharge, the water concentration, the reservoir water level. The deposition in the reservoir were measure every year at same period.

Figure 2. Location of monitoring system and equipment.

The results show some characteristic of empty flushing operating listed follow.

First, with the comparison of the rainfall intense and inflow concentration, the occurrence of the high concentration flow needs 2-3 hour continued rainfall.

Second, the reservoir water level is the key factor during operating. The transportation of the high concentration flow would be cut with the high reservoir water level. The suitable value is suggesting as EL30M.

From these characteristic, some suggestion could feedback to improve the effect of the empty flushing operating.

REFERENCE

Southern Region Water Resources Office. 2009–2014. Report of Sediment observation and efficiency evaluation for desiltation of Agongdian Reservoir by using empty flushing, Tainan, R.O.C. (in Chinese)

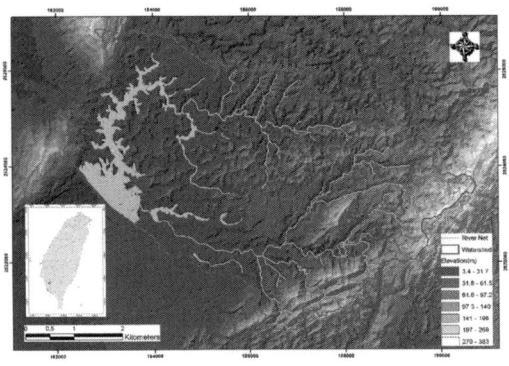

Figure 1. Location of Agondian reservoir.

Study on sediment desilting operation mode and structure layout of Pakistan Karot hydropower project

J. Zhao, X.N. Liu, B. Fan, G. Wei, M. Wang & Z. Jin
Yangtze River Scientific Research Institute, Wuhan, Hubei Province, China

ABSTRACT

The establishment and operation of one hydropower complex will change the movement law of water flow and sediment of the river way, and sediment deposition is difficult to avoid. Due to the characteristics of "large amount of sediment, small storage capacity of reservoir", the desilting operation mode and structure layout design is one of the major technical issues for proposed Karot Hydropower Project, which is the fourth level of the five cascade hydropower plants planned in the Jhelum River within Pakistan.

This paper applies a combing method of one-dimensional hydrodynamic &sediment model and a physical model test of dam area to study the movement law of flow and sediment in reservoir and dam site area, then find the rational flushing operation mode and desilting structure layout. The 1-D mathematical model results show that the appropriate reservoir operation mode is "desilting operation water level of 446 m, desilting startup flow of 2100 m^3/s", which can be ensured that the 20-year-operation reservoir will basically reach the scouring and silting balance and the reservoir will be able to regulate the capacity, according to comparison and analysis of sedimentation amount and pattern of various proposals.

The 1:100 sediment overall model test results show that the desilting facilities and the desilting dispatching method can ensure the "clean inlet" at the water inlet

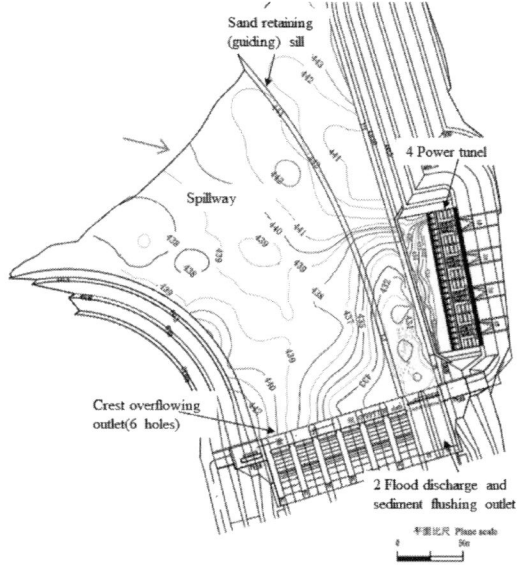

Figure 2. Siltation contour line distribution chart of spillway diversion canal at the end of 20 years.

of power station and also the normal water diversion and power generation, and for which are consisted of 6 flood discharge surface orifices and 2 flood discharge and desilting orifices. The results of this study can be used for providing scientific bases for Pakistan Karot Hydropower Project design.

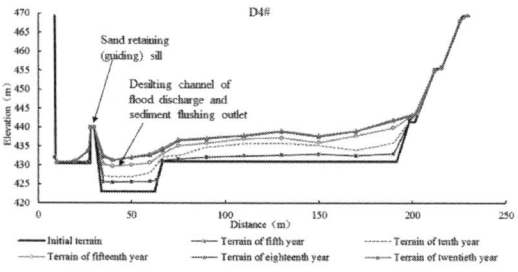

Figure 1. One section siltation change charts of spillway diversion channel.

REFERENCES

Cai, S., Rao, G., Frangaki, M., et al. 2015. Pakistan KAROT Hydropower Project Level 1 Design Report, Chapter 6 Objectives and Scale of Project. Wuhan: Changjiang Survey, Planning, Design and Research Co., Ltd.

Wan, Y., Yang, X., Liu, X., et al. 2014. Pakistan Karot Hydropower Project Sediment Specific Study. Wuhan: Changjiang Institute of Survey, Planning, Design and Research.

F. Social, economic and political aspects of sediment management

Sediment management at Sukkur Barrage – How competing needs and uses of the structure impact the design

S. Aziz
Atkins, Epsom, UK

M. Roca-Collell
HR Wallingford, Wallingford, UK

I. Heijne
Atkins, Epsom, UK

ABSTRACT

Sukkur Barrage was completed in 1932 and is located on the River Indus in Pakistan. The barrage has suffered from sediment management issues since its construction and as a result the so called 'approach/tail channel' was installed on the right bank to induce secondary currents with the aim to reduce sediment intake into the right pocket and the canals. An island has built up between the tail channel and the right pocket which has led to the closure of 10 gates. For some years this approach was perceived as successful but over time the canals on the right bank have experienced significant sediment build-up. In addition to this there is evidence of significant sedimentation in the river channel, both upstream and down- stream of the barrage, and the pockets.

The sedimentation problem was investigated as part of a wider project adding significant complexity to the possible solutions and improvement measures. One of the other major concerns at the barrage was flood risk. The original barrage configuration had 56 main river spans. This number was reduced to 36 main river gate to allow the construction of the approach/tail channel with 8 gates and closure of 10 gates as part of the sediment management improvements. This has led to concerns about the flood capacity in particular following the 2010 floods with calls to reopen bays to increase flood capacity.

At the same time irrigation demands at the barrage have increased over time leading to undersized pockets which increase the sediment management issues at the barrage. The sediment management of the barrage in light of irrigation requirements and the flood capacity can therefore be perceived as competing concerns.

The barrage is of great significance to the agriculture in Sindh Province supplying irrigation water to over 8 million acres on both banks of the Indus. The command area of the barrage is the largest of the three barrages in Sindh Province. As such, the safety of the barrage and the continued performance attracts huge public interest. Both irrigation supply and flood capacity are highly political issues.

Since the construction of the barrage significant further development has taken place on the Indus upstream with construction of a large number of dams and barrages on the Indus and its tributaries. This has significantly altered the flow arriving at the barrage and has also influenced the sediment transport in the river system.

Furthermore, climate change is likely to have played a part in the reduced dry season flows experienced particularly between 2002 and 2011.

The paper sets out how the project addresses the risks to the barrage from droughts, flooding and climate change and how the team have approached the identification of solutions in the context of the competing requirements.

The challenges faced on the project provide lessons learnt for other projects and demonstrates the need for engineers and scientists to understand the wider picture beyond the science of sediment management.

Overlooked costs of dams: Barrier to sustainability

M. George & R. Hotchkiss
Brigham Young University, Provo, Utah, USA

ABSTRACT

With an ever-increasing global population, mounting demand exists for a more sustainable water supply system. Despite this demand, worldwide water storage capacity is relentlessly diminishing due to reservoir sedimentation. A warning in the Reservoir Sedimentation Handbook states that "sudden loss of the world's reservoir capacity would be a catastrophe of unprecedented magnitude, yet their gradual loss due to sedimentation receive little attention or corrective action." Action must be taken to improve the sustainability of reservoirs and meet the increasing demand for water.

Dam construction disturbs the natural sediment equilibrium present in typical open channel flow, while also creating a valuable reserve of water. The resulting reservoir traps sediment due to the decreased channel velocity upon entrance. Over time, the deposition of sediment promulgates significant damages both upstream and downstream of the dam, in addition to loss of storage space within the reservoir. Upstream aggradation can result in clogged intake structures, decreased navigational clearance, and increased flood frequency, while downstream scour can lead to abandoned intake structures, compromised channel stability, and damaged bridge piers and abutments. The loss of storage space within the reservoir itself contributes to a reduction in all project benefits as well as delta starvation at the coast. Even after only a small percentage of storage capacity is lost, severe problems related to sedimentation can appear.

The true costs of such damages are often overlooked and, thus, not included in cost-benefit analyses when designing dams. This results in underestimating costs associated with project benefits (hydropower, irrigation, flood control, etc.) True project costs would be better represented if damages associated with reservoir sedimentation were acknowledged and included in the economic evaluation. Such an omission has resulted in the construction of thousands of dams that are not sustainable.

It is understood that the elimination of sedimentation is neither viable nor possible. As such, sediment must be managed and preventative measures must be taken in order to alleviate the catastrophic loss of reservoir storage space. A sustainable approach would both consider the damages caused by dam construction and operation and address the mitigation of sediment, theoretically leading to an indefinite lifespan. As is, most dams do not have the necessary facilities for such a task.

In order to promote long-term economic viability, dam owners (e.g., hydropower companies) and legislative bodies are encouraged to reconsider the traditional, short-sighted reservoir design life approach in favor of a life-cycle management plan which incorporates sediment management. By incorporating overlooked costs into economic analyses and implementing a life-cycle management plan, reservoir lifespans will be more sustainable, profits will be extended indefinitely, and the economic burdens placed upon future generations by ours will be lessened.

REFERENCES

Annandale, G. 2013. Quenching the Thirst: Sustainable Water Supply and Climate Change. North Charleston: Create Space Independent Publishing Platform.

Hotchkiss, R.H. & Bollman, F.H. 1996. Socioeconomic Analysis of Reservoir Sedimentation. In: Proceedings of the International Conference on Reservoir Sedimentation, Vol. 1. Ft. Collins, CO: September 9–13. 52–32 to 52–50.

Juracek, K.E. 2014. The Aging of America's Reservoirs: In-Reservoir and Downstream Physical Changes and Habitat Implications. Journal of the American Water Resources Association 51(1): 168–184.

Morris, G.L. & Fan, J. 1998. Reservoir Sedimentation Handbook. New York: McGraw-Hill Book.

Vanoni, V.A., (ed.) 1975. Sedimentation Engineering (ASCE Manuals and Reports on Engineering Practice-No. 54). New York: American Society of Civil Engineers.

River sand and gravel mining in Iran

S. Norouzi
River and Coastal Engineering Bureau, Iran Water Resources Management Company, Tehran, Iran

J. Habibi
Conservation and Operation Assistant Director, Iran Water Resources Management Company, Tehran, Iran

E. Zrdchoghai & A. Zardchoghai
Expert, Private Company, Tehran, Iran

ABSTRACT

Not only are rivers counted as the main sources of water supply, but also they are accounted for one of the sources of sediment production and transport. The needs of developing countries, like Iran, to infrastructure establishments, such as dams and highways, easy accessibility to high quality of sand and gravel materials have caused excessive river mining to be under-taken. This is where river mining phenomena, either deliberately or unwittingly, exceeds carrying capacity of the river (Guideline 2006).

If sand mining activities don't perform regarding to technical instructions, it will cause negative impact on the natural balance of the river resulting in destruction of the surrounding land, the aquatic environment as birds and animal's habitats, decrease the slope and degradation of the river bed which threats the piers of bridges, inverted siphon and intakes. Consequently, flood damages impacts may be intensified to culminate on river cross-structures damages (Soltani et al. in press).

In recent years, particularly due to the lack of technical standards and criteria as well as illegal sand and gravel mining without any control, a lot of damages happened to rivers. But in recent years sand and gravel mining is monitored and controlled by technical guidelines, description of services, and related legal regulations.

In this research, sand and gravel mining in Iran were initially investigated by comparing the extracted

Figure 1. Bridge piers scouring due to illegal sand and gravel mining.

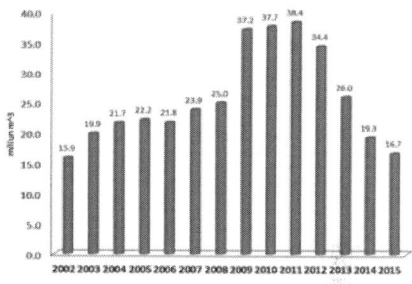

Figure 2. Sand and gravel mining in Iran between 2002 and 2015.

volume of sand and gravel in various years (Norouzi et al. 2013). In the following, the focus of the investigations has been on the analysis of the negative impacts of mining by concentrating on some important rivers where illegal mining was observed (Soltani et al. in press). In this concern, with predicted results of flow and sediment modeling of these rivers is discussed some issues as distance of mining pits from hydraulic structure, determination of allowable volume and depth of mining and refilling of previous mining pits.

Thereafter, the following issues will be addressed: recent management actions in sand mining supervision, control volume of sand mining, prevent unauthorized withdrawals and dealing with offenders in the country.

Finally, some managerial and technical suggestions for proper river mining will be presented while reclamation needs and negative impacts are minimized.

REFERENCES

Norouzi, S., Jafari, J & Meybodi, J. 2013. Study of legal challenges of river sand and gravel mining, 9th International River Engineering Conference, Shahid Chamran University

Guideline on sand and gravel mining from rivers 2006, issue No. 336, management and planning organization.

Soltani, Sh., Jafari, Gh. & Norouzi, S. (in press). River sand and gravel mining

SS 1 Hydropower and sediment management

Challenges facing Atbara Dam Complex (ADC) operation management

A. Abdelsalam Ahmed
DG of UNESCO Chair in Water Resources, Omdurman Islamic University, Khartoum, Sudan

ABSTRACT

Atbara River is one of the tributaries of the Nile River. It is the most northerly tributary of the Nile flowing from the Ethiopian Highlands. It has two branches Setit River (called in Ethiopia Tekeze) and Upper Atbara River. The total catchment area of Setit and Upper Atbara rivers is about 97000 km^2 (68000 km^2 and 29000 km^2 respectively).

Atbara is a seasonal river with annual average flow amounts to 12000 Mm^3. Khashm ElGirba dam (KED) was constructed on Atbara River in 1964 with a storage capacity of 1300 Mm^3. Since then the KED lost about 70% of its storage capacity due to sedimentation. Currently there are two dams (DCUA) under construction, each on one of the two branches (Setit, Upper Atbara) with total storage capacity about 3700 Mm^3 (2520 Mm^3 for Burdana reservoir on Setit river and 1180 Mm^3 for Rumela one on Upper Atbara River). The two reservoirs are linked by a channel flowing from Burdana to Rumela.

Several recommendations are provided by a Consultant to operate the Atbara Dam Complex (ADC) – (the two reservoirs plus KED). The main objective of the project isirrigation, drinking water supply and hydropower generation.

In this paper the experience gained from the reservoirs management in Sudan, in particular KED will be reviewed and discussed. Sediment management will be the core of the operation rules which will govern the ADC. This means that the sustainability of the reservoir to serve its purposes, on which the feasibility study is conducted, mainly depends on how the ADC will be operated.

The impact of the upstream development on the ADC operation (i.e. Ethiopia Tekeze Dam – called T5) will be examined. The paper concludes with several recommendations to maximize the benefits out of the ADC reservoir.

Field calibration of bedload monitoring system in a sediment bypass tunnel: Swiss plate geophone

I. Albayrak, M. Hagmann, C.R. Wyss & R.M. Boes
Laboratory of Hydraulics, Hydrology and Glaciology (VAW), ETH Zurich, Switzerland

ABSTRACT

Under the strong impact of climate change sediment transport in melting water draining from glacier basins, rivers and waterways, and reservoir sedi-mentation have strongly increased in both the Alpine regions and worldwide. As a consequence, three main problems arise with increasing sedimentation: the loss of storage volume for energy production, flood protection, water supply and irrigation; (2) increased hydro-abrasion at turbines and hydraulic structures; and (3) negative environmental impacts as the sediment transfer downstream is prohibited. (1) and (2) result directly in a decrease of energy production and in an increase of maintenance costs. An effective and holistic countermeasure against reservoir sedimentation is to route sediment around a dam by using a sediment bypass tunnel (SBT). A major problem affecting nearly all SBTs is severe hydro-abrasion on the tunnel invert due to the high bed load transport rates in combination with high flow velocities. Depending on site-specific operating conditions and sediment properties, i.e. size, hardness and shape, invert abrasion can cause considerable refurbishment costs.

For optimized design and operation of SBTs with respect to sustainable sediment management and cost efficiency, there is an increasing need for continuous real-time monitoring of bed load transport. Bed load transport can be monitored indirectly by using a passive sensor like geophone or hydrophone. In the present study, the so-called Swiss plate geophone system (Fig. 1) has been implemented at the outlet of Solis

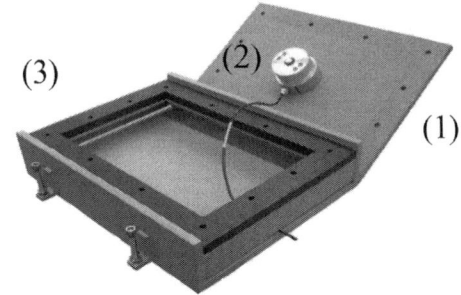

Figure 2. Swiss plate geophone system (before assembly) with steel plate (1), geophone sensor encased by an aluminum hous- ing (2), and elastomer bearing (3) on the steel casing (VAW).

SBT located in Grisons in the Swiss Alps. The geophones with the sampling rate of 10 kHz are placed across the whole tunnel width of 4.40 m and have an inclination of 10° against the invert slope. Geophone sensors mounted beneath the steel plate register the oscillations caused by impacting bedload particles (Fig. 2). The number of impulses computed from the signal registered by the Swiss plate geophone correlates well with the transported bed load mass. However, this relation depends on site-specific conditions like flow velocity, particle-size and shape and hence a field calibration is required.

This study deals with the field calibration of the Solis SBT geophone system. The calibration procedure consists of: (I) introducing 10 m^3 of bedload material for each particle-size class inside the tunnel, (II) running the SBT with surplus inflow after a flood event when discharge is high, thereby keeping a high reservoir level to avoid bed load transport from the reservoir into the SBT inlet, and (III) recording and analyzing of the raw geophone signals. Three particle-size classes are chosen: 16–32 mm, 32–63 mm and 100–200 mm. In this paper, the details of the geophone system at Solis SBT and the results of the field calibration are presented and discussed.

Figure 1. Implemented geophone system at the outlet of the Solis SBT (VAW).

Measuring sediment fluxes in periglacial reservoirs using water samples, LISST and ADCP

D. Ehrbar, L. Schmocker, D.F. Vetsch & R.M. Boes
Laboratory of Hydraulics, Hydrology and Glaciology (VAW), ETH Zürich, Switzerland

M. Döring
Zurich University of Applied Sciences (ZHAW), Wädenswil, Switzerland

ABSTRACT

The expected glacier retreat due to atmospheric warming will offer new perspectives for the construction of reservoirs and hydropower plants in newly formed landscapes in front or just beyond the outer limits of glaciers (periglacial environment). However, the immediate proximity to the glacial environment will pose challenges in terms of construction, operation and maintenance, as the temporal evolution of glacier runoff and sediment transport have to be considered.

Especially reservoir sedimentation is of major importance in the periglacial environment. Several reservoirs in the Swiss Alps face severe sedimentation rates, whereas others are observing accelerating sedimentation processes that are possibly linked to glacier retreat. So far, the governing processes of reservoir sedimentation in highly glaciated catchments are not fully understood. The remaining questions are (amongst others):

- Particle size distribution: Which particles are transported in suspension, which fractions settle down, are transported in turbidity currents or remain in suspension throughout the whole reservoir ("washload")?
- Sediment concentrations: Which are typical concentration profiles in periglacial reservoirs?
- Velocities: What are typical flow velocities of turbidity currents and how can they be determined?

In order to answer some of these questions, an extensive field measuring campaign was conducted in summer 2015.

Prototype measurements were carried out at three different reservoirs in Switzerland. The volume V of the reservoir, the full supply level z, the watershed area A and the glaciation P of the catchments are given in Table 1.

A combination of the following measuring methods was applied: (1) Secci disk; (2) Niskin bottle sampler; (3) Van Veen Grab sampler; (4) Acoustic Doppler Current Profiler (ADCP); (5) Laser In-Situ Scattering and Transmissiometry (LISST); (6) Remote Sensing (Landsat satellite images). The aim was to link different methods with each other, e.g. to compare ADCP and LISST measurements and to calibrate and validate them with bottle samples. Furthermore, limitations and boundary conditions of different measuring techniques and their combinations have been established.

Table 1. Characteristics of investigated reservoirs.

Reservoir	V [hm^3]	z [m a.s.l.]	A [km^2]	P [%]
Griessee	18	2385	10	61
Lac de Mauvoisin	180	1961	114	46
Gebidem	9	1436	198	64

Figure 1. Griessee on 2 July 2015.

In this contribution, a set of different concentration profiles will be used to discuss governing processes of sediment transport in periglacial reservoirs. By comparing them with bottle samples, application ranges of different measuring techniques can be derived. Due to construction works at the intake, the water level at Griessee had to be lowered significantly in summer 2015 (Figure 1). This provided the opportunity to realize measuring campaigns at intermediate and full supply level. These two campaigns will be explained in detail.

Sensitivity analysis of measured sediment fluxes in a reservoir

S. Haun
Institute for Modelling Hydraulic and Environmental Systems, University of Stuttgart, Stuttgart, Germany

L. Lizano
Ministry of Environment and Energy, San José, Costa Rica

ABSTRACT

Reservoir sedimentation is an enormous problem nowadays and reduces the available storage capacity worldwide in the range of about 1% per year. However, due to climate change and other factors, such as geology and land use management this rate can reach values of up to 5% per year. Particularly in areas where people depend on the stored water, these depositions may have a negative influence on the society. However, due to the fact that each reservoir is unique with respect to the occurring flow situation and the behavior of sediments (e.g. flocculation or consolidation) it is not trivial to develop a sustainable and ecological friendly reservoir management. Accurate information regarding sediment transport within the reservoir, with a high spatial and temporal resolution, is therefore necessary.

A combination of two indirect measurement methods is presented in this paper, which were used to evaluate the suspended sediment transport in the Peñas Blancas hydropower reservoir in Costa Rica. A LISST-SL device (Sequioa Scientific Inc., 2011) was used to obtain a time series of suspended sediment concentrations (SSC) and particle size distributions (PSD) in several locations within the reservoir. These measurements were in a first step used to conduct a comprehensive data analysis with respect to natural fluctuations of the SSCs. In a second step a correlation function of the measured suspended sediment concentrations and the acoustic backscatter signal strength (ABS), obtained by measurements with an Acoustic Doppler Current Profiler (ADCP) (Teledyne RD Instruments, 2012) was developed. With the information from the ABS of the moving ADCP measurements it is feasible to evaluate sediment fluxes in pre-defined transects along the reservoir and subsequently to calculate the trapping efficiency of the reservoir.

The aim of the presented study is twofold, first the sensitivity of the evaluated sediment fluxes regarding the input data (SSC measurements) was investigated.

The correlation function, used for calibration of the backscatter intensity, was therefore obtained by taking into account the natural fluctuations which occur during the LISST-SL measurements. The point measurements were conducted for 60 seconds with a frequency of 0.5 Hz. The results showed that differences in the measured sediment fluxes of up to 42% occur by taking these natural fluctuations into account. In a second step the sediment transport within the near bed blanking distance was examined. Due to the ADCP measurement methodology no information, neither on flow velocities nor on acoustic backscatter data, is available close to the reservoir bed due to the ringing effect. Especially in deep reservoirs this near bed blanking distance can reach values of up to several meters. Where generally an extrapolation with information from the last three valid cells is used, within this study the information of the last two, three, four and five valid cells was taken into account for the extrapolation. The extrapolated values from the ADCP measurements were then compared with measured values by the LISST-SL. From the results could be seen that instead of a fixed number of cells (cell size may depend on the used ADCP), a value similar to the blanking distance height was used and more accurate results could be obtained. The error in the calculated sediment fluxes could for this study be reduced by up to 16%, depending on the height of the blanking zone.

This study proved in addition the advantage of combining these two measurement methods because a high spatial distribution can be achieved and time and costs can be reduced.

REFERENCES

Sequoia Scientific Inc. 2011. LISST-SL User's Guide V 2.1. Bellevue, USA.
Teledyne RD Instruments 2012. RiverRay Operation Manual. Publication Number 95B-6063-00, Teledyne RD Instruments, Poway, CA, 28.

HPP Vrhovo operation under reservoir sediment management

L. Javornik
Institute for Water of the Republic of Slovenia, Ljubljana, Slovenia

A. Kryžanowski & M. Mikoš
Faculty of Civil and Geodetic Engineering, University of Ljubljana, Ljubljana, Slovenia

ABSTRACT

The HPP Vrhovo and its 7 km long reservoir was built in 1993 as the first run-of-the-river HPP in the HPP chain on the lower Sava River, which is currently still under construction. Being the first HPP in the chain, it also intercepts the majority of sediment inflow from the upper reaches, causing significant changes in the Sava river sediment transport dynamics.

Operation procedures in the case of low, normal, high, and extreme flows are defined in an operation procedure manual, which in the present state, determines appropriate reservoir levels only in a relation to upstream incoming flows from the gauging stations Hrastnik (on the Sava River) and Veliko Širje (on the Savinja River, a major left Sava tributary) and as such doesn't consider effects of increasing flow levels on the reservoir morphologic conditions. The main goal of this study was to determine new, optimised operating rule curve (ORC) at the HPP Vrhovo dam gates with regard to sediment effects on the reservoir bathymetry.

The basis to determine an optimised, amended ORC, was developed using sediment transport simulations in the HPP Vrhovo reservoir for the case of high and extreme flows. An adequate flood event in the recent past had to be chosen to conduct such simulations, like the September 2010 event, when its peak discharge of 3600 m³/s reached Q_{100}. According to the comparison of two reservoir bathymetry surveys, this single flood event caused reservoir capacity increase of 7% due to sediment flushing The reservoir capacity before the flood event was 4.6 million m³.

Two simulation scenarios were carried out using the CCHE2D sediment transport model, differing in shares of individual sediment size classes at the HPP Vrhovo reservoir input. The first scenario assumed evenly distributed smaller fractions (medium and coarse silt), whereas the second scenario assumed fine sands as the dominant fraction at the model entry. Simulation results of both scenarios were similar; two

Figure 1. HPP Vrhovo dam.

main peaks of sediment concentrations within the flood event and before the flood peak discharge were observed – the first one as a consequence of flushing of recent sediment deposits from the reservoir starting at discharges as low as 0.5 Q_1, and the second one as a consequence of incoming sediments.

The reservoir drawdown has to commence at the time of sediment concentration peak which is a key to amend the ORC. Three different ORC scenarios were developed, each one predicting different turbine operation levels, from the first ORC scenario where all three turbines are working optimally, to the least favourable ORC scenario, when all of the three turbines are inoperable. Consequently, different water levels were obtained for such conditions due to gate and flaps manipulation.

Sediment management for sustainable hydropower development

M. Omelan & J. Visscher
Franzius-Institut for Hydraulic, Estuarine and Coastal Engineering, Leibniz University of Hanover, Hanover, Germany

N. Rüther
Department for Hydraulic and Environmental Engineering, NTNU of Trondheim, Trondheim, Norway

S. Stokseth
Statkraft AS, Oslo, Norway

ABSTRACT

Reservoirs have conventionally been designed to operate for periods between 50 and 100 years without adverse effect by sedimentation. Nowadays, reservoirs are faced with serious sedimentation problems that were ignored by the original designer. Especially, reservoirs in Middle East and China have an annual storage volume loss due to sedimentation of 1.5% and 2.3%, respectively. Reservoir areas, with a high sediment yield will affect the power generation, irrigation and flood control due to the reduction of available storage volume by siltation. Today, to maintain the usage of the reservoirs, it is necessary to predict the reservoir sediment yield in order to identify a sediment management strategy for sustainable hydro power development. For this reason, Statkraft has initialized a research and development project to sustainably handle future sediment issues at their hydro power plants. The presented study is conducted within this R&D project and will contribute to the knowledge of sediment yield calculation and how this data can be used from a sediment management point of view.

The overall goal of the presented study is to evaluate potential sediment management techniques for Banja Dam both, from the technical and economic point of view. The Banja Dam is the most downstream power plant of the Devoll hydro power project owned by Statkraft in the Devoll River catchment in Albania. The Dam is currently under construction and starts operating in Summer 2016.

Within this goal, the first purpose of the study was to calculate the spatial distribution of annual soil loss and the expected amount of sediment at Devoll River catchment using the "Revised Universal Soil Loss Equation" (RUSLE 2014) model. The results illustrate a relatively high average annual soil loss rate over an area of about $3,121$ km^2. Moreover, the sediment delivery ratio (SDR) was utilized for calculation of the total annual sediment yield. The annual sediment yield for the whole catchment was calculated by ArcGIS software and gave similar values as the observed sediment yield which was determined by the average sediment yield of the years 1955–1983.

The second purpose of the study was to find a technically feasible sediment management strategy that maximizes the economic benefits of the Banja Dam by using the Reservoir Conservation (RESCON) model (2003 a&b). The RESCON model examines and compares some sediment removal techniques (flushing, hydro suction sediment removal, dredging, and trucking) both economically and hydraulically. In addition, several sensitivity analyses are carried out in order to investigate the program for variable conditions.

REFERENCES

Palmieri, A., Shah, F., Annandale, G., W. & Dinar, A. 2003a. Reservoir Conservation, Volume I: The RESCON Approach, The World Bank.
Kawashima, S., Johndrow, T.B., Annandale, G.W. & Shah, F. 2003b. Reservoir Conservation, Volume II: The RESCON Model and User Manual, The World Bank, 2003.
United States Department of Agriculture – Agricultural Research Service. 2014. Revised Universal Soil Loss Equation (RUSLE) – Welcome to RUSLE 1 and RUSLE 2

Flow field and sediment flux measurements at alpine desanding facilities

C. Paschmann, J.N. Fernandes, D.F. Vetsch & R.M. Boes
Laboratory of Hydraulics, Hydrology and Glaciology (VAW), ETH Zürich, Zurich, Switzerland

ABSTRACT

Operating high-head hydroelectric power plants under alpine conditions may expose facility components to hydro abrasion due to mineral suspended sediments in the turbine water. Particularly, turbines can be affected by wear, leading to a considerable efficiency decline affiliated to power and financial losses. Therefore, high-head hydroelectric power plants are commonly equipped with desanding facilities to reduce the amount of suspended sediments. Nowadays, climate change causing glacier meltdown entails increasing sediment yield from glaciated catchment areas into alpine waters. Additionally, experiences show that the settling efficiency of existent desanding facilities often is below expectations, frequently due to shortcomings of the geometrical design and the unsatisfactory results of the existing empirical formulations (Ortmanns, 2006). Thus, the geometric optimization of existing and proposed facilities is of great importance.

To cope with these issues a new experimental campaign in three desanding facilities was conducted. The facilities comprehend different geometries and inlet and outlet conditions. The experimental setup was designed to allow for the characterization of the flow velocity and turbulence, the turbidity, the sediment concentration and the water density in the main chamber of the facilities (cf. Fig. 1). The measuring mesh typically consisted of about 750 points. Special care (e.g. finer measuring mesh) was taken in the inlet section.

The equipment installed comprised acoustic Doppler velocimeters (ADV), turbidimeters and a Coriolis flow- and density-meter. A mobile pump was used to collect water samples at numerous positions, spatially and temporally synchronized with the ADV measurements. Amongst other features, the gravimetric sediment concentration of the water samples was determined in the laboratory and correlated with the measured turbidity and density. That procedure allowed for determining the sediment concentration distribution over the whole area of the facility. Additionally, continuous measurements of the turbidity in the inlet and outlet cross sections were performed. The particular desanding effect of each facility was identified by analyzing and comparing these data.

Conclusions regarding the connection of turbidity and sedimentological composition of the suspended particles were drawn. Samples of the deposited sediments were taken at scattered positions. The grain size distributions were determined in order to gain an insight of the specific deposition patterns.

In the present paper the main findings from the experimental campaign are presented. The flow field in each facility is characterized by taking the effects of the geometric shape of the chamber and of the approach flow and inlet conditions into account. Qualitative conclusions are made regarding the influence of the present flow fields onto the particle settling in the investigated facilities.

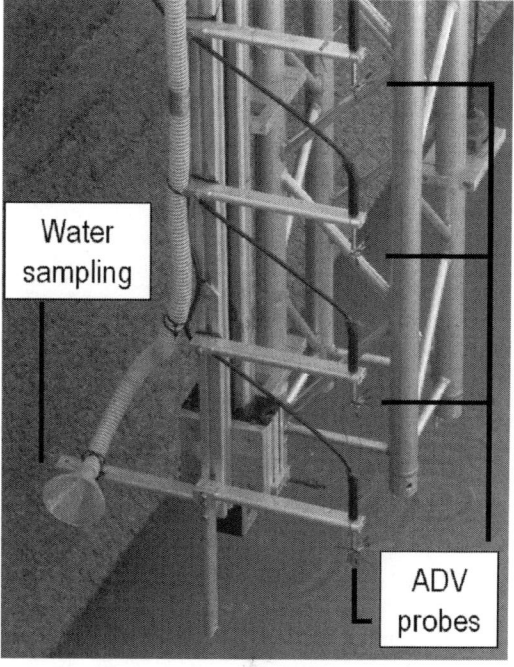

Figure 1. Selected detail view of experimental setup.

REFERENCE

Ortmanns, C., 2006. Entsander von Wasserkraftanlagen ('Desanding facilities at hydropower plants'). *VAW-Mitteilung* 193 (H.-E. Minor, ed.), VAW, ETH Zürich, Switzerland

Ensuring sediment continuity through a reservoir: Challenges and methodology applied to define favorable hydraulic scenarios in the case study of the Champagneux run-of-river dam on the Rhône River, France

C. Peteuil, D. Alliau, T. Frétaud, M. Decachard, S. Roux & S. Reynaud
Compagnie Nationale du Rhône. Lyon, France

N. Boisson, A. Vollant & Y. Baux
Optifluides. Villeurbanne, France

ABSTRACT

By decreasing the flow velocity and turbulence, reservoirs controlled by dam are likely to force inflowing sediments to settle. This process can be more or less temporal and intensive, depending on particle size and reservoir characteristics (Morris & Fan 1997). To limit the reservoir sedimentation and prevent a disruption of sediment continuity, one possible option for dam operators is to recover favorable flow conditions either for routing inflowing sediments or remobilize previous deposits (Kantoush & Sumi 2010). In the case of cohesive sediments, one of the main challenges to deal with is that deposition and erosion thresholds are often radically different as a result of deposit consolidation (van Rijn 1993). The purpose of this communication is to present the methodology followed to define those conditions in the case of the Champagneux dam.

First, field and laboratory investigations performed are described. So as to ensure the representativeness of samples and evaluate the heterogeneity of deposits, preliminary acoustic measurements using side scan sonar and sub-bottom profiler have been performed (Fig. 1). Subaquatic video recording of the river bottom have been also carried out, as well as direct sampling works with a dredge scraping the deposit surface and a core sampler. Erosion tests on undisturbed subsurface samples were performed in a lab flume to evaluate the mechanical resistance of deposits, in particular through parameters like the critical shear stress and velocity for erosion (Briaud & Chen 2005).

The evaluation of flow conditions throughout the reservoir and for different hydrological and operating conditions relies on a hybrid approach combining (1) a 3D free surface numerical model of the whole reservoir, (2) a small scale physical model limited to the downstream part of the reservoir and (3) a CFD numerical model of the dam area. Even if those 3 models have been initially deployed in the frame of a more general project, this experience has shown that such comprehensive approach is required to obtain, with a

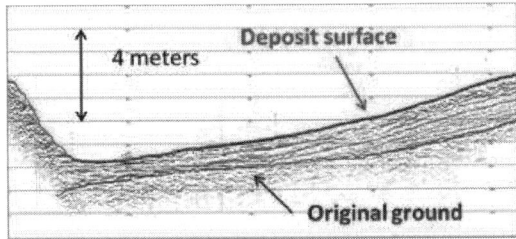

Figure 1. Structure of sub-surface layers of deposits according to acoustic data collected with the sub-bottom profiler deployed in the frame of the study.

high resolution and in the entire reservoir, relevant values of the shear stress and flow velocity close to the river bottom. The erosion and transport potentialities are finally evaluated by a comparison of simulation results with critical thresholds resulting from lab tests.

For the dam operator, such comprehensive survey provides useful information to enhance the management of the reservoir and feedbacks obtained will contribute to an optimization of the methodology for other similar cases.

REFERENCES

Briaud, J.-L. & Chen H.-C. 2005. The EFA, Erosion Function Apparatus: An Overview, Proceedings of the International Conference on Soil Mechanics and Geotechnical Engineering, Osaka, Japan, September 2005.

Kantoush, S.A. & Sumi, T. 2010. River morphology and sediment management strategies for sustainable reservoir in Japan and Europe Alps, Annuals of Disast. Prev. Res. Inst., Kyoto Univ., No. 53B

Morris, G.L. & Fan, J. 1997. Reservoir Sedimentation Handbook: Design and Management of Dams, Reservoirs and Watersheds for Sustainable Use. New York: McGraw Hill.

van Rijn, L.C. 1993. Principles of Sediment Transport in Rivers, Estuaries and Coastal Seas. Aqua Publications, 386 pp.

Experimental analysis of the interaction between hydroelectric sluice gates and sediment transport

G.R. Pisaturo, M. Righetti, F. Amante & E. Bigliotti
Department of Civil, Environmental and Mechanical Engineering, Mesiano, Trento, Italy

ABSTRACT

The design of withdrawal structures for run-of-the-river hydroelectric plants is strongly influenced by the presence of sediment transport. Usually, weirs cause an upstream water depth increase and therefore a flow velocity decrease. This hydraulic effect implicate an alteration of sediment transport dynamics in the river, with accumulation of sediments and morphology alterations in the upstream reach. Sediment trapping caused by weirs can lead to twofold problems: the reduction on withdrawal functioning due to clogging at intake structures and, more important, a significant impact on the downstream reach.

One of the most used management methods to allow the downstream sediment transport is sediment flushing. This technique involves removing sediment by erosion, thanks to the opening sluice gates outlets.

For this research, an experimental model developed at the Laboratory of Hydraulics, University of Trento is used. The kinematic parameters of the model are derived by Froude similarity, whereas the sediment transport parameters are derived by Shields similarity.

The first step is study the clogging process caused by the weir (synthetic resin $d_{50} = 0.7$ mm; $\rho_s = 1130$ kg/m^3).

The second step consists in the opening of the control gate at the left side of the model such that the flow depth is maintained at the weir crest level and -at the same time- the full discharge flows through the gate.

The tests show that the sediment erosion is related to the hydraulic conditions imposed by the opening operations of the bottom outlets.

The maintenance of the water depth at the weir crest level, strongly influences, the erosion process. In this case, in fact, for small flow rates ($Q = 15$ l/s) the erosion of the sediment is limited to a very small portion and close to the used gate (Fig. 1). The scour length in upstream direction is about 0.5 m and in lateral direction is about 0.55 m, comparable with gate dimension.

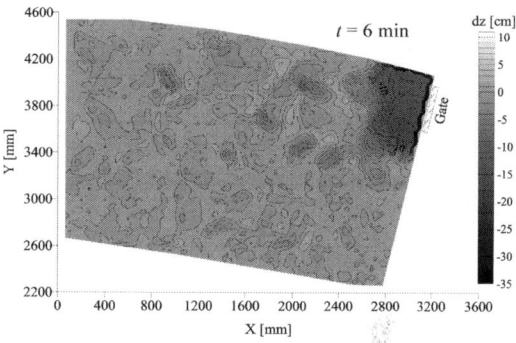

Figure 1. Bed erosion map after 6 minutes of sediment flushing from left gate. ($Q = 15$ l/s).

In this case there is a formation of a phenomenon of erosion called flushing cone (Shen 1999). The size of the flushing cone is dependent on the flow rate from the gate and on the geometric characteristics of the gate itself, particularly in the case of low flow rates. For higher flow rates ($Q = 30-40$ l/s) is possible to observe an increase of the scour dimension in lateral direction.

To understand the fluid dynamic phenomena that causes the erosion located in the proximity of the weir, velocity field is analyzed in the scour area.

It is used both the ADV and the Particle Tracking Velocimetry technique to reconstruct the fluid dynamic field.

It is possible to observe that flow field is strongly 3D. In particular, the formation of horizontal and vertical axis circulations implicate the scour extension and the lateral erosion process.

REFERENCE

Shen, H.W. 1999. Flushing sediment through reservoirs. Journal of Hydraulic Research, vol. 37, n. 6: 743–757.

3D fully coupled numerical modelling of local sediment flushing scour at dam bottom outlets for sustainable hydropower operation

O. Sawadogo & G.R. Basson
Department of Civil Engineering, University of Stellenbosch, Matieland, South Africa

ABSTRACT

Reservoir sedimentation is not only a serious threat to water storage capacity but it could also undermine some hydraulic structures such as hydropower intakes and dam bottom outlets. Reservoir sedimentation affects the long term sustainability of dams. When sediment deposits near the hydropower intakes, the sediment transport capacity could become high enough to transport fine sand into the turbines, with resultant damage and cost implications. To protect hydropower intakes, sediment flushing tunnels/conduits could be installed below or near the power intakes, such as were implemented at the Xiaolangdi Dam on the Yellow River in China. These outlets operate under pressure and are located relatively high above the reservoir bed near the invert levels of the hydropower intakes. Their effectiveness to keep coarse sediment away from the hydropower intakes depends on the three dimensional flow patterns near the intakes.

Bottom outlets at dams can also be used to flush sediment locally near intakes, under pressure or with free outflow conditions. In the latter case much more sediment can be removed from the reservoir during floods, but excess water should be available to allow considerable draw-down of the water level in the reservoir.

To protect hydropower intakes a three dimensional hydrodynamic CFD model can be used to simulate the stream lines, sediment transport and local scour accurately upstream of a bottom outlet or sediment flushing tunnel. This paper focuses on the simulation of flow and sediment transport patterns upstream of a dam bottom outlet.

This paper first describes the development of a coupled fully three-dimensional (3D) numerical model for the prediction of the local pressured sediment flushing scour upstream of the bottom outlet for non-cohesive sediments. The presented numerical model solves Navier-Stokes equations in conjunction with the $k - \varepsilon$ turbulent model which include both sediment transport and hydrodynamic parameters. The proposed coupled fully 3D numerical model is later used to simulate the experimental tests based on non-cohesive sediments. Results from these simulations are in good agreement with the measurements in terms of the geometric features of the scour hole as well as the velocity field upstream of the bottom outlet.

Integrative monitoring approaches for the sediment management in alpine reservoirs: Case study Gepatsch (HPP Kaunertal, Tyrol)

M. Schletterer, B. Hofer, R. Obendorfer & A. Hammer
TIWAG – Tiroler Wasserkraft AG, Innsbruck, Austria

M. Hubmann & R. Schwarzenberger
ARGE Limnologie GesmbH, Innsbruck, Austria

M. Boschi
droneproject.at, Innsbruck, Austria

S. Haun
University of Stuttgart, Institute for Modelling Hydraulic and Environmental Systems, Stuttgart, Germany

M. Haimann, P. Holzapfel, H. Habersack, B. Brock, B. Schmalzer & C. Hauer
BOKU – University of Natural Resources and Life Sciences Vienna, Institute for Water Management, Hydrology and Hydraulic Engineering, Department for Water – Atmosphere – Environment, Vienna, Austria

ABSTRACT

Sedimentation and the management of hydropower reservoirs is one of the big global issues concerning future hydropower use. Exemplarily for Asia a reduction of 80 % of the potential storage volume is predicted by 2035. For Europe those thresholds in sedimentation are predicted by 2080. Related consequences – beside a reduction of the storage volume and the restriction in hydropower production – are effects on aquatic ecology linked to artificial flushing of deposited fines. Impacts on aquatic ecology, however, can be manifold, especially in a change in habitat quality with possible clogging of the gravel matrix and an increase of stress due to an increased turbidity. What is missing up to now, are integrative studies on different scales including catchment-, reach -, and point scale assessments with an integrative evaluation of the biota (e.g. fish and macroinveretebrates).

In principle it is essential to distinguish reservoir flushing from a controlled drawdown of a reservoir. While reservoir flushing aims the removal of sediments and takes place at high flow rates for short time periods (i.e. suspended sediment concentrations are high – similar to natural flood events), a controlled drawdown of a reservoir is often carried out in winter period with low flows (i.e. suspended sediment concentrations are relatively low, but the time of the emission is longer [depending on reservoir size]).

The presented case study is dealing with a lowering of the water surface level of the reservoir Gepatsch (Tyrol) beyond the operational level. This was a controlled drawdown of the reservoir, which has to be distinguishes from reservoir flushing. However, also from the controlled drawdown an increased load of suspended sediments was expected. Based on the awareness of possible negative ecological consequences a complex set of measures and an integrative monitoring design was been developed. Monitoring is based on detailed event based quantification of eroded sediments. High resolution turbidity data are available for the entire Inn river in Tyrol. Moreover, we analysed the biological quality elements macrozoobenthos and fish at selected stretches. In addition, freeze-core samples were taken before and after the lowering of the reservoir volume, in combination with cocooning of brown trout during the spawning period. This case study brought up an extensive data set, however for future monitoring activities in similar projects we suggest to concentrate on a limited number of monitoring sites, including a reference station as well as a station to assess the emission. Downstream effects could be assessed numerically, but it has to be considered that additional stations provide the possibility for a detailed process study, i.e. the analyses of processes that are causing a natural increase of suspended sediments.

This case study brought up an extensive data set, however for future monitoring activities in similar projects we suggest to concentrate on a limited number of monitoring sites, including a reference station as well as a station to assess the emission. Downstream effects could be assessed numerically, but it has to be considered that additional stations provide the possibility for a detailed process study, i.e. the analyses of processes that are causing a natural increase of suspended sediments.

Sediment management of reservoirs: Sediment discharge in dependence on the suspended load concentration in the run-off water – Theoretical foundations and practical experiences

F. Sollerer & Gottfried Gökler
Vorarlberger Illwerke AG, Vandans, Austria

ABSTRACT

Generation companies of hydropower plants often have difficulties to realize an ecological and sustainable as well as economic sediment management of their reservoirs. Because the most common methods (reservoir flushing and machine excavations) do not meet these requirements, the Vorarlberger Illwerke AG has implemented a system for one of their reservoirs that achieves both the environmental and economic requirements.

Currently approx. 350.000 m³ of sediments are located in the three reservoirs of Rodund/Vorarlberg – Austria. These reservoirs are used by the power plants Rodund I and II as lower reservoirs as well as upper reservoirs by the power plant Walgau (WAW). The loss of volume which is caused by the sediments approximates 16% of the original storage volume of 2,1 Mio m³. To allow an optimal usage of the reservoirs it is necessary to restore their original volume within the next few years. Therefore several options have been investigated to find an answer which one is acceptable in ecological belongings as well as in technical and economic aspects. Because of the reservoir's location any solution and road haulage would be due to the quantity of sediments not reasonable for the neighboring country.

The plan is now to lift the sediments in one of the reservoirs by suction excavator and to add them to the head race tunnel of the power plant Walgau, which leads the sediments into the river "Ill". The Ill is at least the source of these sediments and due to the existence of reservoirs low in suspended load. The planned continuous discharge of sediments should be done only with acceptable impact on the environment, depending on the simultaneously measured stream flow conditions (discharge and amount of suspended load) of the diverted waters. This ensures a sustainable return of sediments into the river and that the suspended load concentrations are similar to an undisturbed natural river.

Figure 1. Image of the three reservoirs of Rodund.

For setting up this system investigations had to be done for the questions:

- How much was the suspended load in the river Ill without the reservoirs?
- What can be understood under acceptable impact to the limnology and biology?
- Is there any impact on the communication between the river and the groundwater (role of the silted layer)?
- Technical realization regarding the measurements in the river and the dredging unit
- Potential influences on the power plant Walgau

Now since May 2013 sediments are returned by the dredging unit into the "Ill" and faced us with the differences between the theory and practice. These have been specially the hydrological parameters as well as the influences on the power plant Walgau.

SS 2 Navigation and river morphology

Analysis of sedimentation of the Yangtze Estuary channel, China

X.P. Dou
Nanjing Hydraulic Research Institute, Nanjing, China
Key Lab of Port, Waterway and Sediementation Engineering of MOT, Nanjing, China

Z.X. Jiao, X.Y. Gao, L. Ding & J. Jiao
Nanjing Hydraulic Research Institute, Nanjing, China

ABSTRACT

The Yangtze River Estuary, flowing into the East China Sea, presents a pattern of "Three-level Branching and Four-outlet", namely, Chongming Island divides the estuary into the North and South Branches; Changxing Island and Hengsha Island subdivide the South Branch into the North and South Channels; Jiuduansha Shoal subdivides the South Channel into the North and South Passages (Figure 1).

Since 1998, DCRP has been built at South Channel and North Passage by regulation structures and dredging. In the first two phases (1998–2005), a pair of training jetties are respectively 48 and 49 km and 19 groins with a total length of 30 km. The elevation of the deepwater channel increased from 7 m to 8.5 m after Phase I and to 10 m after Phase II. The channel is only dredged and the elevation to 12.5 m in Phase III (from 2006 to 2010).

During the 12.5 m deepwater channel has entered the maintenance period, the siltation is large and far exceeds 100 million m³ (Table 1). There are some changes including diversion ratios of ebb and suspended load of North Passage, vertical distribution of sediment concentration, velocity and siltation along the channel, concentration and siltation along the channel. However, runoff and sediment quantity from the Yangtze River into the sea have been decreased,

Table 1. Sediment siltation in 12.5 m navigation channel.

Year	2010	2011	2012	2013	2014
Siltation	80.15	85.46	100.85	81.76	76.21

especially in sediment have been greatly reduced. According to the field data, the relationship among velocity, concentration and siltation are analyzed.

With the groins construction, the North Passage has been scoured and became narrow-deep. The volume under 8 m isobaths of NP was expanded. The deposition between groins has increased, the average sedimentation thickness 2.7 m, most of the elevation of the area had reached 0 m. The deposition material was suspended periodically under the action of tide currents and wind waves, which led to high concentration. In addition, due to the increase of the difference of height between the groins area and the channel, the sediment is much easy to enter the channel with density flow or gravity flow. The sediment on the Hengsha shallow shoal and Jiuduansha beside the North Passage is suspended by the action of wave and tide, and flow into the NP with the currents. Furthermore, there are 4 mud storages and 1 mud-dumping area between North Jetty and South Jetty. Some of the dredging mud will go back to the deep-water channel. All the sediment is the source of siltation of the deepwater channel.

Based on the above analysis, some measures to decrease siltation of the deepwater channel are studied.

REFERENCES

Dou XP, Li TL, Dou GR, Numerical model of total sediment transport in the Yangtze Estuary, CHINA OCEAN ENGINEERING, No.13, 277–286, 1999.

Jin Liu, Yu Zhiying, He Qing, Study on River Regime Control and Siltation in Deepwater Channel, China Harbour Engineering, Feb., 2012, Total 178, No.1.

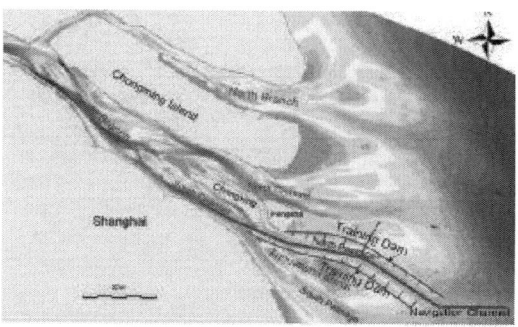

Figure 1. Pattern of Yangtze river estuary.

Evolution characteristics of the north branch of the Yangtze Estuary

X.Y. Gao, X.P. Dou, L. Ding, Z.R. Gao & J. Jiao
Nanjing Hydraulic Research Institute, Nanjing, China
State Key Laboratory of Hydrology-Water Resources and Hydraulic Engineering, Nanjing, China
Key Lab of Port, Waterway and Sedimentation Engineering of MOT, Nanjing, China

ABSTRACT

The planar pattern of the Yangtze River estuary is fan-shaped.It has three-level branching and fourth mouths. As the first-level branch, the North Branch of the Yangtze River estuary flows through Chongming County, Shanghai and Haimen and Qidong, Jiangsu Province. The length between Chongtou and Lianxing Port is 78.8 km (Fig. 1). Study on the evolution rule of the river regime of the North.

There are many factors affecting river regime evolution, such as hydro-sediment conditions, planer pattern, geological conditions and human activities. The North Branch was the main channel flowing into the sea from the Yangtze River in history. Since the 18th century, main stream of the Yangtze River has been diverted to the South Branch, and the runoff of the North Branch has been gradually reduced. In 1915, the diversion ratio of the North Branch was about 25%. Since 1958, Xuliujing node of the Yangtze River estuary has gradually taken shape, playing a positive role in stabilizing the river regime of the South Branch. Since 1970, Xuliujing section has been narrowed down to 5.7 km. the upper channel of the North Branch and main stream of the South Branch is about 90°. It's diversion ratio is reduced to 3%~4%. The hydrodynamic of the North Branch is mainly determined by tidal current. Since 1950s, the economy of the Yangtze River Delta has rapidly developed. Human activities have increased frequently. Large-scale reclamation has been carried out in the North Branch and its entrance region, with an enclosed area up to 440 km². The Three Gorges Project and South-to-North Water Diversion Project constructed in the upper reaches of the Yangtze River have changed hydro-sediment conditions of the Yangtze River estuary. In recent years, bridges and lots of ports and piers have been built in the downstream section of the North Branch. As indicated by former research and data analysis, the river regime evolution of the North Branch is greatly affected by human activities. At present, the upper mouth of the North Branch is characterized by poor inflow and reduced diversion ratio. That leads to gradual siltation and narrowing of river beds and the gradual decreasing of channel-fill volume on the whole (Table 1). From 1984 to 2011, the river width was narrowed down by 36.5% on average. The upper and middle channels of the main stream line of the North Branch are frequently varied.the lower channel is stable. In the developing and utilizing of the upper, middle and lower channels of the North Branch, it is appropriate to regulate them by sections according to the evolution rule. The strategy of utilizing deep water in deep areas and utilizing shallow water in shallow areas are adopted. Engineering methods are taken to maintain and improve hydrosediment conditions of the upper mouth and reduce the influence of tidal current and dredging method is also used to regulate the North Branch.

Table 1. Volume changes at different elevations of the whole reach of the North Branch.

Year	Channel volume/0.1 billion m³		
	0 m	−2 m	−5 m
1984	18.84	10.67	2.25
1991	15.41	8.7	1.61
1998	14.48	8.01	1.64
2001	14.71	8.78	2.58
2003	15.1	9.35	2.9
2005	13.89	8.14	2.2
2008	12.94	7.8	2.23
2011	12.12	7.45	2.31

Figure 1. Pattern of Yangtze river estuary.

Design of bank protection for inland waterways with GBBSoft+

C. Gesing & B. Söhngen
Federal Waterways Engineering and Research Institute, Karlsruhe, Germany (BAW)

K. Kauppert
SWIFT Engineering GmbH, Karlsruhe, Germany

ABSTRACT

In 2008 the GBBSoft software for designing bank and bed protection for inland waterways was introduced in the scope of business of the *Federal Waterways and Shipping Administration (WSV)* especially for application cases deviating from the standard building methods of canals. Its calculation methods and design concepts are based on the newsletter No. 88 *Principles for the Design of Bank and Bottom Protection for Inland Waterways (GBB)* issued by the *Federal Waterways Engineering and Research Institute (BAW)* in 2005 which was comprehensively revised and updated in 2011 (BAW 2010). The principles as well as the software are continuously enhanced and modified by an interdepartmental working group of the BAW. Since 2015 it is also possible to design technical-biological bank protection measures with the software, renamed in GBBSoft+. For this purpose the basic principles of the advisory guideline DWA-M 519 on technical-biological bank protection, which will be published in the beginning of 2016, have been implemented (DWA 2015). This paper, however, focuses on technical revetments which are prescribed in canal sections particularly relevant to safety as well as in highly frequented waterways. GBBSoft+ ascertains the loading on banks resulting from the primary and secondary wave field of typical inland navigation vessels (e.g. motorized freight vessels, sport boats, etc.) in stationary movement parallel to the banks in a prismatic, trapezoidal cross-section through the waterway. The basis for calculating the ship-induced loads is mainly taken from relevant literature. Basic equations of the one-dimensional canal theory and the simplified potential theory are adapted to the particular boundary conditions of inland waterways. Effects taken into account are e.g. shallow water, boundary layers and eccentricity (Fig. 1). This leads to parameter-based, semi-empirical approaches which are calibrated by model and field test results.

The hydraulic actions constitute the input parameters for the design of bank protection which comprises

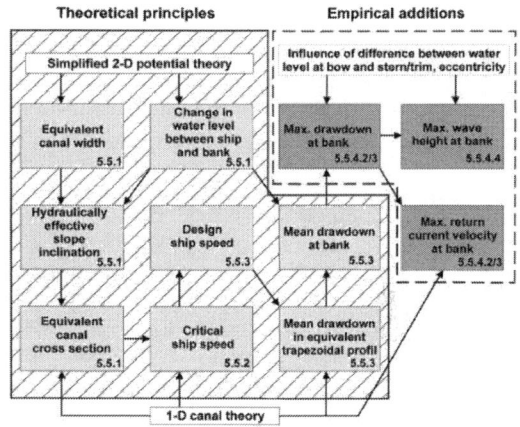

Figure 1. Procedure for determining the hydraulic parameters when subject to e.g. shallow water and boundary layer effects.

a hydraulic, a geotechnical and when required a technical-biological component. GBBSoft+ differentiates between the design of a classical technical rip-rap and a technical-biological bank protection measure. These two design cases have to be carried out separately.

In addition to the principles of the GBB (BAW 2010) the paper summarizes recent developments regarding the software such as the possibility of analyzing multiple water levels at the same time and the implementation of a two-dimensional numerical model.

REFERENCES

BAW. 2010. Principles for the Design of Bank and Bottom Protection for Inland Waterways (GBB). BAW Code of Practice. Bundesanstalt für Wasserbau. Karlsruhe.
DWA. 2015. Technisch-biologische Ufersicherungen an großen und schiffbaren Gewässern. DWA-M 519. DWA.

Tension between bridge and waterway in the middle of Yangtze River with its countermeasures

D. Li & L. Chen
State Key Laboratory of Water Resources and Hydropower Engineering Science, Wuhan University, Wuhan, China

ABSTRACT

Relying on the golden waterway, the construction of the Yangtze River Economic Zones Project has been raised to the national core strategy of China. Nevertheless, many bridges being built or having been built severely limit the development Yangtze River's Golden Waterway. Riverbed evolutions in bridge reaches tend to affect navigation channel conditions. Changes of flow and sediment lead to dramatic riverbed evolutions after the impoundment of Three Gorges Project, which makes navigation conditions in bridge reach difficult and hazardous for the settled navigation spans. In this paper, the authors firstly review the tensions between bridges and waterway in the middle of Yangtze River and navigation problem researches in bridge areas. Then, Taipingkou waterway of Jingzhou Yangtze River Bridge is taken as a case study to analyze its channel variations and navigation-obstruction characteristics by field observation data. Also, based on fixed physical model experiment of Taipingkou, we study the flow characteristics and waterway regulation projects effects.

The middle of Yangtze River starts from Yi-chang, where dams of the Three Gorges Project (TGP) and Gezhouba Project are located. The main stream reach from Yichang to Hankou is about 590 km in length. Below Yichang, the Yangtze River enters a flood plain and the riverbed on this reach is alluvial sediment in contrast to the bedrock of the upper stream (Yuan et al. 2012). The middle of Yangtze River also has a typical meandering river pattern and the variations in hydrology and sediment in this reach are very complex. After the impoundment of TGP, large amounts of sediment are being deposited in the reservoir and the released clear water is causing even more complex and river regimes, which impact the navigation channel conditions in bridge area severely. In the middle Yangtze River from Yichang to Hankou, more than 15 bridges have been built and around 10 bridges are in design or construction. The most typical navigation-obstruction bridges in this reach are Jingzhou Yangtze River Highway Bridge and Wuhan Yangtze River Bridge. Around

Table 1. Diversion ratio before and after regulation projects.

Discharge (m^3/s)	Sanba central bar			
	Before		After	
	North	South	North	South
6000	12.6	87.4	25.4	74.6
12245	21.5	78.5	30.4	69.6

70 accidents of ship against bridges took place since the operation of the Wuhan Bridge and more than ten of them caused a financial loss over one million RMB (0.157 million $) (Dai et al. 2002).

Jingzhou Bridge is taken as a case study to analyze its channel variations and navigation-obstruction characteristics. Based on fixed physical model experiment, we study the flow characteristics and waterway regulation projects effects.

Experimental conditions: five typical kinds of discharges are used in the experiment and they are 6000, 12,245, 18,367, 25,773 and 36,082 m^3/s.

The main designed navigation span of the Jingzhou Bridge is located in the north of the Sanba central bar. Therefore, increase of the hydrodynamic conditions and diversion ratio in the north branch helps to improve navigation conditions of the bridge. The main experiment results are shown in Table 1.

REFERENCES

Dai, T.Y., Nie, W., Liu, Y.J. & Wang, L.P. 2002. Statistical analysis of ship collisions with bridges in china waterway. *Journal of Marine Science & Application*, 1(2), 28–32.

Yuan, W., Yin, D., Finlayson, B. & Chen, Z. 2012. Assessing the potential for change in the middle yangtze river channel following impoundment of the three gorges dam. *Geomorphology*, s 147–148(8), 27–34.

Adaptability of numerical model for siltation in the Yangtze Estuary channel

T.L. Li, L.M. Chen, X.Z. Zhang, W.Y. Zhang & X.Y. Gao
Nanjing Hydraulic Research Institute, Nanjing, China
State Key Laboratory of Hydrology-Water Resources and Hydraulic Engineering, Nanjing, China

ABSTRACT

The Yangtze Estuary Deepwater Channel Project was started from 1998 and carried out in three phases. The Phase-I Project reached 8.5 m bed elevation in September 2002, the Phase-II Project 10 m in November 2005, and the Phase-III 12.5 m in March 2010, finally forming a tow-way channel of 92.9 km long and 350–400 m bottom wide. Before the implementation of the Project, the numerical model for sediment siltation was applied to predict the back-silting quantity of each phase, showing that, under the normal hydrological condition, the back-silting quantity of all phases respectively are 15 million m³, 21 million m³ and 32 million m³. In consideration of runoff changing in different years, the variable amplitude of the back-silting quantity is 20%.

The back-silting quantity of the channel was measured after the treatment implementation of the Project. The paper is to research the adaptability of the numerical model by comparing with the back-silting quantity of the channel measured at each phase and the value calculated by the model.

The numerical model is 2D tidal current and sediment transport. A formula of sediment transport capacity is used to calculate the back-silting quantity which considers the combined action of tidal currents and wind waves. The model has verified flow velocities, flow directions and sediment concentrations of spring tide, moderate tide and neap tide in the flood season and dry season, topographical change in two years. Then the back-silting quantity at different phases is predicted by the model.

Table 1 is the comparison on silting quantity in 1 year after the completion of the Phase-I Project. We also compare the calculating and measuring silting quantity during the maintenance period of the Phase I and Phase II and the silting quantity of the Phase III. It can be found that the calculating values are in good agreement with the measuring ones except the siltation quantity after Phase III. The measured siltation is larger than the predicted values when the channel has been maintained 12.5 m.

Table 1. Comparison on siltation of the deepwater channel after phase I implementation.

Duration	Measured silt quantity/(10⁴ m³)	Calculated silt quantity/(10⁴ m³)
March, 22–Nov. 12, 2000	879	818
Nov. 18–Apr. 26, 2000	693	690

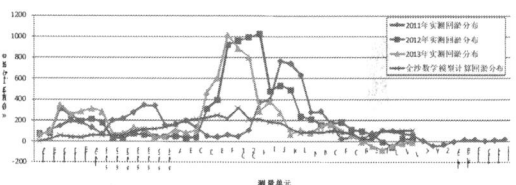

Figure 1. Comparison the silting quantity on the calculated value with measured value of phase-III.

The sediment carrying capacity is one of the most important factors in the mathematical model for sediment siltation. In order to reflecting the back-silting of the 12.5 m deep-water channel via the full load mathematical model for the Yangtze Estuary deep-water channel, the paper focuses on analysis the silt carrying capacity affecting the sediment back-silting distribution and quantity (Figure 1).

REFERENCES

Dou Guoren, Dong Fengwu, Dou Xibing,1995, Sediment Transport Capacity of Tidal Currents and Waves[J]. Chinese Science Bulletin, 40(13): 1096–1101.

Dou Xiping, Li Tilai, Dou Gnoren,1999. Numerical Model of Total Sediment in the Yangtze Estuary[J]. China Ocean Engineering, 13(3): 277–286.

Jin Liu, Yu Zhiying, He Qing, 2013. Response of Deep-water Channel in North Passage to Water and Sediment Exchange between Channel and Shoal in the Yangtze Estuary. Port & Waterway Engineering. pp 101–108 (in Chinese).

The potential of alternative technical-biological bank protection measures on federal waterways – an applied research approach

K. Schmitt & L. Symmank
Federal Institute of Hydrology, Koblenz, Germany

ABSTRACT

The increase of navigation and thus intense maintenance of German federal waterways caused large ecological impacts on river ecosystems. These impacts result in the loss of biodiversity and constraints of ecosystem functionality within river bodies, their banks and adjacent floodplains. According to specifications of the European Water Framework Directive, introduced in 2000, a major part of federal waterways in Germany are currently classified as heavily modified.

To achieve an ecologically sustainable development in terms of bank protection with contemporary securing of navigation processes a joint research project of the German Federal Waterways Engineering and Research Institute (BAW) and the German Federal Institute of Hydrology (BfG) has been established in 2004. Within the framework of the research project "Technical-Biological Bank Protection Measures", test stretches e.g. at the river Weser (Stolzenau) and at the river Rhine (Lampertheim) were implemented to investigate and evaluate different technical-biological measures of bank protection under natural conditions like hydraulic stress and repeated flooding. Moreover single alternative bank protection measures individually implemented by departments of the Federal Waterways and Shipping Administration (WSV) are monitored by the BfG in terms of flora and fauna, whereas bank stability parameters are monitored by the BAW. In addition, a national survey on bank protection within the WSV was conducted which contribute valuable information for future investigations.

Measures of technical-biological bank protection lead commonly to modifications or even entire replacement of the rip rap zone through the implementation of native and site-typical plants or other structural elements. With the ecologic objectives to increase structural diversity and the promotion of natural succession, measures like willow branch cuttings, hedge layers, living fascines or inert wood trunks with roots have been tested and investigated in different water levels. Furthermore, to create shallow water zones including protection against hydraulic impacts, stone walls can be implemented in front of the bank, which creates potential habitats for macrophytes, fish and macrozoobenthos. All tested measures need to meet the requirements of bank stability under intense shipping conditions to avoid bank erosion and at the same time need to support a development of near-natural bank structures.

First monitoring results show that site typical riverbank vegetation has established and that natural succession has extensively set in. Inserted plants have generally resisted diverse stressors and single losses in vitality were predominately compensated. In shallow bank zones first potamogeton species emerged. Further, fixing structural elements turned out to ensure survival of intense fluvial and climate impacts. In conclusion it can be said that various alternative measures can basically be applied successfully on federal waterways leading to a broad increase of biodiversity and ecological functionality of riparian zones.

In close collaboration with other research and development projects at the BfG, we aim to investigate further insights into functionality and ecological potential of alternative bank protection measures with special regard to biodiversity and ecosystem services.

Hence, as a future perspective, evaluation criteria for the WSV will be derived to facilitate the implementation of alternative bank protection on federal waterways to meet the requirements of the European Water Framework Directive. This development is affirmed by tools, such as profiles of individual alternative measures and concepts to transfer them to small waterways implying different challenges (e.g. water tourism) in the future.

Study on sediment transport of silt coast by wave and tidal current

J. Mu & C. Yin
Zhejiang Institute of Hydraulics & Estuary, Hangzhou, China

ABSTRACT

The wave and tidal current are the important factors for the sediment transport of silt coast(Sonu C J 1972). Some coastal zones are main influenced by tidal current, some are by wave. But for most coast zones, the main factor for sediment transport and coastal evolution is the joint action of wave and tidal current, and it is difficult to describe the coastal sediment movement if only considering one factor.

United the third generation wave model SWAN for shallow water, this paper approximated the wave motion as time-averaged wave flow distribution field. And based on the wave flow radiation stress, a 2D mathematical model for sediment transport by wave and tidal current has been established (Jinbing Mu, 2013). Based on triangle grid and finite volume method, The Roe scheme based on Riemann solver is used to compute the flux in the paper. The control volume is CC type grid (Fig. 1). The result shows that the wave is the main dynamics factor for the bedload movement. Wave-current interaction has a significant impact of sediment transport. The suspended sediment concentration increased more than 5 times when the typhoon landfalling. And it the main cause of scour for the shoal area during typhoon.

The model has been verified by the topographical data of trial dredging area at Cangnan thermal power plant. And it replayed the sudden sedimentation caused by typhoon. Fig. 2 shows the comparison between

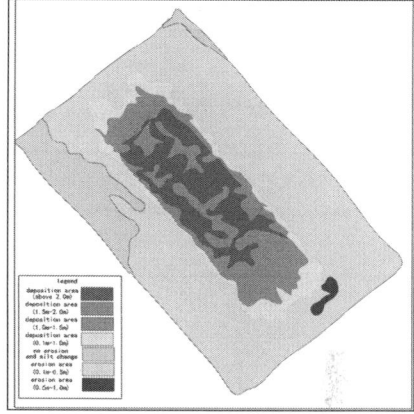

Figure 2a. Measured morphological change.

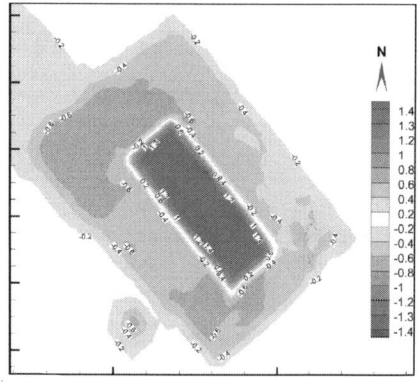

Figure 2b. Simulation morphological change.

simulated and measured morphological change. The research is very significance for the sediment transport during typhoon at silt coast.

REFERENCES

Jinbing Mu, Shichang Huang. The influence of the estuarine large-scale reclamation project on the surrounding hydrodynamic environment. Journal of Sichuan University; 2013, 45(1):61–66.

Sonu C. J. Field observation of nearshore circulation and meandering current. Journal of geophysical Research, 1972, 77(18):3232–3247.

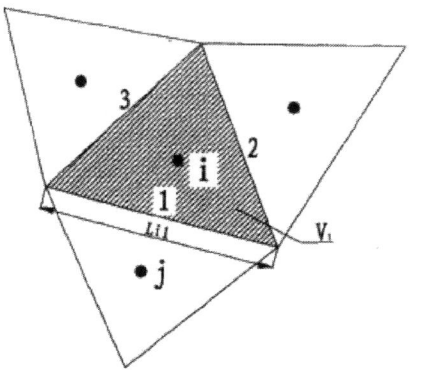

Figure 1. The sketch of control volume for 2-D shallow flow equations.

Back siltation in Bach Dang navigation channel, Nam Trieu Estuary, Vietnam

V.T. Nguyen
Faculty of Civil Engineering, University of Transport and Communications, Hanoi, Vietnam

M.D. Do
Faculty of Civil Engineering, University of Transport and Communications, Hanoi, Vietnam
Vietnam Maritime Administration, Hanoi, Vietnam

M.T. Vu
SEATECH, UTLN, Toulon, France

ABSTRACT

Hai Phong navigation channel has total length of 42.8 km from buoy No. 0 to Berth No. 11 of Hoang Dieu Port (Figure 1). It is a major life line of the Vietnam National maritime navigations. The vessel from Vietnam East Sea entrance to LachHuyen estuary then goes to Ha Nam navigation channel and next to Nam Trieu estuary, and afterwards to Cam River navigation channel to Hai Phong Ports. Bach Dang navigation channel is located in Nam Trieu estuary from Ha Nam navigation channel to the gate of Dinh Vu channel. The navigation channel has length of 8.2 km and width of 80 m. The natural elevation of the navigation channel ranges from −6.4 m to −48.0 m (CD).

The morphological evolution of the Bach Dang navigation channel is very complicated due to the complexity of the hydrodynamic regime and sediment transport in Nam Trieu estuary. The navigation channel needs to dredge five to six times a year, and a significant budget is allocated to maintain the navigation depth. Sometimes, silt in the navigation is deposited immediately after dredging. Based on measurement data in Bach Dang navigation channel from 2010–2014, the back siltation process in the navigation channel is presented. The result shows that, the back siltation in the navigation channel ranges from 212,266 m³ to 1,803,200 m³. The back siltation volumes in the flood season are higher than in the dry season about 26%. The flood current is main cause induced the back siltation, a huge of sediment is transported from the upper course of the rivers system to the navigation channel where the current velocity reduces due to enlarged of the river mouth. Especially, due

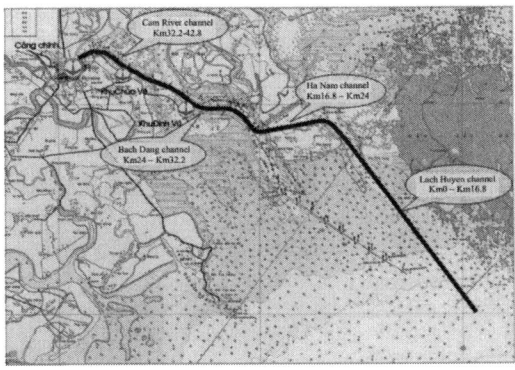

Figure 1. General layout of the navigation channel entrance Hai Phong Port, Vietnam (Thanh et al. 2015).

to the Con Son storm caused the high floods in the Northern of Vietnam in 2010, the intensity of back siltation reaches to 367,467 m³/month. To keep the channel design depth, the local authorities continued to perform dredging in every year.

REFERENCES

Thanh, L.Q. et al. 2014. The final report – Option selected pilot project contractors perform maintenance dredging shipping lanes Haiphong and Saigon – Vung Tau under open tendering mechanism for package deal. JSC Consulting Maritime Construction, Hanoi, Vietnam.

Thanh, N.V., Tuan, N. & Back, A. 2015. Siltation in Ha Nam navigation channel, Hai Phong, Vietnam, Vietnam – Japan Workshop on Estuaries, Coasts and Rivers 2015, 7–8th September, CKT, Hoi An, Vietnam.

Scour geometry and flow velocities induced by an experimental ship propeller jet

F. Núñez-González & K. Koll
Leichtweiß-Institut für Wasserbau, Technische Universität Braunschweig, Braunschweig, Germany

B. Söhngen & D. Spitzer
Bundesanstalt für Wasserbau, Germany

ABSTRACT

The size and power of propellers used to drive inland vessels have considerably increased during the last decades, bringing about beneficial impact on the efficiency of commercial transportation. This notwithstanding, new vessels also increase adverse effects over the environment and morphology of waterways. Among such effects, scouring action generated by the wash of a ship propeller, may cause severe damages to streambeds and quay structures. Assessment of the scouring induced by propeller jets is thus required, but there is a lack of a general criterion to quantify the characteristics of the scour holes generated by different types of ships propellers, under a wide range of hydraulic conditions and for different bed materials.

The main difficulty in developing a general approach for calculating scouring induced by ship jets, are the large amount of variables involved in the phenomena. For instance, LWI (2013) counted 54 relevant variables, which were reduced to 19 by applying dimensional analysis. Some authors have presented relations for scour depth prediction (e.g., Verhey 1983, Hamill 1988), but they do not consider all of these variables, and they are mainly based on experiments with free propellers (i.e., without a ship-body).

In this work we present results of experiments performed with a model of an inland vessel (scale 1:16) (Fig. 1). The experiments are intended for the development of a criterion for determination of scour induced

Figure 1. View of the experimental basin and stern model.

Figure 2. Scour developed after 7200 s over a sand bed.

by ship propellers. Unlike previous studies, the model used in this work gives a closer representation to real vessels, since it includes a ship stern. We show results about the effect on the scour development of the elevation of the propeller axis above the bed level (Fig. 2), for a maneuver situation, i.e., when the ship is standing with the motor on. Besides, we also present results regarding the effect of these two parameters on the velocity field near the bed surface. The velocity field was investigated, by performing additional experiments over a fixed rough bed.

Analysis of experimental results will provide further evidence to conclude whether existing relations for estimating the distribution of flow velocities induced by ship propellers, are useful to be considered in the determination of scouring development, under a wide range of conditions.

REFERENCES

Hamill, G.A. 1988. The scouring action of the propeller jet produced by a slowly manoeuvring ship. PIANC Bulletin 62, 85–110.

Leichtweiß-Institut für Wasserbau, 2013. Modellversuche zur Kolkbildung infolge eines Schraubenstrahlangriffs in einer Manöversituation. Bericht 1041, TU Braunschweig, in German, unpublished.

Verhey, H.J. 1983. The stability of bottom and banks subjected to the velocities in the propeller jet behind ships. 8th Int. Harbour Congress, Antwerp, Belgium. VBD.

German guidelines for designing alternative bank protection measures

B. Söhngen & P. Fleischer
Federal Waterways Engineering and Research Institute, Karlsruhe, Germany

H. Liebenstein
German Federal Institute of Hydrology, Koblenz, Germany

ABSTRACT

Facing i.a. the European Water Framework Directive, more environmental friendly bank protection measures will be demanded for, not only for small waters, but also for inland waterways. These measures should contain as much as possible living or at least dead plants, combined with technical building materials, if necessary, to avoid erosion from the natural and vessel-induced flow and wave field. These measures will be called "technical-biological" and shall replace technical protection methods as riprap or sheet piling, if applicable.

Numerous experiences and corresponding guidelines for such alternative measures are available for waters without navigation. For large and generally navigable inland waters as those in German waterways, reliable design rules are lacking up to now, especially if living plants shall take over the protection function.

For this reason the German Association for Water, Wastewater and Waste (DWA) founded in 2008 a working group named "Alternative Bank Protection Measures". The aim was to collect and condense the existing knowledge in this field, to evaluate the ecological value and efficiency of these measures and to assess the construction and maintenance expenses, especially of measures with living plants. The main idea was to transfer existing experiences from waters without navigation to those with significant impact from vessel-induced waves and currents and to account for first results of an ongoing mutual research project of the German Federal Waterways Engineering and Research Institute (BAW) and the German Federal Institute of Hydrology (BfG) concerning the same subject for waterways only. The results should be published in a design guideline for technical-biological bank protection measures to make it usable for planners of waterway infrastructure. The guidelines are available now.

Parallel to this the DWA founded another working group named "Bioengineering Measures in Water Construction". This working group focusses on planning, construction, tendering, realization and success monitoring of these measures. The corresponding guidelines shall be published in 2016 too.

The following paper demonstrates the way how to use the new guidelines for designing technical-biological bank protections, considering relevant local boundary conditions, decisive design aspects, the hydraulic and geotechnical impacts in comparison to load thresholds, the demands of bioengineering constructions and ecological aspects, to find out an optimal solution from a list of 10 recommended constructions. The latter concern to measures to protect the bank slopes, not to reduce impacts.

A more general design code, considering e.g. a possible reduction of flow and wave impact with measures in front of the bank, basing on international experiences of implemented measures, shall be worked out in the next years in a renewed PIANC (World Association for Waterborne Transport Infrastructure) working group with the same topic. Also some information about the present state of work of this working group will be dealt with in this paper.

It should be mentioned that the present state of knowledge concerning waterways is still restricted. Therefore, the present design guidelines recommend measures which are generally on the safe side, especially concerning the resistance to excess pore water pressure from vessel-induced water level drawdown. Nevertheless, it makes sense to publish a design code as it was done as soon as possible in order to promote the usage of alternative bank protections for waterways and therefore to be able to have more experiences with such measures in the next years. So, the new DWA-guidelines which are discussed in this paper have to be modified, basing on extended experiences, in just a few years' time if necessary.

Sediment budget of the Rhine River as basis for optimizing navigation along the Mittelrhein waterway

S. Vollmer, G. Hillebrand, T. Hoffmann & S. Schriever
Federal Institute of Hydrology, Koblenz

ABSTRACT

Knowledge about sediment transport in rivers is needed by engineers, geomorphologists, ecologists and river managers for a wide range of purposes, such as channel design, maintenance, sediment management etc. The Rhine River is one of the most important waterways in Europe and therefore a sediment budget analysis is one basis for maintenance and optimization of the waterway. The Rhine drainage basin covers 197,000 km^2, is heavily populated with 58 million inhabitants and is connecting the port of Rotterdam to the industrial areas in the hinterland (IKSR 2005).

In order to improve navigation conditions, 19th century river engineers decided to narrow and straighten the river. This increased the water depth, energy slope and bed-shear stress and therefore led to an increase in transport capacity. In the 20th century, the sediment supply from the hinterland was strongly diminished by the building of dams in the river and its tributaries. The Rhine is still reacting to these impacts by entraining sediments from the river bed leading to net erosion. Despite the general bed erosion, some parts of the Rhine are subject to sedimentation, which causes navigational problems, both.

Embankments alongside the river protect from flooding. The river itself has been deepened and narrowed, with groynes and bank revetments preventing bank erosion. In order to allow year-round navigability continuous dredging operations are carried out. Dredged sediments are re-allocated to the river downstream. Furthermore, large amounts of sediments are artificially supplied to the river to stop bed degradation and to fill scour holes.

The Federal Waterways and Shipping Administration carries out systematic measurements of the sediment transport in the Rhine since 1965 (Frings et al. 2014) using the BfG bedload sampler (Dröge et al. 1992) and suspended sediment samples. This unique dataset with thousands of bed-load and suspended-load measurements was used to quantify the fluxes of gravel, sand, silt and clay through the northern Upper Rhine Graben and the Rhenish Massif. Sediment transport rates were found to change in the downstream direction: silt and clay loads increase due to tributary supply; sand loads increase due to erosion of sand from the bed; and gravel loads decrease due to a reduced sediment mobility caused by the base-level control exerted by the uplifting Rhenish Massif. Sand being eroded from the bed is primarily washed away in suspension, indicating a rapid supply of sand to the Rhine delta. In contrast to previous studies, this sediment budget includes the effects of tributaries, floodplain deposition, abrasion and anthropogenic sediment fluxes. Bed degradation and artificial sediment feeding represent the major sources of sand and gravel to the Rhine River; only small amounts of sediment are supplied naturally from upstream or by tributaries. Sediment sinks include dredging, abrasion and the sediment output to the downstream area. Sand was found to be the main morphological agent. The budget analysis represents one of the fundamentals for the optimization of the Mittelrhein waterway. The results of this sediment budget may also serve as calibration dataset for numerical models.

REFERENCES

Dröge, B., Nicodemus, U. & Schemmer, H. 1992. Instruction for bedload and suspended material sampling. Report BfG-1718. Bundesanstalt für Gewässerkunde, Koblenz.

Frings, R.M., Gehres N., Promny M., Middelkoop H., Schüttrumpf H. & Vollmer S. 2014. Today's sediment budget of the Rhine River channel, Upper Rhine Graben and Rhenish Massif. Geomorphology 204, pp 573–587.

IKSR, 2005. Merkmale, Überprüfung der Umweltauswirkungen menschlicher Tätigkeiten und wirtschaftliche Analyse der Wassernutzung. Koordinierungskomitee Rhein/ Internationale Kommission zum Schutz des Rheins, Koblenz.

River Rhine between Mainz and Bingen – Morphodynamic analysis of a navigational bottleneck

S. Wurms
Federal Waterways Engineering and Research Institute, Karlsruhe, Germany

ABSTRACT

Looking back on the history of navigation on River Rhine, the morphologic conditions within the stretch between Mainz and Bingen (Rhine-km 493–529) continuously caused a great effort to achieve reasonable fairway conditions.

Located at the lower end of the Upper Rhine, the stretch shows the least slope of the free flowing Upper Rhine with cross section widths up to 900 m and mean water depths during the effective discharge concerning morphodynamics of about 4 to 5 m. Sediment supplied from the upstream part of the Upper Rhine mainly tends to deposit within the stretch. Until the beginning of river regulation, these conditions led to a dynamic, braided river morphology between Mainz and Bingen, coming along with moving shoals and obstacles hindering navigation.

Against this background, the article exemplifies past, current and potential future river engineering measures such as river regulation and sediment management within the stretch, aiming at the improvement of navigational conditions. By means of morphodynamic analysis of the stretch, the morphodynamic responses to the mentioned anthropogenic activities and other changing conditions such as potential climate induced changes of hydrological conditions are shown. The findings serve as a base for the adaptation and the development of further measures concerning sediment management and river engineering.

An indispensable element of sediment management within the stretch is the bed load trap located nearby Mainz Weisenau (Rhine-km 494.3–494.46). The bedload trap was implemented in 1989 with a width of 250 m, a length of 160 m and a depth of 1.5 m related to the surrounding bed level. The measure aims at the reduction of bedforms such as dunes passing the stretch between Mainz and Bingen, as bedforms are dynamic obstacles within the fairway and thus difficult to locate in order to be dredged. Up to the year 2014, the bedload trap was emptied 28 times with a dredging volume of about 1.99 million m^3 of sediment. 27% of the material was supplied downstream of the bedload trap, whereas 73% was removed from the system. The reduction of dunes occurring within the stretch could be reached successfully by this measure. Nevertheless, comparisons of echo soundings of the last two decades show a trend towards bed degradation downstream of the bedload trap. The current magnitude of degradation indicates a modification of the dredging practice aiming at an increased sediment input into the downstream stretch. Ongoing investigations are focusing on the impact of modified dredging rules and geometries of the bedload trap on the further development of bed levels downstream of the measure.

Furthermore, examples of current morphological phenomena occurring locally between Mainz and Bingen are presented. One phenomenon of interest is a scour occurring in the right-hand branch of the bifurcation between Eltville and Ostrich (Rhine-km 512.5). The coarse upper layer with a thickness of about 0.2 m up to 0.3 m, mainly consisting of gravel and cobbles is eroded locally, exposing the subjacent Tertiary clay. Detected for the first time in the year 2004, the scour dimension increased continuously up to a volume of about 2300 m^3 and a depth up to 1.7 m below the surrounding bed level in 2014. Potential countermeasures to stabilize the affected riverbed are subject of current investigations.

Looking from present to future, the potential development of bed levels up to the year 2100 within the stretch between Mainz and Bingen is shown. The underlying studies, based on 2D-morphodynamic-numerical long-term simulations were conducted to quantify the potential effort necessary to maintain the today's fairway conditions even under changed climatic conditions.

Turbulence based approach for the transported particle size concerning ship induced propulsion flux

R. Zimmermann & J. Stamm
IWD, Institute of Hydraulic Engineering and Technical Hydromechanics, TU Dresden, Germany

T. Beck & B. Söhngen
BAW, Federal Waterways Engineering and Research Institute, Karlsruhe, Germany

ABSTRACT

In order to estimate the grain size of channel embankments, which can withstand ship induced propulsion flux (e.g. bow thruster), a new turbulence based approach is introduced. The theoretical foundation can be found in Führer & Römisch (1972) as well as Führer & Römisch (1974), who were investigating on propulsion flux propagations and velocity distributions. Not only the induced propulsion velocity but also the velocity at the bed and embankment of channels was part of their physical modeling research.

However, the above stated literature only delivers a basis for the consideration of turbulent flux processes. At this point Söhngen developed a modified, turbulence related approach for a dimensioning equation of the grain size, which can withstand a certain, known strain at channel embankments, according to the basic structure also found in BAW (2004).

This contribution will summarize the state of work in numerical modeling of ship induced flux processes related to the usage of a bow thruster, which were done by the authors so far. The implementation of the above mentioned approach by Söhngen claimed a good understanding and knowledge of the wall surface near flow velocities at the embankment. Therefore a viewpoint based routine was programmed, in order to access velocity data at any needed point and time in the numerical model. This model was built and set up in OpenFOAM (version 2.3.0), itself an open source CFD software package. Simplifying the model was set up to be a one phase system, where the free surface border of the water level can be identified to be the upper boundary. Instead of a ship movement, the water body and the channel borders were moved with the corresponding ship velocity. This circumstance enabled the possibility to pass on a moving or morphing mesh around the ship hull in the model but simultaneously forced a time oriented selection of data between multiple viewpoints to gain a strain-time relation for one realistic point on the embankment. The basic method behind the time related selection of data and their usage in order to obtain all needed information for the turbulence related approach of Söhngen will be shown.

With focus on selected parameters (e.g. ship velocity, capacity bow thruster, bank distance etc.) different scenarios were investigated in the numerical model. The test cases were chosen to match the scale model measurements, carried out in Duisburg at the Development Centre for Ship Technology and Transport Systems (DST) commissioned by the BAW in 2014.

The measurements were based on a physical model with ship and channel in the scale of 1:16. Besides propulsion trials with a stationary ship, studies of the amount of moved particles with one and two ships in the channel were performed. Unlike the parameters mentioned above, the grain size and the slope angle were constant during the measurements. A matrix of three ship velocities, three bank distances and two draughts have been analyzed.

Pointing out the key information about the scale modeling this article will also provide a comparison and judgement between the results of the physical and numerical modeling processes. An empirical relation between the numbers of particle movement and the selected parameters mentioned above will be shown.

REFERENCES

BAW 2004. Grundlagen zur Bemessung von Böschungs- und Sohlensicherungen an Binnenwasserstraßen. Mitteilungsblatt der Bundesanstalt für Wasserbau Nr. 87. Karlsruhe.

Führer, W. & Römisch, K. 1972. Abschlussbericht zur Forschungs- und Entwicklungsarbeit „Wirkung des Propellerstrahls auf Sohle und Böschung". Berlin.

Führer, W. & Römisch, K. 1974. Abschlussbericht zur Forschungs- und Entwicklungsarbeit „Erosionserscheinungen bei Schiffsmanövern". Berlin

SS 3 Innovative measurement techniques

Bathymetry of Zipingpu Reservoir by earthquake and flood induced turbidity currents

A. Ruidong, L. Jia & Y. Zhongluan
State Key Laboratory of Hydraulics and Mountain River Engineering, Sichuan University, China

ABSTRACT

A field measurement to assess the bathymetry was conducted in Zipingpu Reservoir from 2007 to 2013, Min River, China. The Wenchuan (Ms = 8.0) earthquake (2008) took place in the upper reach and sedimentation by flood induced turbidity currents changed reservoir bathymetry in a short time.

The Zipingpu Reservoir effectively controls the 98% sediment inflow and 90% torrential rain area of upstream reaches, which the average annual flow is 469 m³/s. It meets irrigation and municipal water supply of Dujiangyan irrigation project and Chengdu, balancing flood control, power generation and environmental protection, etc.

All data obtained from multi-beam echo sounder system (MBES) are used to analyses the changes of bathymetry. It consists of acoustic, auxiliary, data collection and visualization system, which was mounted on a vessel. The similarities of compassion with reservoir section measurement suggest that high resolution (1.25 cm) and superior performance of MBES can accurately detect the characteristics of bathymetry by transmitting fan-shape pulses perpendicular to cover bottom, as shown in Figure 1.

A natural barrier induced by landslide is found in a distance of 6.95 km upstream of the dam, which had been raised for 33.84 m. The volume of the landslide in reservoir is estimated about 5.3×10^6 m³. A flushing cone formed close to the dam because of turbidity

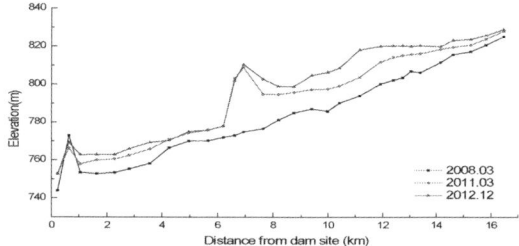

Figure 2. Comparison of thalweg.

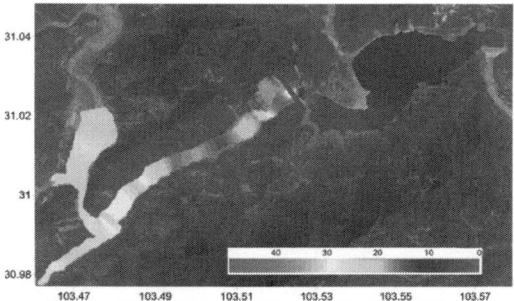

Figure 3. Distribution of section area loss ratio.

currents. Figure 2 shows that the most severe deposition occurs behind the barrier formed, with area loss ratio reaching to 40%.

Triangulated Irregular Network method was used for creating 3D terrain model and analyzed storage capacity. The results show that the storage capacity loss ratios are 17.07% at lowest level and 12.23% at flood limit water level compared to 2008.

REFERENCE

Yin, Y., Wang, F. & Sun, P. 2009. Landslide hazards triggered by the 2008 Wenchuan Earthquake, Landslides. 6: 139–151.

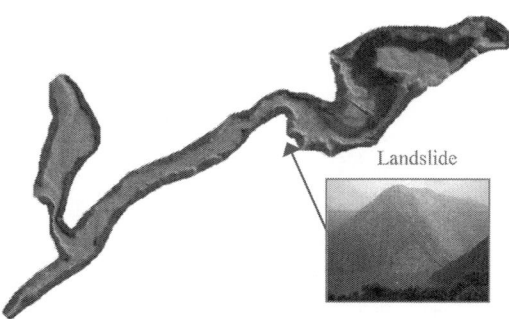

Figure 1. Measurement results by MBES.

Feasibility tests to airborne gravelometry for prealpine rivers

M. Detert
Laboratory of Hydraulics, Hydrology and Glaciology VAW, ETH Zurich, Zürich, Switzerland

L. Kadinski
Deliusstrasse, Aachen, Germany

V. Weitbrecht
Laboratory of Hydraulics, Hydrology and Glaciology VAW, ETH Zurich, Zürich, Switzerland

ABSTRACT

This paper presents basic tests to develop an airborne photogrammetric methodology that derives grain size distributions of pre-alpine gravel bed rivers. The data acquisition was performed by a lightweight action cam *GoPro Hero3+ black edition* with a ~170° fisheye lens operated in video mode with 4096×2160 px^2 and 11.988 fps, and a hand-held digital single lens reflex camera *Nikon D7100*. Image processing comprised the Structure from Motion (SfM) technique to obtain digital elevation models of gravel beds. SfM is increasingly used in terms of surveying river environments and generating georeferenced high resolution digital elevation models (e.g. Javernick et al. 2014).

Laboratory results indicate that the accuracy of digital elevation models based on action cam data is about four to six times lower than data recorded by a laser scanner *Leica Scanstation P15*. However, the model computed via photos taken by the reflex camera is almost of the same order as the mm-precision of the laser data.

Field measurements aimed at testing the applicability of SfM-technique in combination with manual field sampling at riverine outdoor conditions. The measurement campaign has been carried out on October 22, 2015, at River Töss, Switzerland, located at N 47.464 E 8.728. The action cam was mounted to a low-cost quadrocopter, while the reflex camera was operated in hand-held mode. Figure 1 illustrates an exemplary resulting view to a 1×1 m^2 cut-out of a 3D gravel patch computed by SfM based on *Nikon*photos. Further analyses in comparison to manual grain samplings indicate that the measurement combination used is able to generate data from that characteristic grain size parameters can be estimated.

As consequence for forthcoming field measurements, surveying flights on reach length scale may be conducted with a fisheye action cam mounted to a quadrocopter. However, for detailed analysis of roughness heights it is advised to use a handheld

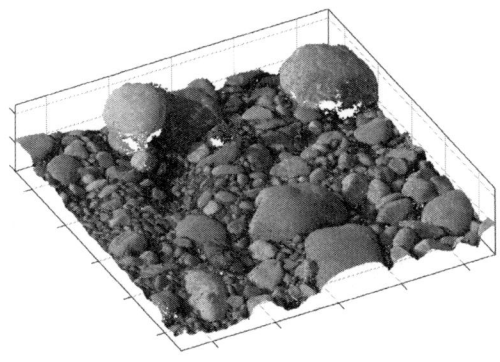

Figure 1. 3D-view to a dense point cloud covering a 1×1 m^2 cut-out of a gravel patch, generated with *PhotoScan* (Agisoft 2015) based on 25 photographs taken with a *Nikon D7100*.

reflex camera, if the field conditions allow for physical access to the gravel bed to be measured.

Photogrammetric techniques have a high potential to improve gravelometric field measurements. The next step should combine the benefits of SfM technique and 2D image based object detection tools as BASEGRAIN (Detert & Weitbrecht 2012). The overarching vision is to automatically generate surface maps of gravel beds, with known geo-location and size of each individual grain.

REFERENCES

Agisoft 2015. *Agisoft PhotoScan Professional Edition* (version 1.1.6). www.agisoft.com.

Javernick, L., Brasington, J. & Caruso, B. 2014. Modeling the topography of shallow braided rivers using Structure-from-Motion photogrammetry. *Geomorphology* 213: 166–182.

Detert, M. & Weitbrecht, V. 2012. Automatic Object Detection to Analyze the Geometry of Gravel Grains – a free stand-alone tool. *River Flow 2012*, R.M. Muños (Ed.), Taylor & Francis Group, London, ISBN 978-0-415-62129-8, 595–600.

Combining *in-situ* laser diffraction (LISST) and vibrating tube densimetry to measure low and high suspended sediment concentrations

D. Felix, I. Albayrak & R.M. Boes
Laboratory of Hydraulics, Hydrology and Glaciology (VAW), ETH Zürich, Switzerland

ABSTRACT

Field measurements of suspended sediment mass concentration (SSC) and particle size distribution (PSD) with high temporal resolution and good accuracy are required for a better understanding and management of many fine-sediment related processes. Although many techniques for suspended sediment monitoring are available and described in literature (Wren et al. 2000; Bishwakarma and Støle 2008), it is not evident how to design monitoring systems capable of measuring SSC and PSD in relevant ranges with reasonable accuracy, reliability and cost.

This paper deals with the recent field study on the monitoring of SSC and PSD in the turbine water at the high-head hydropower plant (HPP) Fieschertal in the Swiss Alps. To mitigate turbine abrasion, real-time measurements of high SSCs are of particular interest. The following techniques and devices used in the field study are treated in the paper:

- A turbidimeter measuring at a free falling jet (*AquaScat* from Sigrist Photometer),
- A 'Laser in-situ Scattering and Transmissiometry' device, LISST-100X (from Sequoia Scientific) with an optical path length reduced to 5 mm,
- Vibrating tube densimetry using a Coriolis flow and density meter (CFDM, *Promass 83F* DN15, from Endress + Hauser),
- Automatic bottle sampler (*Isco 3700* from Isco Teledyne).

The raw field data were converted to SSCs based on reference SSCs gravimetrically determined from bottle samples. Furthermore, the PSDs were obtained by LISST every minute.

The results showed that the SSC measuring range of the LISST device without dilution chamber is limited to about 1 g/l of medium silt, as expected from the manufacturer's information and earlier laboratory investigations (Felix et al. 2013a, b). With the CFDM however, it was possible to measure higher SSCs with a good accuracy, i.e. well in line with gravimetrical SSCs (Fig. 1).

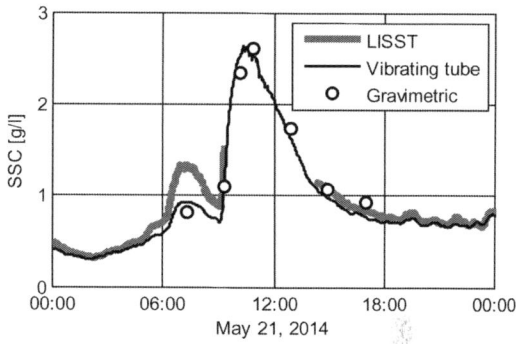

Figure 1. Example of SSC time series obtained from vibrating tube densimetry (Coriolis flow- and density meter) and from a LISST device (without dilution chamber), compared to gravimetrically determined SSCs of bottle samples.

Based on these findings, the combination of a LISST device without dilution chamber and CFDM, supported by bottle sampling, appears to be an interesting option for SSCs measurements from a few mg/l to several 10 g/l (e.g. during floods) at relatively low cost and uncertainty.

REFERENCES

Bishwakarma, M.B. & Støle, H. 2008. Realtime sediment monitoring in hydro-power plants. *J. of Hydr. Research*, 46(2): 282–288.

Felix, D., Albayrak, I., Boes, R.M. 2013a. Monitoring of Suspended Sediment – Laboratory Tests and Case Study in the Swiss Alps. In *Advances in River Sediment Research*, Fukuoka S., Nakagawa H., Sumi T., Zhang H. (eds.), Taylor & Francis Group, London: 1757–1766.

Felix, D., Albayrak, I., Boes, R.M. 2013b. Laboratory investigation on measuring suspended sediment by portable laser diffractometer (LISST) focusing on particle shape. *Geo-marine Letters*, 33(6): 485–498.

Wren, D., Barkdoll, B., Kuhnle, R., Derrow, R. 2000. Field Techniques for Suspended-Sediment Measurement. *J. of Hydraul. Eng.*, 126(2): 97–104.

UAV based determination of grain size distribution at River Jachen, Germany

C. Haas & P. Thumser
I AM HYDRO, St. Georgen, Germany

L. Seitz
Institute for Modeling Hydraulic and Environmental Systems, University of Stuttgart, Stuttgart, Germany

ABSTRACT

Unmanned aerial vehicle (UAV) based data ascertainment of riverine systems is strongly increasing, as it is a cost effective way to gain data. It be-came state of the art in various remote sensing applications in diverse engineering fields over the last years. In this work we present a study of techniques for grain size determination of the top layer above and under the water surface, using UAV based aerial images. The results are compared with different conventional methods, such as sieving and pebble counts. The proposed methods enable the creation of detailed substrate maps, based on high precision orthomosaics with reference grain size distribution as an input for various modeling applications.

The study area is at the River Jachen, which is a 23 km long river in South Germany. The Jachen is the natural outflow of Lake Walchensee, with an altered flow regime from hydropower production. The only hydraulic structure in the river is a weir, approx. 1.5 km above the outlet of River Jachen into River Isar. The structure divides the river in an in-take of a hydropower plant and the old riverbed, supplied with a minimum flow. A custom made hex copter with a DJI NAZA-M V2 Multirotor Autopilot is used for the data ascertainment. The maximum payload of the hexacopter is approx. 2 kg. The hexa-copter carries a Sony Nex 6 camera with a 16 Mega-pixel APX-C Sensor with a 30 mm fixed focal length lens. In addition a DJI Phantom 3 Professional quad-copter was used for further aerial images. The quad-copter carries an integrated camera with a 12 Mega-pixel Sony EXMOR 1/2.3" Sensor and a 20 mm fixed focal length lens on a on a 3 axis. The maximum flight time of the setup is approx. 23 min with a 4480 mAh lithium polymer battery. The full setup of the quadcopter including camera and battery is 1.3 kg.

Aerial pictures of the full stretch are taken at 50 m altitude and lower, to ensure sufficient ground resolution of the photos. Additionally, 6 colored frames (0.5 m * 0.5 m) were placed in areas with different grainsizes below and above the water surface. These frames will be captured on the photos and serve as reference areas, where different approaches of grain-size distribution are tested including pebble counts and sampling of the top layer after all aerial data is ascertained. Water depth at each underwater reference area is measured in addition.

Photosieving can give a good impression about the grain size distribution, especially when the dominating substrate is gravel. Different algorithms of image processing will be used in this study and compared to each other.

In addition, all taken photos are merged to a georeferenced aerial picture. UAV based data in a birds eye view has the advantage of a better perspective compared to mapping in the field. The generated orthomosaic unfortunately is not sufficient for photo sieving purposes, as the full file is too big to handle with current software and in addition a ground resolution of 2 cm per pixel is not sufficient when it comes to areas with smaller fractions. Nevertheless a detailed substrate class map can be generated out of this orthomosaic, referenced based on photo sieving of its raw photos.

Photo sieving with aerial pictures works satisfyingly good. Especially the DJI System generated good footage as the camera was stabilized by a 3-axis gimbal. As the sensor of this setup is not per-forming as good as the one of the Sony NEX6 cam-era a lower flying altitude especially in the areas of finer fractions is necessary to produce a good dataset for photo sieving.

REFERENCES

Eisenbeiss, H. & Sauerbier, M. 2011. Investigation of UAV systems and flight modes for photogrammetric applications, Photogramm. Rec., vol. 26, no. 136: 400–421,

Nex, F. & Remondino, F. 2013. UAV for 3D mapping applications: a review. Appl. Geomat., vol. 6, no. 1: 1–15.

Experimental study on development and migration of sand waves in a flume

C. Liu, W.H. Cao, L. Xu, J. Lu & L. Liu
State Key Laboratory of Simulation and Regulation of Water Cycle in River Basin, Beijing, China
Key Laboratory of Hydraulic and Sediment Science and River Harnesting of the Ministry of Water Resources, Beijing, China
Department of Sediment Research, China Institute of Water Resources and Hydropower Research, Beijing, China

ABSTRACT

The initiation and development of the bedforms in alluvial rivers have dramatic effect on the flow resistance and sediment transport. Therefore, precisely measuring the size and shape of the bedforms is one of the bases for further study on the mechanism of water and sediment interactions. Due to the technical limitation, measurements on the bed forms of the river were mostly a static prototypes bed, could not reflect the development of the shape of the bed caused by their movement. On the other hand, experiment in a flume usually measure no more than a single point, which do not reveal spatial distribution of the bed form, especially for a three dimensional sand wave. In this research, a measurement system for monitoring the dynamic movement of bedforms in a flume was developed. The system consists of a hardware system and a data acquisition software system which collecting flume bed elevation data automatically at a certain interval. Ten independent underwater ultrasonic distance sensors with a diameter about 10 mm were employed in the system. Ten sensors were arranged as a T-shape or Triangle shape as a whole, as shown in Figure 1, nearest distance between two sensors were 30 mm. The data of ten different positon of the bed elevation can be measurement synchronously. Because of the novel arrangement of our sensors group, temporal and spatial variations of the bedforms, especially the small scale change at different position, can be analyzed conveniently. The bed form measurement system would provide more precise, reliable and abundant data for the study of bed form morphology over different flow conditions.

Three runs of experiments have been conducted in a 50 m long, 1 m wide and 1.2 m height flume. uniform sand with 0.9 mm diameter were employed in the experiment. The elevation variation of the mobile bed were measured by the monitoring system. Three runs of test, representing the nearly immobile bed, micro sand waves bed and large number of successive dunes, respectively. The validity of the measurement system was verified at different experiment conditions. The results show that the development and characteristic parameters of bedforms are different under different flow conditions. The height, length, shape of the sand waves and their migration velocities, were determined and analyzed based on the measured data. These key parameters would provide invaluable insight for the research of the effect of bedforms on the flow and the sediment transport.

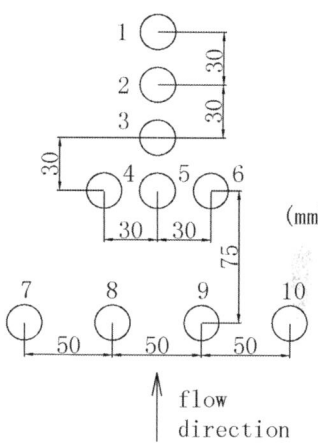

Figure 1. Sensor arrangement.

Comprehensive measurement techniques of water flow, bedload and suspended sediment in large river using Acoustic Doppler Current Profiler

S. Okada
National Institute of Technology, Kochi College, Kochi, Japan

A. Yorozuya, H. Koseki & S. Kudo
International Centre for Water Hazard and Risk Management under the auspices of UNESCO (ICHARM), Ibaraki, Japan

K. Muraoka
Bangladesh Water Development Board (BWDB), Dhaka, Bangladesh

ABSTRACT

In order to elucidate sediment processes in a large river, comprehensive and efficient measurement techniques are necessary. However, they require a number of direct sampling at multiple point in a cross section. Therefore, these works are time-consuming and commonly temporal and spatial resolution are restricted.

Recently, measurement techniques using acoustic Doppler current profiler (ADCP) were developed for not only flow but also sediment discharge measurement. For example, bed-load velocity are calculated from the instantaneous position difference at each point measured by GPS and bottom track function; thereafter, bed-load discharge were estimated by author's method (Yorozuya et al. 2010), applying the bed-load velocity, the shear velocity estimated with vertical water flow distribution, thickness layer and shear stress relationship originally proposed by Egashira et al. (2005).

$$q_B = \int_0^{h_s} c \cdot u \cdot dz \cong v_s \cdot h_s \cdot c_s \quad (1)$$

where q_B = unit bedload rate, c = sediment concentration, u = velocity of sediment, h_s = sediment layer. Also v_s = vertical averaged velocity of sediment, and c_s = vertical averaged sediment concentration. In here, we assumed that water depth and bedload velocity were determined by ADCP based on the difference of density (i.e. scattering strength), and thickness of sediment layer h_s was determined by Egashira's formula related to non-dimensional shear stress.

Also, suspended-sediment concentration (SSC) was estimated using acoustic backscatter obtained from ADCP measurement proposed by Kitsuda et al. (2006). Procedure of field measurement and data analysis of suspended sediment are 1) Measure the vertical distribution of turbidity using a turbidity meter and pressure gage at representative points during cross-sectional flow observation, 2) Take water samples to find out the grain size and obtain the turbidity-SSC relation, 3) Determine the variables in the equation (2) by using measurement setting and observed results as the calibration data. 4) Estimate the suspended sediment discharge using velocity distribution and turbidity-SSC relation.

$$\log M(r) = S\{dB + 2r(\alpha_w + \alpha_s)\} + K_s \quad (2)$$

where $M(r)$ = SSC at a distance r from the transducer, S = backscattering coefficient dB = backscatter intensity after the diffusion correction α_w = absorption coefficient of the water α_s = absorption coefficient of the particle K_s = constant number related to transducer.

Flood flow observation using tethered boat equipped with ADCP and RTK-GPS, turbidity measurement and water sampling were conducted in Brahmaputra River, Bangladesh. River width and maximum depth were 4.8 km and 15 m respectively and maximum discharge during observation was 63,000 m³/s (almost averaged maximum discharge in a year), representative grain size of bed material d_{50} was 0.15 mm.

As the results, maximum bedload velocity exceeded 1.3 m/s and high positive correlation between bedload velocity measured using RTK-GPS and depth-averaged velocity can be confirmed.

Ratio of bedload discharge to suspended sediment discharge was the range from 10 times to 25 times. Therefore, it is clear that suspended sediment is dominant factor of sediment transportation in this river.

REFERENCES

Kitsuda, T. et al. 2006. Issues of River flow measurement method and Applied observation result using acoustic Doppler current profiler, Advances in River Engineering, JSCE, Vol.12, 133–138.

Egashira, S. Miyamoto, K & Itoh, T. 1997. Constitutive Equations of Debris-Flow and Their Applicability, Proc. of 1st International Conference on Debris-Flow Hazards Mitigation, C. L. Chen (Eds.), ASCE: New York; 340–349.

Yorozuya, A., Okada, S., Kanno, Y. & Fukami, K. 2010. Bed-load discharge measurement by ADCP in actual rivers,River Flow 2010, 1687–1692.

Estimation of sediment deposition in Koyna Reservoir by integrated bathymetric survey

R.A. Patil & R.V. Shetkar
Government College of Engineering, Aurangabad, Maharashtra, India

ABSTRACT

In southern part of India, most of the precipitation occurs due to south-west monsoon during the months of June to September. Storage reservoirs are constructed to store the run-off volume and utilize it for long term requirements in the command such as irrigation, power generation, domestic water supply etc. The sediment deposition in the reservoir is a continuous and complex process. Sediment in a river which originates from the land erosion process in the catchment is propagated along with the river flow and finally gets deposited in the reservoir. This has adverse effect such as reduction in storage capacity and increase in backwater levels. For the long term and yearly planning of efficient utilization of reservoir capacity, estimation of sediment deposit is not only sufficient but also estimation of trapped sediment at different levels in the reservoir is required. Differential Global Positioning System (DGPS) based Integrated Bathymetric Survey System for hydrographic survey is used for estimation of sediment deposition pattern in the Koyna reservoir. In this method three or more satellite based GPS receivers are used in differential mode for positioning. Ecosounder is used for depth measurements. GPS reference receiver is located at a known latitude and longitude position and GPS mobile receiver is located on the survey boat. Both the GPS operate in differential mode through a communication link. From this study (2004) it is observed that sedimentation rate in Koyna reservoir is more than 6.5 Ham/100 sq.km/year anticipated in 1949.

REFERENCES

de Vente, J., Poesen, J. & Verstraeten, G. Evaluation of reservoir sedimentation as a methodology for sediment yield assessment in the Mediterranean: challenges and limitations, Soil Conservation and Protection for Europe Sedimentation Study Report of Shivajisagar Lake of Koyna Dam 1980, 1986, October 2004.

Densitometric probe based on non-differential pressure: A monitoring technique for high suspended sediment concentrations

D. Petrovic, A. Marescaux, J.-P. Vanderborght & M.A. Verbanck
Department of Water Pollution Control, Université Libre de Bruxelles (ULB), Belgium

ABSTRACT

Presently, there exist many different surrogate methods for determination of suspended matter loads. However, all those methods suffer for certain limitations and there exists no universal technique for sediment monitoring at all flow and sediment transport conditions. Particularly, difficulties are explored in documenting data under intensive storms events when actually the largest fraction of the annual particle flux is delivered.

In this paper, first results of application of an innovative densimetric technique based on precise absolute pressure measurements for determining suspended matter load concentrations in slurries are presented. The method is intended to monitor semi-continuously or continuously the amount of suspended particles in any highly-concentrated open-channel flows.

The method relies on a combination of pressure, temperature and water level sensors for collecting environmental variables which are mounted in such an assembly that once processed, their combined signals can be converted into the pursued concentration information.

Table 1. Maximal SSC in g/l at experimental test catchment Laval (Draix).

Date	Filtration	Densitometric Probe
14 Jun 1989	407	440
31 Jul 1990	378	370
13 Aug 1997	798	890

Determination of suspended sediment volumetric concentration in static conditions was tested in the laboratory. Six different, known, volumetric concentrations of dry matter (density 2710 kg/m^3) were observed using absolute pressure measurements with relative error less than 9%, as shown in Fig. 1. Due to relatively low volumetric concentrations, results are influenced by accuracy of the pressure sensor and of the depth-level monitoring device.

In order to test the method in natural environment, the method has been applied to the preexisting historical data collected at the monitoring test site Laval located in the strongly eroding Black Marls area of Draix (French Southern Alps) which has the great merit of displaying hyperconcentrated flow conditions during the most intense rain events. Although there was a problem with data resolution (since data was produced with a different aim), it is demonstrated that the method is able to reproduce suspended sediment concentrations during short and intense rain event when SSC was reaching up to 800 g/l, Table 1.

Note: A one-liter bucket of hyperconcentrated marly slurry sampled taken on 13 Aug 1997 carries 0.8 kg of solids, transported as combined suspended load and bed-load.

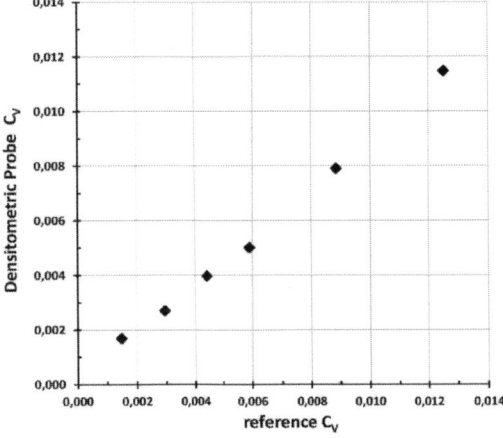

Figure 1. Observed versus reference suspended sediment volumetric concentration in static conditions in a tank.

Suspended sediment measurements with multi-frequency backscatter acoustics

J. Skripalle
HydroVision GmbH, Kaufbeuren, Germany

T. Hies & H.H. Nguyen
HydroVision Asia Pte Ltd, Singapore, Singapore

ABSTRACT

Monitoring of sediment load in rivers is an essential task for a sustainable management of rivers and reservoirs (Basson 2009). In particular sedimentation related effects like erosion, transport and deposition are having an increasing environmental impact (Wood & Armitage 1997)

Reasonable reliable measurements of the Total Suspended Sediments (TSS) are mostly achieved by taking water sampler. However this procedure does not provide high spatial and/or temporal resolution of TSS in a river or reservoir system.

Acoustic backscatter technology is a promising technology to conduct spatial and temporal measurements of suspended sediment loads in water (Landers 2010). However such "acoustic surrogate metrics" still require taking water samples to calibrate data and instruments. Due to changing particle size distributions, re-calibration of the instruments can only be done subsequently. This causes the risk that actual data are over or under estimating the amount of total suspended sediments.

A multi-frequency backscatter model has been developed to utilize ultrasonic frequencies in the range from 0.5 to 4 MHz. Higher ultrasonic frequencies are beneficial when smaller sediment grain sizes have to be considered (Carpenter et al. 2014).

The detailed model is based on the typical SONAR equation and allows to determine the sediment concentration of non-cohesive suspended sediments without calibration steps. The model runs in real-time and calculates the total suspended sediment concentration (TSS), the mean grain size and the slope of the particle size distribution (PSD) simultaneously. The mean size and slope are used to estimate the particle size distribution (PSD) (Skripalle et al. 2014).

The instrument has been intensively tested under controlled laboratory conditions showing a high performance of ±10% for TSS concentrations from 50 mg/l up to 50 g/l and grain sizes ranging from 20 μm to 1000 μm (Hies et al. 2015).

Tests have been conducted in rivers and reservoirs in India and Vietnam comparing the measurement data with laboratory analysis of the TSS in accordance with APHA 2540D. The non-calibrated measurement data achieve an average deviation of ±22% of the TSS compared to the 10 water samples analyzed in the laboratory.

The implemented model is being fine-tuned to further increase the accuracy and reliability.

REFERENCES

Basson, G.R. 2009. Sedimentation and Sustainable Use of Reservoirs and River Systems, Draft ICOLD

Carpenter, Jr. W.O., Goodwiller, B.T., Chambers, J.P. & Wren, D.G. 2014. Acoustic Measurement of Suspensions of Clay and Silt Particles Using Single Frequency Attenuation and Backscatter. Applied Acoustics 85 (2014): 123–129.

Hies, T., Nguyen, H.H. & Skripalle, J. 2015. Analysis of Multi-Frequency Backscattering Signals for Sediment Concentration Measurements. Proceedings International Conference on Hydropower for sustainable development 05–07 Feb 2015: 95–104.

Landers M.N. 2010. Review of methods to estimate fluvial suspended sediment characteristics from acoustic surrogate metrics. Joint Federal Interagency Conf 2010; 2:1–2.

Skripalle J., Hies T. & Nguyen H.H. 2014. Real-Time Concentration and Grain Size Measurement of Suspended Sediment Using Multi-Frequency Backscattering Techniques. 10th International conference on hydraulic efficiency measurement, 16th – 19th September 2014, Brazil.

Wood P.J. & Armitage, P.D. 1997. Biological effects of fine sediment in the lotic environment. Environmental Management 21: 203–217.

Continuous grid monitoring to optimize sedimentation management

T. Van Hoestenberghe & R. Vanthillo
Fluves, Ghent, Belgium

M. De Paepe
Hexa Studios, Nevele, Belgium

N. Dezillie & N. Van Ransbeeck
VMM, Brussels, Belgium

ABSTRACT

The efficiency of sediment management techniques often increases when closely monitored. Examples are shore face nourishments and sediment traps. Shore face nourishment has become more popular in Europe at the expense of beach nourishment because of their (cost) efficiency (Van Leeuwen et al. 2007). Sediment traps as installed in the Elbe estuary near Hamburg facilitate dredging by concentrating sedimentation (HPA 2013).

The efficiency of both type of measures are nowadays monitored by bathymetric measurements. Due to shipping traffic, permanent measurement installations are not possible in the water column. Large area echo-soundings and sediment sampling are used as key measurement techniques, but can't close sediment budgets due to the large gaps in the data.

With the recent developments in fiber optic measurement techniques, solid-matter dynamics can be monitored in a very detailed way both spatially and temporally. A continuous grid of the sea- or riverbed morphology can continuously be measured. The measurement system is not present in the water column, but is a cable installed below the sedimentation layer to be monitored. Depending on the shipping traffic, dredging intensity and thickness of the sediment layer, different fiber optic cables and processing algorithms are appropriate.

Fluves was awarded in 2015 a project by the Flemish Environmental Agency (VMM) to monitor continuously during one year the erosion and sedimentation of a sediment trap with dimensions 200 × 20 m, by means of fiber optic techniques. Spatial resolution of the measurement grid is every m^2 with a daily temporal resolution. A detailed digital terrain model (DTM) of the sediment trap as shown in Figure 1 can hence be transmitted to the client on a daily basis. With the changes in DTM and sedimentation layer, morphological evolution and efficiency of the sediment trap are monitored. With this knowledge, operational decisions can be taken (i.e. when to dredge the trap) and the design of future traps can be optimized.

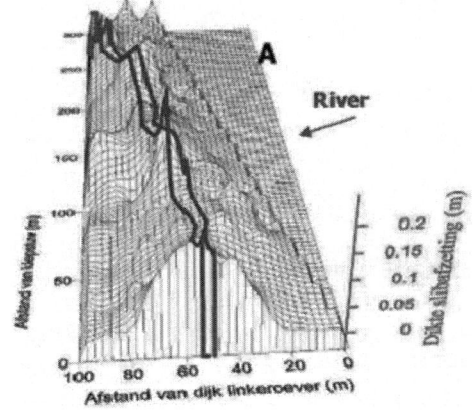

Figure 1. Figure 1. Overview of a sediment trap in Flanders with thickness of the sedimentation layer (in meter) between two grid measurements shown in the Z-axis. The double blue line in the middle of the sediment trap is the river flowing through the trap (Van Hoestenberghe 2006).

When applied to shore face nourishments, this continuous grid monitoring system can drastically increase the insights in sediment dynamics over large areas at the shallow coastal areas.

REFERENCES

HPA. 2013. Pilot Project: 'Evaluation of the sediment trap near Wedel'. Measure analysis in the framework of the Interreg IVB project TIDE. Measure 5. Hamburg Port Authority (HPA), Hamburg, Germany.

Van Hoestenberghe, T. 2006. Proceedings symposium 'Watersysteemkennis', studiedag 'Water en Sediment', Aalst, 16 november 2016, Meten en karakteriseren van slibafzetting.' Water 26, 5 pp.

Van Leeuwen, S., Dodd, N., Calvete, S. & Falqués, A. 2007. Linear evolution of a shoreface nourishment. Coastal Engineering, 54, 417–431

SS 4 SEDITRANS – Sediment transport in fluvial, estuarine and coastal environment

Failure by overtopping of earth dams: Novel methods to determine the breach effluent hydrograph

S. Amaral & T. Viseu
Laboratório Nacional de Engenharia Civil, Portugal

J.E. Santos, A. Lopes & A.M. Bento
Instituto Superior Técnico, Universidade de Lisboa, Portugal

L. Caldeira
Laboratório Nacional de Engenharia Civil, Portugal

R. Cardoso & R.M.L. Ferreira
CERIS, Instituto Superior Técnico, Universidade de Lisboa, Portugal

ABSTRACT

Overtopping is the most common cause of failure of earth dams. Yet, there is still lack of phenomenological insights on the processes that occur during breach growth, which ultimately determines key features of those hydrographs. Evidently, to advance knowledge on the causes of breach hydrograph, one needs methods to determine it experimentally. This entails theoretical and practical difficulties. On a theoretical level, an unambiguous definition of breach hydrograph is yet to be offered – the flow over a breaching dam is unsteady and spatially varied, varying the shape of the cross-section where it is evaluated. Practical issues, in fact the difficulty to measure flowrates over an evolving breach, have nullified this discussion. The breach hydrograph has been traditionally presented as the result of a mass balance in the upstream reservoir or the flowrate measure in a downstream reference cross-section. This paper addresses the practical issues involved in determining non-intrusively the breach hydrograph. Its goal is to present exploratory results of novel techniques based Large-scale PIV (LSPIV), optical techniques and acoustic instrumentation. The resulting breach hydrograph is determined for a crosssection located over the initial crest of the dam. Hence, it should respond immediately to changes in the breach morphology such as those originated by sudden mass failure.

Following Amaral et al. (2014), two laboratory tests were performed in a medium scale facility at the Portuguese National Laboratory of Civil Engineering (LNEC) under hydraulic and geotechnical controlled conditions. Tests main characteristics were maintained, namely in terms of grain-size distribution curves, water content (ω), dry volumetric weight of the soil (γ_d) and degree of compaction, being the initial breach geometry the only variable different among tests (Fig. 1).

The variables monitored during the experiments were: i) the flow discharge at the reservoir inlet; ii) the

Figure 1. Intial Breach geometries.

temporal variations of the water level inside the reservoir, near the breach and downstream the dam; iii) the geometric evolution of the breach during the failure event and iv) the surface velocity maps in the approach and at a cross-section over the initial crest. To measure iii) and iv) a 6 W laser sheet was deployed over the dam crest. Acoustic probes measured the variations of the water surface in the reservoir. Two high-speed video systems were employed to register the evolution of the cross-section and surface distribution of free-surface tracers for LSPIV data. Two novel non-intrusive methods for the breach hydrograph were determined: A) direct integration of the velocity profile on the breach cross-sections (only available to low flow discharges) and B) outflow hydrograph from the energy balance equation within a control volume around the breach.

Additionally, traditional indirect estimates based on the mass balance equation within the reservoir and on the rating curve of a downstream spillway were determined.

Key features of the estimated hydrographs are discussed and correlated with the breach evolution. It was observed that there is a delayed response of the flow field and was affected by mass instability events during breach enlargement. Hence the increase of the outflow volume rate is, initially, due to geometric enlargement. This strongly points to the need of estimating breach effluent hydrographs at the breach site in order to understand the relative effect of hydraulic and geotechnical phenomena involved in breach evolution.

A particle counter prototype and video imaging techniques for calculation of bedload fluxes

F. Antico, P. Sanches & L. Mendes
Instituto Superior Técnico, Universidade de Lisboa, Lisbon, Portugal

R. Aleixo
GHT Photonics, Padova, Italy

R.M.L. Ferreira
CERIS, Instituto Superior Técnico, Universidade de Lisboa, Lisbon, Portugal

ABSTRACT

Bedload transport is an important process in fluvial hydraulics as determines fluvial morphology. Classical approaches describing bedload transport processes were focused on the macroscopic aspects of the granular-fluid flows, with no consideration of the mechanics of sediment motion at grain-scale. Better predictions of bedload fluxes can be achieved through fundamental research on grain mechanics for which detailed measurements of bedload fluxes at grain-scale are instrumental. In controlled laboratory environments, different bedload measurement techniques are employed, such as devices based on a metallic plate and accelerometers to count the particles impact on the plate (Heyman et al. 2013) or high speed-video recording to count individual particles (Roseberry et al. 2012, Ballio et al. 2014, Aleixo et al. 2015). The latter technique has been increasingly used; however, the need for optical access and the large amount of data generated, due to long periods of acquisitions, are important drawbacks associated to this method.

To overcome some drawbacks related to bedload measurement methods described in the literature, the present paper proposes a novel impact-based particle prototype particle counter. This device, a second generation of the apparatus described in Aleixo et al. 2015, works by detecting the impacts of sediment particles on a sensitive surface. Digital signal processing and pattern recognition techniques are employed to detect impacts and count them.

Its performances have been tested comparing the results obtained with real reference data collected with high-speed video recordings.

The advantages, monitoring bedload for long periods, small data footprint, no need for optical access, real-time analysis, adaptability to different channel widths, and the limitations of the impact surface technique are discussed.

The validation of the prototype was conducted in 12.5 m long and 0.405 m wide recirculating flume of the Laboratory of Hydraulics of Instituto Superior Técnico, Lisbon. Experimental tests were carried out under conditions of weak bedload transport; 5 mm glass beads were used as bed material. For a sampling time of 90 s, it was observed a good overall agreement in what concerns the number of detections. The lateral distribution of bedload showed a slight disagreement, mainly due to false particle impacts detected by the particle counter prototype and included in the analysis. It was also found that both the systems were able to capture the expected temporal stochastic variation of bedload and both leaded to similar autocorrelation functions and energy spectra.

This work was partially supported by Project SEDITRANS funded by the European Commission under the 7th Framework Programme and by FED-ER, program COMPETE, and by national funds through Portuguese Foundation for Science and Technology (FCT) project RECI/ECM-HID/0371/2012.

REFERENCES

Aleixo, R., L. Mendes, F. Antico, P. Sanches, F. Alegria, & R. M. L. Ferreira (2015). Assessment of video imaging and particle counting techniques for calculation of bedload fluxes. In E-proceedings of the 36th IAHR World Congress 28 June, 3 July, 2015, The Hague, the Netherlands.

Ballio, F., V. Nikora, & S. E. Coleman (2014). On the definition of solid discharge in hydro-environment research and applications. Journal of Hydraulic Research 52(2), 173–184.

Heyman, J., F. Mettra, H. Ma, & C. Ancey (2013). Statistics of bedload transport over steep slopes: Separation of time scales and collective motion. Geophysical Research Letters 40(1), 128–133.

Roseberry, J.C., M.W. Schmeeckle, & D.J. Furbish (2012). A probabilistic description of the bed load sediment flux: 2. Particle activity and motions. Journal of Geophysical Research: Earth Surface (2003–2012), 117(F3).

Coupling of large eddy simulations with the level-set method for flow with moving boundaries

F. Kyrousi
Idrostudi s.r.l., Trieste, Italy
Department of Engineering and Architecture, University of Trieste, Trieste, Italy

A. Leonardi & F. Zanello
Idrostudi s.r.l., Trieste, Italy

V. Armenio
Department of Engineering and Architecture, University of Trieste, Trieste, Italy

ABSTRACT

The sediment transport generated by some environmental flow is intense enough to have a deep influence on the flow itself, due to the modifications occurring to the erodible interface. This process is challenging to reproduce in a numerical environment, due to the simultaneous presence of moving boundaries and turbulence structures. Motivated by the need for a three-dimensional model able to deal with the erosion generated by gravity currents, the coupling of a turbulent flow with a mobile bed is developed.

The flow is numerically simulated by means of the large eddy simulation (Armenio & Piomelli 2000), and the evolution of the interface between the bed and the fluid flow is tracked using the level-set method (Osher & Sethian, 1988). The implementation of the level-set method using both the upwind and the essentially non-oscillatory schemes is presented. The treatment of the moving boundary is handled using an immersed boundary technique (Roman et al. 2009) over a fixed Cartesian grid. The level-set method provides a simple and efficient way to manage the time evolution of the geometry and is used to handle the complexity of the moving immersed interface. A channel flow over a sinusoidal bed is simulated for validation purposes, and the numerical results concerning the bed deformation due to the flow shear are discussed. Furthermore, the influence of the motion of the sinusoidal bed on the flow is studied as a separate test case.

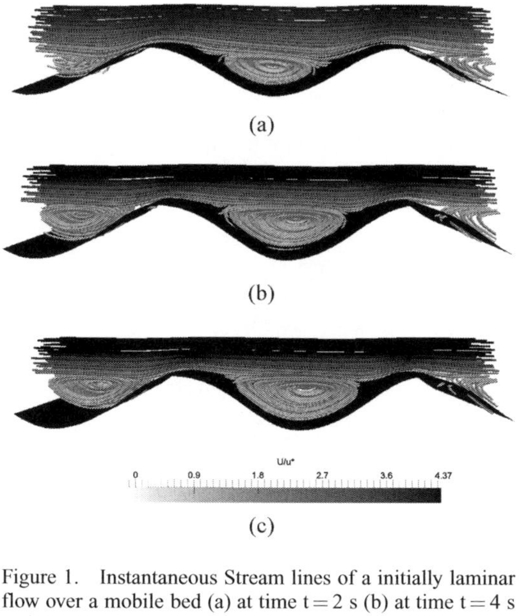

Figure 1. Instantaneous Stream lines of a initially laminar flow over a mobile bed (a) at time $t=2$ s (b) at time $t=4$ s and (c) at time $t=6$ s.

REFERENCES

Osher, S.J. & Sethian, A., 1988. Fronts propagating with curvature-dependent speed: algorithms based on Hamilton-Jacobi formulations. Journal of computational physics, 79(1), 12–49.

Roman, F., Napoli, E., Milici, B. & Armenio, V. 2009. An improved immersed boundary method for curvilinear grids. Computers & Fluids, 38(8), 1510–1527.

Armenio, V. & Piomelli, U. 2000. A lagrangian mixed subgid-scale model in generalized coordinates. Flow, Turbulence and Combustion, 65(1), 51–81.

River morphodynamics under the effect of flow variability

B. Oliveira & R. Maia
Faculty of Engineering of the University of Porto, Portugal

ABSTRACT

A great variety of techniques has been developed over the years for representing the relationship and interaction between river's flow magnitudes and its erosive potential, transport capacity and morphodynamic behavior in general. These techniques include a simple function-based relationship between streamflow and sediment transport (often expressed as a power-law function), or a broad set of mostly semi-empiric expressions relating the flow's erosive capacity (represented, e.g., by an estimate of bed shear stress) and sediment mean/representative diameters.

The relationship between streamflow and sediment transport (and its representations) is relevant to all morphodynamic studies, from forecasting morphological river change or reservoir sedimentation to the vast majority of numerical simulation of open channel morphodynamics. These applications are based on a set of assumptions, of which the gradually varying nature of streamflow is one of the most uncertain for any generic application.

This study aims to analyze the morphodynamic behavior of rivers when subjected to the effects of hysteresis which result from the naturally changing flow. Common approaches in engineering (particularly in more practice-oriented environments or when a simplified approach is necessary) often rely on the assumption of a gradually varied flows which do not account for hysteresis, without the associated error being quantified. In this paper, an attempt was performed at assessing this error. This task will be performed by way of multivariate analysis (viz., direct comparison, data clustering and curve fitting) of the extensive database available in the Portuguese National Water Resources Information System (SNIRH in the Portuguese acronym), with a total of 77 dual hydrometric/sedimentological data collection stations.

The evolution of morphodynamic processes over time is compared and related with the succession of flow values in the series and the hysteresis which results from this same succession of flow values. Likewise, sediment transport, both as suspended and as bed load (as measured directly from the river) was related, not just with flow magnitudes, but also with the naturally varying streamflow (i.e., hysteresis) which has proven to be a relevant portion of the natural forcing exercised by the flow on the river bed. Whilst the suspended fraction of sediment transport displays the expected behavior (with a directly proportional relationship with streamflow magnitude, in locations where there is sufficient sediment availability), bed load presents a non-linear behavior which is dependent on the succession of streamflow values.

A new form for sediment rating curves, albeit presently under study for further validation, is experimented with, producing significant improvements over the traditional sediment rating curves.

Common approaches to defining bed load transport involve the Meyer-Peter and Müller (modified by Wong & Parker, 2006) or the Wilcock and Crowe formula (Wilcock & Crowe, 2003) which, directly or indirectly, directly relate streamflow and bed load sediment transport. This paper shows that the use of these formulas is questionable, particularly when quickly varying flows are involved.

REFERENCES

Wilcock, P.R. and Crowe, J.C., 2003. Surface-based Transport Model for Mixed-Size Sediment. Journal of Hydraulic Engineering, Volume 129, pp. 120–128.

Wong, M. and Parker, G., 2006. Reanalysis and Correction of Bed-Load Relation of Meyer-Peter and Müller Using Their Own Database. Journal of Hydraulic Engineering, Volume 132, pp. 1159–1168.

Numerical investigation on the effect of suspended sediment load on flow field around a cylinder

T. Paone, R.M.L. Ferreira & A.H. Cardoso
CEHIDRO, Instituto Superior Técnico, Universidade de Lisboa, Lisbon, Portugal

V. Armenio
Dipartimento di Ingegneria e Architettura, Università degli Studi di Trieste, Trieste, Italy

ABSTRACT

The presence of suspended sediments in the flow has influence on its fluid dynamics properties such as the velocity distribution and turbulent characteristics. The flow around cylindrical obstructions is characterized by a complex geometry and dominated by energetic coherent structures. Such vortices interact with the suspended sediments endangering the stability and undermining the reliability of hydraulic structures (e.g. bridge piers).

The study of the flow field around cylinders has benefited from numerical simulations as step beyond potential flow approximations. In recent times, large eddy simulations (LES) were used to predict the flow field, bed evolution and sediment entrainment mechanisms around piers (e.g. Kirkil et al. 2008).

The purpose of this paper is to analyze the effect of the suspended sediment load on the flow field around a cylinder by performing numerical simulations. A numerical code, which includes a suspended sediment transport module, is used. The code is an Euler-Euler single-phase approach involving LES with sub-grid-scale model to solve the 3D Navier-Stokes equations, whereas an advection-diffusion equation is for suspended sediment transport (Dallali and Armenio 2014). Furthermore, it considers the Boussinesq approximation of the Navier- Stokes equations to take into account the effect of suspended sediment concentration, C, on flow field (sediments buoyancy effects). In this work, numerical results with ($C = 0.01$) and without ($C = 0.0$) the suspended sediment are compared in terms of the mean velocity, vorticity, shear stress, turbulent intensities and Strouhal number. The numerical results for the clear water case are also compared with existing experimental data to analyze the flow field downstream of the cylinder (Ricardo et al. 2014a, b). Small differences in first and second-order moments of velocity fluctuations are identified and discussed.

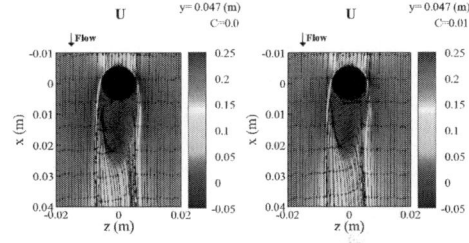

Figure 1. Mean stream-wise velocity contour map in a horizontal plane close the free surface. Comparison of numerical results with (right) and without (left) suspended sediments.

ACKNOWLEDGEMENTS

This work was partially supported by ITN SEDI-TRANS, funded under EU 7th Framework and by FEDER, program COMPETE, and national funds through Portuguese Foundation for Science and Technology (FCT) project RECI/ECM-HID/0371/2012.

REFERENCES

Dallali, M. & V. Armenio (2014). Large eddy simulation of two-way coupling sediment transport. River Flow – the 7th International Conference on Fluvial Hydraulics – at EPFL, Lausanne, Switzerland.

Kirkil, G., S. G. Constantinescu, & R. Ettema (2008). Coherent structures in the flow field round a circular cylinder with scour hole. J. Hydraul. Eng. 134(5), 572–587.

Ricardo, A. M., K. Koll, M. J. Franca, A. J. Schleiss, P. Sanches, & R. M. L. Ferreira (2014a). Vortex interaction in patches of randomly placed emergent cylinders. In: Proceedings of River Flow 2014. Ed: Schleiss et al., pp. 63–69, Taylor & Francis Group, London, ISBN 978-1-138-02674-2.

Ricardo, A. M., K. Koll, M. J. Franca, A. J. Schleiss, & R. M. L. Ferreira (2014b). The terms of the turbulent kinetic energy budget within reaches of emergent vegetation. In: Water Resour. 50, doi: 10.1002/2013WR014596.

Impact of placer mining on suspended sediments in rivers of the Kamchatka Peninsula (Russian Federation) and the Selenga River basin (Mongolia) and its modeling

E. Promakhova & N.I. Alexeevsky
Lomonosov Moscow State University, Moscow, Russian Federation

ABSTRACT

Placer mining are associated with river valleys. Most of these regions are situated in remote, sparsely populated areas, where the stream gauging network is underdeveloped, thus wise the study of placer mining impact remains challenging. The fieldwork was carried out in the Selenga (Mongolia) and the Vyvenka river (the North of Kamchatka, Russia) basins near the placer mining.

It was found that placer mining operations leads to an increase of the suspended sediment concentration (*SSC*) and additional input of fine particles in river waters. These processes are amplified during the river flood. During this period the excess of *SSC* with anthropogenic genesis is 8.8 times higher than the baseline values for minor rivers in the Selenga basin, and 3.3 times higher in the North of Kamchatka. For the Tuul river (the tributary of the Selenga river) the increase of *SSC* depends on the water phase and reaches 1.6–2.7 times over the baseline. Downstream from the mining the *SSC* does not change along the stream of minor rivers because of the high concentration of dust, silt and clay in river waters, which forms a sediment transfer stream. The anthropogenic *SSC* decreases by 1.7 times independently from the stream flow state along the flow of the Tuul closer to its mouth.

The lack of full-rate hydrometeorological data complicates a large-scale modeling and the estimation of anthropogenic impact on the environment in general. The analytical solution for equation of turbulent diffusion for suspended particles (Graf 1984 and Makkaveev 1931) was obtained under some assumptions and taking into account the rate of *SSC* along the flow depth and width:

$$SSC(x) = SSC_0 \exp\left(-\frac{2mC}{gh}\left(\frac{\omega}{V}\right)^2 x\right), \quad (1)$$

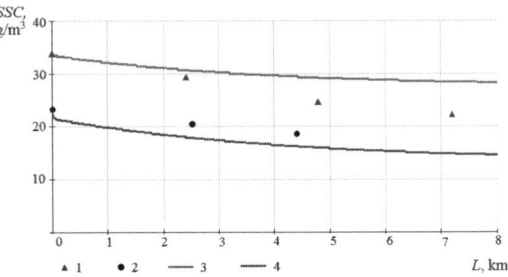

Figure 1. The real SSC (1, 2) and the modeled SSC (3, 4) along the Levtyrinyvayam river (the tributary of the Vyvenka river) for the conditions of best performance of industrial object (1, 3) and reduced platina production (2, 4).

where SSC_0 – the *SSC* in the initial cross section, m – the Bazen-Bussinesk coefficient, C – the Chezy-coefficient, g – the acceleration of gravity, V – the average stream flow, h – the total depth, ω –particle fall velocity, x– longitudinal spatial axe.

The model analytical solution of cross section averaged *SSC* demonstrates the accuracy in a range of 6–31% (Fig. 1). A solution of the longitudinal *SSC* transport equation for the unsteady state is a subject of further research.

In memoriam of my Teacher Nikolay Alexeevsky, Professor and Head of Hydrology Department at Faculty of Geography, Lomonosov Moscow State University.

REFERENCES

Graf, W.H. 1984. *Hydraulics of Sediment Transport*. New York: Water Res. Publ.

Makkaveev, V.M. 1931. On the theory of turbulent regime and weighing deposits. *State Hydrological Institute Bulletin* 32: 5–26.

SS 5 Sustainable land management

Research-praxis integration in South China – the rocky road to implement strategies for sustainable rubber cultivation in the Mekong Region

Thomas Aenis, Jue Wang & Susanne Hofmann-Souki
Humboldt-Universität zu Berlin

Tang Lixia
China Agricultural University, Beijing

Gerhard Langenberger, Georg Cadisch, Konrad Martin & Marc Cotter
University Hohenheim

Manuel Krauss
University Stuttgart

Herrmann Waibel
Leibniz-University Hanover

ABSTRACT

The Sino-German project "SURUMER: Sustainable rubber cultivation in the Mekong region" is looking for "...an integrative, applicable, and stakeholder-validated concept for sustainable rubber cultivation" in Xishuangbanna region, South-west China. In order to at least partially implement solutions for problems of the complex rubber-dominated land use situation, project partners are testing a holistic approach to research-practice-integration which combines profound scientific analyses with a range of activities to enhance interaction and collaborative learning amongst stakeholders (i.e., practitioners and reseachers).

In this presentation we discuss the approach to research-practice-integration. Furthermore we will discuss strengths and weaknesses, identify the mayor challenges and draw conclusions for future process management. The effort follows the assumption that the possibility for implementation will be the higher the more "realistic" the concepts are and the better they are communicated with stakeholders. Three key stakeholder groups have been identified so far, namely innovative rubber farmers and village heads, prefecture administration, and provincial politicians. For stakeholder involvement, a pragmatic communication approach has been chosen which consists of a range of discourse instruments such as bilingual information material, newsletters, focus groups, and a series of workshops with key stakeholders. One of its main elements is participatory scenario development including discussions on the economic and ecologic trade-offs of different land use.

Various aspects supporting or hindering implementation have been analysed: Generally, communication amongst researchers but mainly between researchers and practitioners takes more time and resources than initially estimated.

This mainly affects the testing of new approaches. For example, "scenarios" are discussed in a complete different way within the group of researchers than amongst practitioners, which means that there is need for translation not only from English to Chinese and vice versa, but between abstract modeller-thinking and concrete needs of practice. "Demonstration plots" on intercropping with endangered tree species can show only the very initial planting processes; evidence that intercropping is more sustainable than conventional rubber cultivation is limited.

Stakeholders are open towards innovative solutions. Most of them are aware of ecological problems such as water quantity and quality even if they do not necessarily link them with rubber cultivation. More important is the fact that prices for latex have decreased to less than a quarter in recent years and farmers may have lost two-third of their income. Many farmers either search for work in town or they rent their community land to investors. If farmers rent out land to external investors there is a threat that rubber might be replaced by ecologically more adverse cultures such as banana.

Despite the time constraints and limited resources for interaction with stakeholders, SURUMER has shown ways to mutually develop solutions amongst research and practice. Experience shows that communication processes must be managed carefully. It is obvious that significant resources are needed, and that even long research projects of five years can not expect full implementation.

Managing rubber plantations towards improved water protection

G. Langenberger, H. Liu, S. Blagodatskiy & G. Cadisch
Institute of Plant Production & Agroecology in the Tropics and Subtropics, University of Hohenheim, Stuttgart, Germany

M. Krauss
Institute for Sanitary Engineering, Water Quality & Solid Waste Management, University of Stuttgart, Germany

J. Wang & T. Aenis
Albrecht Daniel Thaer-Institute of Agricultural & Horticultural Sciences, Humboldt-Universität zu Berlin, Germany

S. Min & H. Waibel
Institute of Development & Agricultural Economics, Leibniz Universität Hannover, Germany

ABSTRACT

Rubber plantations became a considerable element of Mainland South East Asia, often dominating whole landscapes. The expansion of rubber plantations took place on behalf of diverse, traditional land-use systems and near-natural or old growth forests, especially mountainous and formerly remote areas, which became accessible due to infrastructure development. This land-use transformation inevitably affects Ecosystem Functions and Services, one of it being the provisioning of water in general and drinking water in particular.

In the framework of the interdisciplinary SURUMER-project we studied the socioeconomic framework conditions as well as the consequences of the rubber management practices in Xishuangbanna Prefecture, Yunnan Province, SW China[1] concerning: a/ farmers' perception of the importance of ecosystem services, b/ the use of agro-chemicals and farmers' awareness of its potential negative consequences, c/ the impact of the weeding scheme on erosion, d/ farmers' micro-economic situation. The respective disciplinary methodologies have been applied.

It turned out that 'water' is by far the most important ESS to farmers, reflecting their experience of degrading water sources accompanying rubber expansion. At the same time the use of agro- chemicals and pesticides is ubiquitous, showing the low awareness concerning the potential impacts on water quality and health, respectively.

Based on an integrative analysis of the different aspects we suggest alternative management concepts on plot but also landscape level. These comprise a shift from chemical to mechanical weeding. Since the traditional manual brushing is labor intensive, the alternative option of using brush cutters is currently assessed. Additionally, labor extensive permanent intercropping schemes with economically but also ecologically valuable tree species are suggested to improve soil stability in the rubber-inter rows but also to reduce weeding requirements.

Since water provisioning can hardly be safeguarded on a plot level but needs the consideration of the respective watershed, water protection zones have been suggested, where rubber management needs to be strongly restricted or abandoned. The idea received considerable attention from farmers but also illustrated the complexity of the topic. Thus, the beneficial farmer communities are not necessarily the respective land owners, making agreements challenging.

---

[1] The nature reserve is designed according to the UNESCO Man & Biosphere concept, allowing people to stay in reserves and to manage land in the buffer and development zones.

Dynamics of soil erosion in rubber plantations and its mitigation by herbicide management

H. Liu, S. Blagodatskiy & G. Cadisch
Institute of Plant Production and Agroecology in the Tropics and Subtropics, University of Hohenheim, Stuttgart, Germany

ABSTRACT

Rubber plantation, as a new land use system introduced into Xishuangbanna, Southwest China, increased by 194,151 ha (a gain of 90%) from 1976 to 2003. Erosion may increase from forest-to-rubber due to less understory; but may decrease from agriculture-to-rubber by canopy interception of precipitation, litter cover and roots from trees. Lifespan of rubber plantation ranged from 25 years to 40 years. Plantation development and forming of multi-layer structure comprising of canopy cover, surface cover and root system can control the runoff and sediment yields dynamics. Herbicide application may strongly influences erosion in rubber plantation by reducing understory plant cover. Little research has been done on temporal change of soil protection from rubber plantations with tree growing due to dense canopy and root density; and potential soil conservation improvement by land management. This study aimed at 1) assessing dynamic change erosion in rubber plantations during its development; 2) identifying the most vulnerable plantation development phase to erosion. 3) evaluating the possibility to improve soil conservation of rubber system by reduced herbicide management. Soil erosion was measured in rubber plantations of 4, 12, 18, 25 and 36 year age. Rainfall, soil texture and carbon content at top 5cm, density of fine roots, canopy radius for each age of rubber plantation, understory plant cover and surface cover were determined using Gerlach troughs, soil coring, photography, respectively.

Soil loss of rubber plantation with different ages varied from 52 gm^{-2} to 277 gm^{-2} with highest soil loss of mid-age (12 and 18 years) and lowest of old (25 and 36 year) rubber. Cropping management (C) factor, which represented anti-erosive effectiveness of rubber system, was calculated for different age rubber plantations by applying Universal Soil Loss Equation (USLE) model. Anti-erosive effectiveness of rubber plantation was mainly controlled by surface cover. The mid-age rubber was found to be most vulnerable phase to erosion due to less understory vegetation compared to young rubber and less litter fall compared to old rubber. To search for potential improvement of soil conservation of rubber plantations, different herbicide application schemes were used to assess impact of land management on rubber system's erodiblity. Runoff and sediment yield was measured in 12 year old rubber plantation under three treatments: i) herbicide application according to farmers' practice frequency two times per year; (ii) no herbicide application to keep high understory plant cover; and (iii) bimonthly herbicide application in order to maintain a low understory plant cover. Highest soil loss (425 gm^{-2}) was observed under intensive herbicide load, while soil loss was minimal (50 gm^{-2}) without herbicide application. For mid-age rubber, understory plant cover was the most influential factor and most efficient in reducing soil loss, with no herbicide application. Therefore, it is possible to improve soil conservation function of rubber plantation by herbicide management.

Mitigation of forest to rubber change impact on soil erosion and stream quality by integrated land management

H. Liu, X. Yang, S. Blagodatskiy, C. Marohn & G. Cadisch
Institute of Agricultural Sciences in the Tropics (Hans-Ruthenberg-Institute), University of Hohenheim, Garbenstraße, Stuttgart, Germany

ABSTRACT

Land use in Xishuangbanna, SW China, has been dramatically changed over the past 30 years. Rubber plantation combined with tea cultivation boosted from 1.3% to 11.8%, while deforestation decreased the forest cover from 69% to 45%. The major factor affecting hydrological processes was land cover change overarching the climate change impact. The conversion of tropical rain forest to rubber plantation may markedly affect quantity and quality of surface runoff, which serves as the most important source for drinking water in the mountainous area. Therefore this land use change can further threaten water security of local villages. In order to reduce impact of rubber expansion, the action guidance of Twelfth Five Year Plan of Xishuangbanna government has limited rubber planting area to elevation below 950 m and slope smaller than 25 degree. However, this threshold was proposed by considering growth of trees instead of improving ecosystem services, e.g. soil conservation. This study aims at 1) evaluating the effect of rubber plantations on erosion at plot and watershed level; 2) enhancing ecosystem function of rubber plantations by improving soil conservation at plot level and ameliorating water quality by better landscape planning at watershed level e.g. establishing buffer zone; 3) balancing land use between economic benefic and environmental protection. We evaluated sediment yields and runoff production in different plantation ages within one lifespan: young rubber with open canopy (4 year), mid-age with closed canopy (12 year, 18 year) and old rubber with dense canopy (25 year and 36 year). Driven factors like rainfall intensity, canopy interception, surface cover, soil properties, fine root density were measured simultaneously. Different herbicide applications schemes were applied to assess impact of improved management on decreasing soil loss at plot level. Rubber-dominated and forest-dominated (as reference) watersheds were selected for continuous measurement of water discharge and turbidity in the outlet position to evaluate the water yield and export of suspended solids from rubber plantation at watershed scale. Based on field survey we found that highest soil loss at plot level (277 g m^{-2}) was in mid-age rubber plantation, which implicated potential water pollution (threshold 100 g m^{-2}). Reduced herbicide rate (2 times per year V.S bimonthly application) can efficiently decrease sediment yields to 50 g m^{-2}. Land use change from forest to rubber increases stream turbidity at watershed level from 350 to 1300 nephelometric turbidity units (NTU). Soil loss in rubber dominated watershed was estimated as 0.3 t ha^{-1} y^{-1} while high frequency (over 15% time in rainy season) of medium high turbidity (> 30 NTU) and turbidity over 1200 NTU appeared in storm events still strongly threatened security of local drinking water resource. Measures at plot (reduced herbicide) and watershed level (introduction of riparian buffer zones) should be taken to supply better ecosystem service. Size and location of the buffer zones serving as effective filter strip needs to be determined. Therefore, the Land Use Change Impact Assessment (LUCIA) model is used to simulate integrated land management (reduced herbicide + buffer zone) effect on erosion and sediment transport. LUCIA model simulates the erosion process in a spatially explicit dynamic way, depending on development of plant canopy and terrain properties. Data collected in field are used to parameterize and validate the baseline scenario. Meanwhile, hotspots of deposition and erosion along the stream can be identified. Plot (reduced herbicide) and landscape (buffer zone) controls on erosion are simulated by the model to identify best management practices for improved ecosystem function (better soil conservation and decreased stream turbidity) as well as maintained economic profitability (total latex production in study region). The results can further assist the cost-benefit analysis and decision-making in land management and landscape planning.

Reduction of fine sediment infiltration into rivers by implementing riparian buffer strips in an agricultural dominated area in Southwest China

L. Seitz & S. Wieprecht
Institute for Modelling Hydraulic and Environmental Systems, University of Stuttgart, Stuttgart, Germany

M. Krauss, N. Azizi & H. Steinmetz
Institute for Sanitary Engineering, Water Quality and Solid Waste Management, University of Stuttgart, Stuttgart, Germany

ABSTRACT

One of the negative effects of monoculture is soil erosion and the loss of fertile soil surface. Fine sediment runoff from agricultural land enters the rivers with serious negative impacts on the aquatic environment. The settlement of fine sediments on or even into the river bed clogs the pore space and reduces the living space for juvenile fishes and macroinvertebrates. Furthermore, the natural exchange processes between surface water and groundwater can be disturbed. Nutrients and pesticides, which preferably bind to sediments are also transported to the rivers and may further reduce the water quality. The implementation of riparian buffer strips along river banks can attenuate the effects of fine sediment and nutrient input by retaining them. The aim of this study is to present a concept to reduce the introduction of fine sediments into river beds by implementing riparian buffer strips in a high monoculture dominated area. The riparian buffer strips are planned according to land use, slope as well as soil composition of the surrounded area. Therefore, different scenarios are developed to retain sediments by the buffer strips. An additional scenario is developed with different buffer widths and vegetation types to retain not only sediments but also nitrogen and phosphorous.

The study site is located in Jinghong County, Dai Autonomous Prefecture Xishuangbanna of Yunnan Province (South China). Xishuangbanna is one of Chinas most biodiverse regions and included in the Indo-Burma biodiversity hotspot (Li et al. 2007). With Chinas growing economy the demand of rubber has been increasing continuously in the last decades. Therefore, large areas with tropical seasonal rain forest have been transformed into rubber monocultures. This led to an increase in the cultivation of rubber of more than 20% between 1976 and 2003 and, also influenced by other changes, to a decrease of rain forest by 67% (Li et al. 2007).

Continuously measured turbidity data at the study site show a high increase in turbidity during rain events. Samples of suspended solids are used to correlate them with the turbidity data. Detailed sediment samples withdrawn from the river bed with the so-called freeze-core technique show a high amount of fine sediments within the river bed with a percentage of fines under 2 mm up to 26%.

To quantify the effectiveness of buffer strips, a GIS-based model is used that calculates the reduction of erosion risk when implementing buffer strips. This model is based on the well-known RUSLE approach, which calculates the annual runoff based on different parameters such as land use and soil properties (Chen et al. 2011). In a first step, the model estimates the current annual soil loss and allows for investigating the effect of the different scenarios with buffer strips.

The results show a significant retention of fine sediments within the buffer strips and a reduction of the erosion risk compared to the current state. It can be concluded that riparian buffer strips are an efficient and easy-to-implement measure to reduce the introduction of fine sediments into rivers.

ACKNOWLEDGMENTS

The authors would like to thank the German Federal Ministry of Education and Research for funding the project 01LL0919B. The responsibility for the content of this publication lies with the authors.

REFERENCES

Chen, T., Niu, R.Q., Li, P.X., Zhang, L.P. & Du, B. 2011. Regional soil erosion risk mapping using RUSLE, GIS, and remote sensing: a case study in Miyun Watershed, North China. Environmental Earth Sciences, 63 (3): 533–541.

Li, H., Aide, T.M., Ma, Y., Liu, W. & Cao, M. 2007. Demand for rubber is causing the loss of high diversity rain forest in SW China. Biodivers Conserv. Vol.16: 1731–1745

Author index

Abad, J.D. 136
Abdelrazek, A.M. 45
Abdul Razad, A.Z. 97
Adema, J. 102
Adhikari, B.R. 129
Aenis, T. 273, 274
Ahmad, M.N. 97
Ahmad, Z. 96
Ahmed, A.A. 223
Akoh, R. 119
Albayrak, I. 224, 255
Aleixo, R. 266
Alexeevsky, N.I. 25, 270
Alfredsen, K. 186
Alliau, D. 230
Amante, F. 231
Amaral, S. 265
An, Y. 21
Annandale, G.W. 198
Antico, F. 266
Antonini, A. 57
Arai, R. 46
Archambeau, P. 90, 208
Ariffin, J. 67
Armenio, V. 267, 269
Asaeda, T. 180
Ata, R. 136
Auel, C. 169, 204
Aziz, S. 217
Azizi, N. 277

Babiński, Z. 195
Bakker, M. 47
Banda, M.S. 120
Basirat, Sh. 48
Basson, G.R. 232
Bastolla, V.P. 56
Batalla, R.J. 196, 210
Battisacco, E. 121
Baulig, Y. 171, 175
Baux, Y. 230
Beck, T. 249
Beckers, F. 49
Beevers, L. 71
Belyaev, V. 22
Bento, A.M. 265
Berger, V. 170
Bidorn, B. 23
Bidorn, K. 23
Bigliotti, E. 231
Bizane, M. 50
Blagodatskiy, S. 274, 275, 276

Bockelmann-Evans, B. 182
Boes, R.M. 128, 156, 224, 225, 229, 255
Boisson, N. 230
Borgsmüller, C. 171
Borthwick, A.G.L. 144
Boschi, M. 233
Boyd, P. 199
Bricker, J.D. 129
Brignoli, M.L. 196, 210
Brock, B. 233
Bui, M.D. 87

Cadisch, G. 273, 274, 275, 276
Cai, Y. 36
Caldeira, L. 265
Cao, W.H. 192, 257
Cao, Y. 51
Capapé, S. 52
Cardoso, A.H. 269
Cardoso, R. 265
Cardoso-Landa, G. 122
Castro, J.L. 56
Chalov, R.S. 187
Chalov, S.R. 187
Chen, F. 107
Chen, H. 100
Chen, K.H. 107
Chen, L. 240
Chen, L.M. 241
Chen, Y. 55
Chen, Y.S. 66, 172
Cheng, W. 137, 172
Chiu, Y.Y. 53
Chmiel, O. 197
Cleveringa, J. 102
Cleyet-Merle, P. 200
Colombo, F. 52
Čomaj, M. 179
Cotter, M. 273
Crave, A. 82
Creed, M.J. 144
Crosa, G. 196, 210
Cui, Z.F. 54
Cuthbertson, A. 71

Dai, C.M. 187
de Cuyper, A. 208
de Jong, C. 24
De Paepe, M. 262
Decachard, M. 230
Dehghani, A.A. 83

Deng, A.J. 34
Deng, H. 124
Deng, J. 61, 124
Deng, S.S. 125, 159
Detert, M. 254
Dewals, B. 90, 208
Dezillie, N. 262
Di Silvio, G. 126
Ding, L. 237, 238
Disse, M. 108
Dittrich, A. 120
Do, M.D. 244
Dong, M.J. 100, 127
Donoghue, J.F. 23
Döring, M. 225
Dou, X.P. 237, 238
Duan, G. 95
Duan, H.F. 104
Duangkamol, K. 30

Efthymiou, N. 198
Ehrbar, D. 225
El Kadi Abderrezzak, K. 90, 136
Ergenzinger, P. 24
Erpicum, S. 90, 208
Espa, P. 196, 210

Fan, B. 213
Fang, H.W. 66, 172, 176
Farahani, F. 38
Fazeli, M. 172
Feld, C.K. 170
Felix, D. 255
Fernandes, F. 173
Fernandes, J.N. 229
Ferreira, R.M.L. 265, 266, 269
Fleischer, P. 246
Formiga, K. 65
Franca, M.J. 121
Franzoia, M. 126
Friedl, F. 128
Fröhle, P. 149
Frétaud, T. 230
Fu, X. 55
Fu, Z.M. 107
Fujita, M. 68
Fukuoka, S. 132, 133, 152

Gamaro, P.E. 56
Gao, A. 100
Gao, L. 58
Gao, X.Y. 237, 238, 241
Gao, Z.R. 238

Gautier, E. 123
Geleynse, N. 102
Geng, X. 79
Geng, Y.F. 191
Gentili, G. 196, 210
George, M. 218
Gerbersdorf, S.U. 174, 188
Gesing, C. 239
Gibson, S. 199
Gilet, L. 123
Gintz, D. 175
Giri, S. 129, 148, 200
Gjunsburgs, B. 50
Gob, F. 123
Gökler, G. 234
Goseberg, N. 93
Goto, K. 80
Gotoh, H. 153
Gotoh, T. 133
Grasmeijer, B.T. 102
Gu, L.H. 166
Guan, X. 27
Guerrero, M. 57
Gumpinger, C. 177, 178
Guo, C.S. 192
Guo, X.H. 166
Guo, Y. 58
Guyot, J.L. 82

Haas, C. 256
Habel, M. 195
Habersack, H. 76, 233
Habibi, J. 219
Hagmann, M. 224
Haihua, H. 111
Haimann, M. 233
Hamdi, E. 110
Hammer, A. 233
Han, L.F. 157
Han, L.Q. 77, 100
Hanada, R. 153
Hanmaiahgari, P.R. 87
Hao, S.Y. 59
Hassan, K. 60
Hauer, C. 177, 178, 233
Haun, S. 33, 134, 185, 226, 233
Haynes, H. 60
He, F. 69
He, G.J. 66, 172, 176
He, Y. 61
He, Z. 62
Heijne, I. 217
Herrera-Granados, O. 63
Hidayat, F. 201
Hies, T. 261
Hillebrand, G. 165, 202, 247
Hinkelmann, R. 113, 165
Hirashita, S. 119
Hitz, O. 31
Hofer, B. 233
Hoffmann, T. 165, 202, 247

Höfler, S. 177, 178
Hofmann-Souki, S. 273
Hohermuth, B. 64
Holdefer, A.E. 65
Holubová, K. 143, 179
Holzapfel, P. 233
Homma, T. 211
Hope, J.A. 3
Horiuchi, K. 135
Hotchkiss, R. 218
Hu, C.H. 34, 69
Hu, D.C. 54
Hu, J. 55, 130
Hu, P. 62
Huang, E. 104
Huang, L. 66, 172, 176
Huang, W. 23
Hubmann, M. 233
Hunzinger, L. 31

Ibrahim, S.L. 67
Ikeshima, T. 80
Ishikawa, O. 153
Ismail, S.I.H. 97
Itoh, H. 180
Itoh, T. 68, 80, 105

James, C.S. 131
Järvelä, J. 189
Javornik, L. 227
Ji, Z. 69
Jia, L. 253
Jiang, E. 26, 51
Jiang, Y.M. 36
Jiao, J. 237, 238
Jiao, Z.X. 237
Jin, Z. 213
Juepner, R. 109
Juwono, P.T. 201

Kadinski, L. 254
Kaitsuka, K. 105
Kanda, K. 147
Kaneko, Y. 132
Kantoush, S.A. 203, 204
Karki, P. 198
Kato, S. 133
Kauppert, K. 239
Kawaguchi, H. 147
Kawaike, K. 70, 84
Kean, J. 71
Kemayou Tchamako, E. 181
Kenworthy, J.M. 3
Khorami, N. 39
Kikillus, A. 134
Kim, H.S. 72
Kim, N. 70
Kimura, I. 45, 72, 88, 135
Kish, S.A. 23
Kishimoto, S. 155
Kitaguchi, O. 84
Kitzhofer, J. 76

Kobayashi, S. 169
Koll, K. 245
Koppe, B. 181
Koseki, H. 258
Koshiba, T. 204
Krauss, M. 273, 274, 277
Kryžanowski, A. 227
Kudo, S. 258
Kuksina, L.V. 25
Kumar, A. 96
Kyrousi, F. 267

Lafhaj, Z. 110
Lai, R. 73
Lai, Y.G. 158
Lane, S.N. 47
Lang, U. 206
Langenberger, G. 273, 274
Langendoen, E.J. 136
Legono, D. 201
Leonardi, A. 267
Li, C.C. 205, 212
Li, D. 240
Li, F. 95
Li, G.D. 77
Li, J. 26, 51, 125
Li, S. 137
Li, T.L. 241
Li, Z. 74, 164
Li, Z.W. 108, 138
Liang, D. 113, 139
Liao, C.T. 75, 158
Lichtneger, P. 76
Liebenstein, H. 246
Lin, C.H. 140
Lin, L. 27
Lin, S.H. 140
Link, O. 92
Liu, C. 28, 141, 257
Liu, C.J. 140
Liu, D.B. 192
Liu, G.H. 75
Liu, H. 142, 274, 275, 276
Liu, L. 257
Liu, M.X. 77, 100, 112
Liu, Q. 137
Liu, Q.Q. 21
Liu, S.G. 36, 78, 187
Liu, X.N. 104, 141, 213
Lixia, T. 273
Lizano, L. 226
Lopes, A. 265
Lou, S. 78, 187
Lu, C. 61
Lu, J. 192, 257
Lu, J.Y. 159
Lu, Y. 130, 142
Lukac, M. 143

Ma, G.F. 78
Maeno, S. 119

Mahede, F. 176
Maia, R. 268
Maity, H. 81
Maji, S. 87
Makhinov, A.N. 36
Makhinova, A.F. 36
Malcherek, A. 151, 197
Maldonado, L.H. 56
Maldonado, S. 144
Malyutina, A. 22
Manhart, M. 92
Mao, J.-X. 79
Mao, L. 89
Marescaux, A. 260
Marohn, C. 276
Martin, K. 273
Martín-Vide, J.P. 52
Matsuda, S. 80
Matsumoto, T. 119
Mazumder, B.S. 81
Meddi, M. 29
Mende, M. 145
Mendes, L. 266
Mendoza, A. 136
Mewis, P. 146
Mhashhash, A. 182
Miike, T. 105
Mikoš, M. 227
Min, S. 274
Mirbach, S. 206
Mishra, A. 28
Mishra, K. 207
Miwa, H. 147
Miyamoto, A. 105
Miyata, S. 68
Mizutani, H. 84
Mizuyama, T. 68, 105
Mohamed Mustafa, T. 40
Monteiro, C.S.G. 106
Moonjun, R. 30
Morera, S.B. 82
Moser, H. 165
Movahedi, N. 83
Mravcová, K. 179
Mu, J. 243
Müller, M. 145
Muramatsu, H. 105
Muraoka, K. 258

Nabi, M. 72, 129, 148, 200
Nagayama, T. 68, 80, 105
Nakagawa, H. 70, 84
Naulin, M. 197
Nehlsen, E. 149
Nguyen, H.H. 261
Nguyen, V.T. 244
Ni, J. 109
Niemann, A. 170
Nishi, Y. 80
Noack, M. 49, 91, 185, 202
Nones, M. 126, 183

Norouzi, S. 219
Núñez-González, F. 245

Obendorfer, R. 233
Ochi, T. 147
Ogawa, H. 105
Okada, S. 258
Oliveira, B. 268
Olsen, N.R.B. 150
Omelan, M. 228
Oplatka, M. 145
Ota, K. 46
Ottevanger, W. 148
Özgen, I. 113
Ozturk, M. 85

Pakuwal, S. 86
Pal, D. 87
Palt, S. 198
Pan, S. 182
Panthee, S. 86
Paone, T. 269
Park, M. 72
Paschmann, C. 229
Paterson, D.M. 3
Patil, R.A. 259
Patsinghasanee, S. 88
Pattison, I. 35, 184
Pauli, M. 31
Peltier, Y. 208
Perk, L.M. 102
Perkins, R.J. 114
Pervez, J. 120
Peteuil, C. 230
Petkovšek, G. 209
Petrovic, D. 260
Piberhofer, B. 178
Picco, L. 89
Pillai, B.R.K. 200
Pintz, P. 198
Pirotton, M. 90, 208
Pisaturo, G.R. 231
Poleto, C. 173
Prager, E.A. 76
Prasetyo, A. 42
Promakhova, E. 270
Pujiraharjo, A. 201

Qin, L. 111
Qu, G. 162, 166
Quadroni, S. 196, 210
Quick, I. 171

Rademacher, E. 151
Raee, N. 94
Rainato, R. 89
Ramasamy, J. 28
Reindl, R. 185
Reynaud, S. 230
Rifai, I. 90
Righetti, M. 231

Roca-Collell, M. 217
Römkens, M.J.M. 32
Roucou, R. 184
Roux, S. 230
Ruidong, A. 253
Rutschmann, P. 87
Rüther, N. 57, 186, 228

Saadon, A. 67
Sadat, S.H. 101
Sadid, N. 33
Said, I. 110
Salehi Neyshabouri, S.A.A. 48
Sanches, P. 266
Sanjaya, K. 180
Santos, J.E. 265
Sasaki, T. 152
Sato, T. 46
Sawadogo, O. 232
Schanderl, W. 92
Scheder, C. 178
Schendel, A. 93
Schlabing, D. 91
Schleiss, A.J. 121
Schletterer, M. 185, 233
Schlurmann, T. 93
Schmalzer, B. 233
Schmidt, H. 174, 188
Schmitt, K. 242
Schmocker, L. 225
Schott, T. 98
Schriever, S. 247
Schwanghart, W. 129
Schwarzenberger, R. 233
Schäfer Rodrigues Silva, A. 91
Seitz, L. 134, 185, 256, 277
Senn, G. 185
Shafai-Bejestan, M. 94
Shang, Q.Q. 66, 172, 176
Shen, Z.Z. 37
Shetkar, R.V. 259
Shi, H.L. 34
Shi, Y. 74
Shieh, C.L. 140
Shimizu, Y. 45, 72, 88, 135
Shrestha, D.P. 30
Shu, A. 95
Shukor, M.R. 97
Sieber, P. 145
Simons, F. 113
Sindelar, C. 76
Singh, U.K. 96
Sinnakaudan, S.K. 97, 99
Sisinggih, D. 201
Skripalle, J. 261
Slamet, N.S. 42
Söhngen, B. 239, 245, 246, 249
Sollerer, F. 234
Speckter, T. 165
Spitzer, D. 245
Spreafico, M. 28

Stamm, J. 249
Stedtnitz, W. 98
Steijn, R.C. 102
Steinmetz, H. 277
Stokseth, S. 57, 228
Sugano, M. 180
Sugiyama, M. 105
Suharyanto, A. 201
Sui, J. 109
Sulaiman, M.S. 40, 99
Sumi, T. 169, 203, 204, 211
Sumitomo, K. 161, 190
Sun, D.P. 77, 100, 112
Sun, Y. 100, 112
Symmank, L. 242
Szabo-Meszaros, M. 186

Tai, H.C. 205
Takemon, Y. 169
Takezawa, M. 153
Tan, B. 187
Tan, L. 62
Tanaka, M. 153
Tang, F. 166
Tang, H. 139
Tang, L.Q. 192
Tao, A. 187
Tassi, P. 136
Thapa, P.K. 198
Thom, M. 174, 188
Thulstrup, H.D. 28
Thumser, P. 256
Tian, Q.Q. 34
Tian, S. 163
Tominaga, A. 101
Tomita, K. 211
Tong, S. 55
Toyoda, Y. 46
Tsai, Y.J. 140, 205, 212
Tsutsumi, D. 68, 204
Tuijnder, A.P. 102
Turan, A. 154
Twohig, S. 35

Uno, K. 155
Ursic, M.E. 136
Utsunomiya, R. 68

Van Hoestenberghe, T. 262
Van Ransbeeck, N. 262
van Rijn, L.C. 102
Vanderborght, J.-P. 260
Vanthillo, R. 262
Västilä, K. 189
Verbanck, M.A. 260
Vetsch, D.F. 156, 225, 229

Violeau, D. 90
Virmoux, C. 123
Viseu, T. 265
Visscher, J. 228
Vollant, A. 230
Vollmer, S. 247
von Bargen, U. 181
Vonwiller, L. 156
Vu, M.T. 244

Waibel, H. 273, 274
Waikhom, S.I. 103, 160
Walling, D.E. 28
Wang, C.H. 192
Wang, D. 141
Wang, H. 73, 162
Wang, J. 74, 273, 274
Wang, M. 73, 73, 213
Wang, P.Y. 157
Wang, R. 74
Wang, S. 95
Wang, X. 139
Wang, X.K. 104
Wang, Y. 84
Wang, Y.H. 192
Wang, Z.L. 130, 164, 191
Watabe, H. 105
Watanabe, Y. 161, 190
Wei, G. 213
Weitbrecht, V. 64, 128, 254
Wieprecht, S. 33, 49, 91, 134, 174, 188, 277
Wilson Júnior, G. 106
Work, P.A. 85
Wu, K.W. 75, 158
Wu, P. 109
Wu, Q.F. 36
Wu, T.H. 205
Wu, W. 9
Wurms, S. 248
Wyss, C.R. 224

Xia, J.Q. 115, 125, 159
Xia, Y.F. 59, 107
Xiao, P.Q. 37
Xiao, Y. 55
Xu, H. 59, 107
Xu, L. 257

Yadav, S.M. 103, 160
Yadav, V.K. 160
Yamada, Y. 105
Yamaguchi, S. 161, 190
Yamamoto, T. 153
Yamauchi, S. 180

Yan, X.F. 104
Yang, C.X. 37
Yang, C.Y. 157
Yang, K. 141
Yang, L. 61
Yang, S. 55
Yang, X. 276
Yang, Y. 61
Yao, S. 162
Yao, W.Y. 37
Yazdandoost, F. 38, 39
Yeh, K.C. 53, 75, 158
Yi, Q. 137
Yin, C. 243
Yokohama, H. 190
Yorozuya, A. 258
Yoshida, K. 119
Yu, B. 109
Yu, G.A. 108, 138
Yu, P. 139
Yu, S. 163
Yu, T. 27, 157
Yusoff, N. 40

Zaccara, S. 210
Zahiri, A.R. 83
Zainal Abidin, R. 40, 99
Zanello, F. 267
Zardchoghai, A. 219
Zelleg, M. 110
Zhandi, D. 111
Zhang, C.D. 138, 164
Zhang, H. 70, 84
Zhang, J. 74
Zhang, Q. 51, 165
Zhang, W.Y. 241
Zhang, X. 26, 112
Zhang, X.Z. 241
Zhang, Y. 115
Zhang, Y.F. 192
Zhao, H.M. 172, 192
Zhao, J. 113, 213
Zhong, G.H. 78
Zhongluan, Y. 253
Zhou, H. 109
Zhou, L. 114
Zhou, M.R. 125, 159
Zhou, X. 95
Zhu, M. 41
Zhu, Y.H. 166
Zimmermann, R. 249
Zong, Q.L. 115
Zrdchoghai, E. 219
Zulfan, J. 42
Zuwen, J. 111